Tectonic Consequences of the Earth's Rotation

Tectonic Consequences of the Earth's Rotation

ROBERT C. BOSTROM

UNIVERSITY PRESS

2000

OXFORD
UNIVERSITY PRESS

Oxford   New York
Athens   Auckland   Bangkok   Bogotá   Buenos Aires   Calcutta
Cape Town   Chennai   Dar es Salaam   Delhi   Florence   Hong Kong   Istanbul
Karachi   Kuala Lumpur   Madrid   Melbourne   Mexico City   Mumbai
Nairobi   Paris   São Paulo   Shanghai   Singapore   Taipei   Tokyo   Toronto   Warsaw

and associated companies in
Berlin   Ibadan

Copyright © 2000 by Oxford University Press.

Published by Oxford University Press, Inc.
198 Madison Avenue, New York, New York 10016

Oxford is a registered trademark of Oxford University Press.

All rights reserved. No part of this publication may be reproduced,
stored in a retrieval system, or transmitted, in any form or by any means,
electronic, mechanical, photocopying, recording, or otherwise,
without the prior permission of Oxford University Press.

Library of Congress Cataloging-in-Publication Data
Bostrom, Robert C.
Tectonic consequences of the Earth's rotation / Robert C. Bostrom.
p. cm.
Includes bibliographical references.
ISBN 0-19-509028-4
1. Earth tides.   2. Earth—Rotation.   I. Title.
QC809.E2 B68   2000
551.1'3—dc21   00-055062

1 3 5 7 9 8 6 4 2

Printed in the United States of America
on acid-free paper

This work was occasioned by
the loss of my companion of 40 years
Cecile Hagen Bostrom,
1930–1994,
mother of my sons;
valiant proponent of a U.S. national health service

Teacher, gentle comrade, loving wife,
fellow farer true through life

Preface

> There is no end, but addition.
> T. S. Eliot

A major part of this work could be entitled "The Secular Term." The existence of a nonreversing term in the Earth's tidal distortion, causing *cumulative* displacement, may possibly interest not only the geologist, but colleagues in astronomy concerned with orbital evolution. An incentive to research has been an indignation that investigators have modeled an earth that is geologically so active, as inert and "dead."

For a century investigators have accepted a finding by Sir George Darwin that, aside from the insignificant retarding torque, no nonreversing tidal force system exists; in tectonics, we need not seek a tidal contribution. In geophysics, models have been based on a formulation by the mathematician Augustus Edward Hough Love. Love specified deformation of an earth subjected to a geostationary potential, reversing as by approach and recession of the Moon.

To an unversed geologist, toiling with his hammer in the field, it has seemed that the Moon and Sun are not stationary in the sky, approaching and receding twice daily—but instead rise and set. Relatively, tide-causing bodies move *around* the Earth. This may come as no surprise to astronomers, convinced since Newton that the Earth rotates relative to the Moon's gravity.

The situation has invited exploration of the assumptions supporting Love-type, oscillatory tidal models. Exploration has not been straightforward. For decades the appealing vernacular, "tidal drag," has caused Darwin's minute retarding torque to divert attention from forces that might be significant. As I might have divined earlier, these turn out to include stress generated because the gradient of the potential is not stationary, but instead rotates relative to the Earth. As distinct from the ebb and flow of the marine tides, perhaps dominant in the thoughts of Darwin and Love, the principal bodily tides are nonreversing. M_2, S_2 are in reality not represented by periodic geostationary bulging, but exist as waves some tens of centimeters in amplitude, progressing unimpeded around the Earth. The stress imposed by their passage is eight orders of magnitude greater than that generated by the retarding torque. Nevertheless, this is not "tidal drag"!

Under unidirectional wave passage the vectors describing the stress ellipsoid rotate without reversal in one direction, unlike those in classical models which are irrotational. Under departure from elasticity, the forcing stress leads the elastic restoring force. Departure is intrinsic, due to the existence of stress-dependent entropic processes including flow, fracture, chemical and phase transformation, and differentiation. For this reason, under rotation forcing and restoring stress are perpetually misaligned, inducing ubiquitous nonreversing body couples and *cumulative* distortion.

I have become keenly aware of a remark by Hans Lugt (1983): "A reason for the fascination which radiates from rotating media may be sought in the fact that the human mind—developed and formed in a locally non-rotating environment—cannot interpret motions intuitively in a rotating system." Distortion of a body subjected to a rotating potential takes the form of rotation of parts relative to itself; translation is absent. Von Helmholtz identified the species of motion (vortical) with respect to which there is no velocity potential in 1857. To date (1997), his insight has been disregarded by those of us struck by a tidal energy flux, astronomically well measured, that has seemed to "disappear" within the Earth. For scale, this is at least as great as that released in global seismicity. It may be desirable to examine the effects of actual external gravity, rather than those of a geocentric field that might well be attributed to Ptolemy.

This writer has found it hard to visualize Helmholtz's vortex motion. Is it most convenient to conceptualize motion of individual particles or pairs, elementary volumes or Helmholtz's vortex filaments? For elucidation Sir Horace Lamb attempted a sketch. A major difficulty is that in an extended medium such as the Earth's mantle, vorticity must be endowed with a self-organizing, i.e., cell-forming potential, akin to that identified by Kelvin, Bénard, and Rayleigh with respect to thermal buoyancy. Most-efficient dissipation requires that under continuous change in heat flow, equilibrium switches discontinuously into successive cellular flow states.

Although tidal body couples are elementally as minute as those under the density gradients driving thermal convection, they must likewise be additive within cells benefiting from the huge dimensions of the mantle. The high exponent of the spatial dimension in the Rayleigh number reflects this. Whatever the magnitude of its contribution, vorticity induction under the rotating potential is likely eventually to merge with the flow under buoyancy. Being pervasive at all scales, besides affecting mantle-wide cells the shaping action of induction must affect more localized aggregations of reduced viscosity, such as those underlying the ocean-ridge system and regions of back-arc spreading.

In this book it has been possible to give only passing reference to the possibility of numerical modelling. Long-term forward projection of flow in the mantle is permanently unattainable, for reasons made apparent in chaos theory. Whereas only tidal models incorporating rotational flow can simulate reality, their importance is a function of the elastic quality factor Q at unlimited period. Pertinent $Q_{vorticity}$ is almost unknown. Phase lags suggest that it is only a fraction of Q observed in seismic absorption and decay of the Chandler wobble. Artificially restricted to Q_{seis} as in orthodox models, dissipation is small.

Treading between multidimensional pitfalls, it seems possible that by acknowledging the existence of wave tides, explanations may emerge of data we have been reluctant to accept. Among these are observations made by the Centre International des Marées Terrestres in Brussels that the gravimetric tide is larger and more diverse than in models permitted by convention. It may be inadvisable to have looked askance at worldwide data that contravene fictitious models. In Brussels, the magnitude of the phase lag (determining that of the couples) is the subject of important, difficult, and ongoing research.

Geologically, tectonic phenomena cannot be addressed in terms of the Darwin-Love tidal model because this excludes the secular distortion. The difficulty is compounded, because models have assumed an inert earth. It is obvious that the Earth is not uniform and passive, but heterogeneous and internally dynamic. Its integrity and elasticity are continually destroyed by internal heat generation. The ensuing flow under convection causes the mantle continually to be the locus of critical phenomena, including phase and state transition. A condition of stress-sensitive marginal stability exists in voluminous subduction zones, and beneath the global rift system. It may be the case that the curiously large phase lag observed by Brussels results from the fact that in sweeping through all directions, the tidal stress discovers instability constantly replenished.

Perhaps because of this, volcanic components of the convection have been found to some extent to be synchronous with tide phases. If the "extent" is substantial, representing as it does a temporal modulation of the convection, there must eventually develop a spatial modulation. We might speculate that pervasive vorticity-induction relates to the puzzling "regimentation" of the Atlantic fracture zones. As products of ridge-crest vorticity about a vertical axis, the latter are systematically latitudinal in orientation; their domain extends symmetrically about the Equator, for one third the Earth's circumference. Kindred reasons may explain the gross extension of the co-tidal surface-westward limbs of the tropical Pacific and Atlantic spreading axes, at the expense of counter-tidal east limbs. We might suspect that vorticity induction is responsible for the difference between our steady-state "plate tectonics" and the spasmodic convective overturn on Venus—rotating only slowly, and devoid of a Moon.

I have attempted to present material in a form that permits the reader to make up his or her mind as to its validity. To some it may seem that orthodox models, although omitting the primary tidal action, are sufficiently close to reality. Others may feel that it is worth looking further, comparing major surface features with expected tidal displacement. The situation may conceivably resemble geotectonics in the early part of the century. Du Toit and Southern Hemisphere colleagues became convinced of the break-up of a former continent, Gondwana. Until acknowledgment came of mantle convection, this view could not be sanctioned. Perhaps those favouring fictitious models of the tide will sanction a similar review.

In chapter 6, discussion of tectonic effects of polar wander, essentially rotation of the Earth about an equatorial axis, is no less speculative than the fundamental paper by Goldreich and Toomre (1969) pointing to its existence, and commonly accepted plate reconstructions.

It has been exceedingly hard to write chapter 7, assessing the observation problem at a time when measurement techniques are proliferating. In observing minute secular motion, we at once encounter Hutton's difficulty: extrapolating the effect of processes barely detectable, but cumulative over time without limit. Laskar, Prigogine and confreres have now shown that indeterminacy makes it impossible to forward-model over geological time. Backward projection encounters an identical block; but uniquely the geological record, of history actually followed, endures. Geophysically, there is the problem of a fiduciary or benchmark. In *The Nature of the Physical World*, Eddington pointed out that "The great stumbling-block for a philosophy which denies absolute space is the experimental detection of absolute rotation." We still are frustrated in our efforts to detect "absolute" plate motion. The Hubble orbiter now provides images simultaneously of the effects of space and time. Monitoring rotation, VLBI observes signals from sources so ancient/distant as to have become non-existent before the appearance of observers—and Earth. Yet the geological effect of tides on the accessible Earth is almost unexamined.

I wish to record my gratitude to many colleagues for discussing and disputing aspects of this exercise. Primarily, I wish to thank my mentor and DPh adviser, Brig. Guy Bomford, O.B.E., for tolerating unorthodox thoughts while giving me something of geodesy. My interest in the tides rose as a result of presenting a fiercely contested paper, *Arrangement of convection in the Earth by lunar gravity*, in the rooms of the Royal Society in May 1972. At the time, in seeking an influence of lunar gravity on the pattern of thermal convection, only Darwin's retarding torque was known. In correspondence, Walter Munk swiftly made me aware of its feebleness—even in the presence of endogenous convection. Already at this juncture, Walter Sullivan, for many years science editor of *The New York Times*, was interested in the possible contribution of tides to peculiarities of plate motion. His investigation was halted by impediments presented as comprehensive and unassailable, adduced originally by Sir Harold Jeffreys in his 1920s debate with Joly.

I count myself privileged to have had the moral support of Keith Runcorn in examining vorticity induction. The toroidal plate motion identified by him decades ago is dimensionally akin. I have been encouraged in communications with George Platzman to think it feasible that we should have included vorticity in estimations of the "solid" dissipation. Correspondence and conversation with David Cartwright have encouraged me to think that not all may be settled in respect to the alternative: solely marine dissipation.

In the geological world, Bill Dickinson at Stanford saw the necessity of admitting the asymmetries in the Pacific basin when exploring mantle flow. Unlike many geophysicists, my old acquaintance Seiya Uyeda at Tokyo has been so bold as to tolerate the possibility of tidal input in large-scale tectonics, in particular that of the Pacific world. I have greatly benefitted from crossing foils with my friend Richard Ray at NASA, and from discussing with Nazario Pavoni at ETH the bipolar nature of the tectonic Earth, first identified by him. Harald Schuh at München has discussed the possibility that wave tides may explain peculiarities in the dissipation revealed by VLBI. François Roure has sent me sections from France showing the similarity of surface-west displacement in the Mediterranean world, to that which has been so enigmatic in the Pacific. It must be emphasized that none of these colleagues are culpable in respect to perceived heresies.

In particular I wish to thank Bill Kaula, who (while not believing substantial effects can exist) took the time to provide a frame within which to consider multitudinous interactions of the tides and convection in the mantle. Eduard Berg in Honolulu caused me to follow up the role that stress-diffusion must assume in a lithosphere-asthenophere situation. Dick Rapp at Ohio State has patiently explained to me the parameters of his elevated gravity-gradient map, forecasting the output of an orbital gradiometer. Paul (now Baron) Melchior and his colleague Véronique Dehant have taken time from their activities at the Observatoire in Brussels to discuss the untidiness imposed on polished algorithms by a geological Earth. I must record my gratitude to Melchior for correspondence that now extends over many years. I wish to thank Ole Andersen at Copenhagen for providing me with preprints summarizing Topex-based models of the ocean tides, displaying the underlying assumptions. Lately I have become indebted to Jim Ryan of NASA's Space Geodesy Branch, for allowing me early access to data issuing from the intercontinental VLBI baselines.

For accomodation, I am indebted to the University of Washington for providing me with unlimited facilities and unfailingly friendly office assistance. I owe much gratitude to Professor Dewey at what is now the Department of Earth Sciences at my alma mater for allowing me space and facilities where these are scarce. Renewed proximity to the Bodleian permitted instantaneous access to publications having otherwise to be identified and sent for. I must also record my monopolizing the resources of the library at the Hagen ranch in south Texas in pursuits which must have seemed extraordinary.

In particular, I wish to acknowledge the guidance given to me by my youngest son, Douglas Kaj, separately from his occupation with broadcasting and data management. Without his help and coming forward with specific hardware and software, this work would not have been completed.

Oxford, 1995   Robert Christian Bostrom
Seattle, 1996–1998

Contents

1 Introduction and Overview	**3**
1.1 Tidal Distortion	5
1.2 Polar Wander	10
1.3 Research Field	12
2 Historical Perception: Earth Rotation and Global Tectonics	**13**
2.1 Overview	13
2.2 Passive Earth	14
2.3 Internally Unstable Earth	20
2.4 Status: Classical Models	24
2.5 Present Stage	37
3 Tidal Action in a Uniform, Inert Earth: Static Bulge, Mobile Bulge	**39**
3.1 Overview	39
3.2 Tidal Deformation	40
3.3 Total Tidal Dissipation	42
3.4 Dissipation: Marine Component	42
3.5 Solid-Earth Dissipation: Models that Assume Bulge Is Geostationary	43
3.6 Distortion Under Rotation of the Potential	48
3.7 Interaction, Bodily and Marine Tides	61
3.8 Partition, Marine/Solid-Earth Dissipation	62
3.9 Summary: Tidal Action in Inert, Uniform Earth	64
4 Tidal Action: Pre-Stress Under Convection	**68**
4.1 Overview	68
4.2 Characteristics of the Convection	70

4.3 Dissipation Paths	78
4.4 Dissipation: Material at Critical Point	90
4.5 Dissipation in Pre-Stressed Earth	103

5 Tectonic Record — 106

5.1 Overview	106
5.2 Plate Motion	107
5.3 Data-Indicated Flow	111
5.4 "Absolute" Motion	133
5.5 Lithosphere Mobility	144
5.6 Tectonic Record	154

6 Polar Wander: Figure Adjustment — 155

6.1 Overview	155
6.2 Displacement in Latitude	157
6.3 Geophysical Evidence as to TPW	162
6.4 Figure Adjustment: Requirements of Bulge Migration	166
6.5 Figure Adjustment: Geological Record	167
6.6 Axis Stabilization (Dynamic Inhibition of TPW)	175
6.7 Status: Models of TPW and Figure Adjustment	178

7 The Observation Problem — 181

7.1 Overview	181
7.2 Standards (Base-Line Models)	182
7.3 Validation	190
7.4 Observation Techniques	190
7.5 The Volcanicity Association	210
7.6 Orbital Interactions: Multiple Outcome (Indeterminacy)	214
7.7 Numerical Models	217
7.8 Status of Field	218
7.9 Scope	220

References	*221*
Index	*262*

Tectonic Consequences of the Earth's Rotation

Plate I. Tidal dissipation in a "geological Earth": Marine; Solid-Earth; Interactive

Marine dissipation, accounting for major but uncertain fraction of whole, is thought to take place principally in shallow-sea regions: *colored red*.

Dissipation in the solid Earth takes place under two distinct types of tidal action: (1) an oscillatory deformation, causing such effects as periodic bulging and a twice-daily subsurface pressure pulse; and (2) a progressive induction of vorticity (increments of circulation). With reference to the reversing type, *white plus signs* (+) mark regions underlain by material at the point of transition to denser phases under subduction (continuous pressure-increase). These are susceptible to a great depth to enhanced dissipation during the rising phase of the tidal pressure pulse, and reduced reversion during the pressure decline. *White negative signs* (−) mark regions of mantle-upwelling (steady pressure decrease), susceptible to enhanced transition during the declining part of the pressure cycle. Induced vorticity is dimensionally identical to the flow in mantle convection, combining with it to an unknown extent. Due to the rigid-plate structure of the lithosphere, stress and dissipation under tidal working must be concentrated at plate intersections.

Solid-Earth/marine interaction is illustrated by the ocean-ridge system. The site of upflow in the underlying mantle, the ridge system generates deep-water baroclinic waves, potentially responsible for much of the marine dissipation (Morozov, 1995; Ray and Mitchum, 1996; Munk, 1997). The meridional orientation adopted by the ridge system in low latitudes enhances this effect. Regions such as the Patagonia continental shelf and west Pacific intra-arc seas are potentially the site both of shallow-sea dissipation, and that in underlying subducted material. The principal solid-Earth deformation affecting the ocean tide is likely to be sea-floor "heaving" of unknown magnitude and distribution, not here visible, of the type identified by Hendershott (1972).

The total dissipation is well measured, via astronomic observation of the recession of the Moon and the increasing length of day, and amounts to about 3.37 terawatts.

The image of seafloor topography, based on satellite altimetric radar tied to bathymetric control, is due to Smith and Sandwell (1997), to whom I am indebted. Their projection makes visible the distinctive "regimented" fracture zones of outer Earth (see for instance the Atlantic region), atypical of flow under random thermal convection.

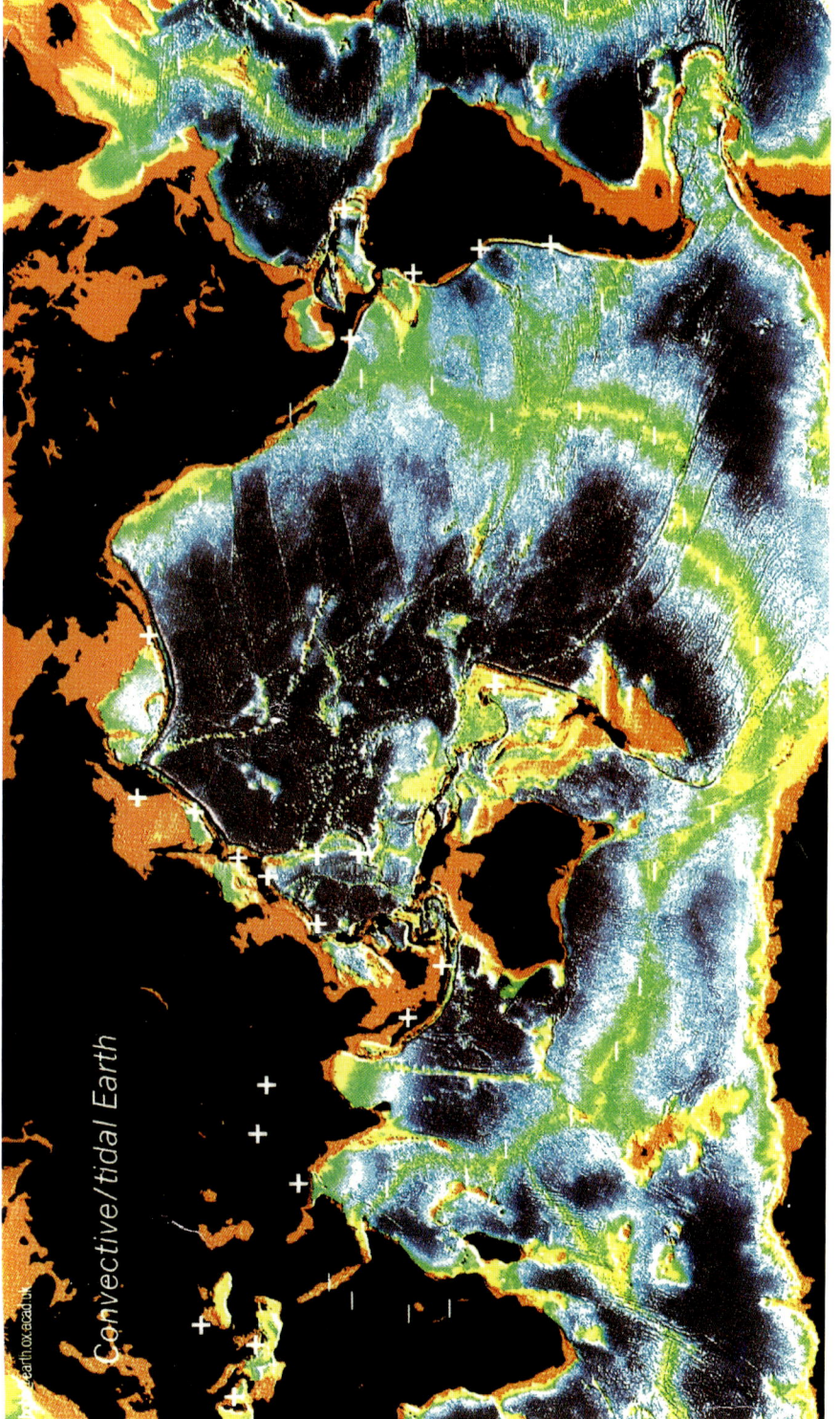

Convective/tidal Earth

1

Introduction and Overview

The Earth's rotation, conspicuous by reason of the Sun's motion and the ocean tides, has been suspected for the past century of playing a role in tectonism.

A connection between rotational forces and crustal displacement was proposed, among others, by G. H. Darwin (1879a,b), Wegener (1915), and Joly (1925, 1928). Joly's conjectures were dismissed by Jeffreys (1929), in papers that are still influential.

At the time, despite remarkable syntheses by Suess (1909) and Taylor (1910a,b), structural maps of the Earth were fragmentary. Seismological and other data were taken to suggest that the oceanic two-thirds of the Earth has the character of foundered continent (Daly 1938) and dates from the Archæan.

In the subsequent 60 years, fractional world maps have been replaced by almost-synoptic images of the entire planet. Marine research has shown that the structure of the ocean floor differs fundamentally from that of the continents. Instead of being ancient and passive, the oceanic portion of the Earth is seen to be a primary site of tectonism. Magnetic and seismologic data have demonstrated that instead of being inert, the Earth's mantle is affected throughout by bulk flow.

Topical factors seem to justify revisiting an ancient subject. The present volume is not intended to "prove" a connection between the Earth's rotation and geotectonics. Instead, it is intended to review evidence that a connection may exist. At an earlier stage, a landmark review, "The Rotation of the Earth," (Munk and MacDonald 1960a) eschewed speculation as to a connection with global tectonics.

With respect to a connection between tectonics and the Earth's rotation, the situation may resemble that concerning mantle convection. In the first half of the century it was perceived by Pekeris (1935) that the mantle may not be inert, but rather subject to convection, of which the effect would be dominant in tectonics.

It was apparent that in the mantle the critical Rayleigh number must be greatly exceeded. Neither the discovery by Vening Meinesz (1948, 1958) of the unique gravity anomalies associated with what are now termed subduction zones, nor the well-founded conjectures of Ampferer (1941) and Holmes (1944) in respect to a mid-Atlantic spreading-axis, could withstand objections to "continental drift," based on "the strength of the earth."

Disregarding deep seismicity detected by Wadati (1928), it was proposed that intense gravity lows signal permanent mantle strength, rather than hydrostatic disequilibrium maintained by flow. Many would identify a turning-point as work by Runcorn (1957) and colleagues, displaying the record provided by rock paleomagnetism, requiring serious discussion as to continental drift.

The connection, if any, between the Earth's rotation and its deformation, has an even more lengthy history. The subject reaches into several disciplines and has generated intense discussion. To disentangle threads, it has been advisable to separate research lines that have proved fruitful from valuable enterprises that have not. For clarity, it has been found necessary (chapter 2) to relate current research to antecedents extending to Newton, Kant, and Helmholtz.

Embracing astronomy and fluid mechanics as well as the geological record, the subject exists as a matrix of simultaneous, conflicting views. As the matrix can only be set down sequentially, it has been necessary to outline the historical development of tidal theory, giving credit to views that have had to be discarded with accession of geophysical data.

To utilize new data, it has been found advisable to incorporate realistic tidal distortion. In his comprehensive review, Lugt (1979) has remarked that, until the time of Helmholtz (1858), the distinction between rotational and irrotational flow was not recognized. Curiously, models of tidal action have maintained this situation (chapters 2 and 3). Among others, those of Darwin (1879 onward; 1908), Love (1909, 1911), Jeffreys (1924), Munk and MacDonald (1960b), Zschau (1978), Molodenskii (1984), Wahr and Bergen (1986), and Dehant et al. (1991) are constructed of an earth undergoing spheroidal oscillatory strain. Loss is then evaluated in terms of the seismologic anelasticity factor Q^{-1} and irrotational motion across equipotential surfaces. Dissipation in this mode encompasses solely that due to variations in the lunar and solar distance, as represented by subordinate tides.

In reality, the tidal bulges move continuously around the Earth, as waves. Having significance with respect to observed dissipation, it has become conspicuous (chapter 2) that distortion then takes place in the mode of rotational, nonpotential flow. Rotational and irrotational flow are distinct entities. Following Helmholtz such figures as Sommerfeld (1950) and standard texts, e.g., Landau and Lifshitz (1959), have established that dissipation in the former cannot be measured in terms of the difference between strain extrema, but only through piecewise integration of the entire flow path.

In this mode, there is furthermore added to orthodox models an increase in the dissipation as function of an elastic quality factor Q pertinent to strains of infinite period (as distinct from a seismologic Q indefinitely larger). Evidence suggests that within materials in the state likely to be present in the mantle, Q decreases with lengthening period of applied stress; stress of infinite period must lead to dissipation difficult to estimate.With respect to tectonics, tidal rotational flow (a circulation), caused by continuous rotation of the stress ellipsoid under east-to-west bulge migration, is cumulative and independent of deformation in spheroidal oscillation. The rotational flow is not derived from the minute tangential stress represented by Darwin's retarding torque, itself derivative from the primary dissipation.

Gravity observations made with great difficulty (Melchior and Ducarme 1991; Melchior 1994a,b) suggest that in scale dissipation in the semidiurnal bodily tide, M_2, exceeds that in seismicity and many flow processes, and is regionally variable. It seems probable that valuable efforts to reconcile the observed tidal gravimetric factor with that computed (Dehant 1992), our chapters 2 and 7, will be unavailing until based on the actual tidal distortion.

It would be desirable to discover whether the energy degradation takes place directly into heat or affects tectonic processes. The stresses produced (chapter 3), additive over great distances, are comparable with those generated by thermal buoyancy. Establishing the extent of the interaction might assist, among other benefits, in deciphering flow processes in the mantle.

It has not been obvious why there seems to exist a latitudinal factor in geotectonic features, of the kind seen in the "quasi-geostrophic" orientation of the Atlantic fracture zones, the grossly extended west limb of the Pacific seafloor spreading, and the western embayment of the Pacific expressed in the displacement of Sundaland. Asymmetries have been perceived for several years by Dewey and Bird (1972), Dickinson (1978), Uyeda and Kanamori (1979), Ricard, Doglioni, and Sabadini (1991), O'Connell, Gable, and Hager (1991), and Lithgow-Bertelloni et al. (1993). It is perhaps significant that the principal, Helmholtzian, tidal distortion, essentially tending to add a preferred circulation component, must affect fluid aggregations or cells at all scales, ranging from regional features believed by geologists to characterize the crestal region of the ocean-ridge system, to cells extending throughout the the mantle.

In what follows, tectonic effects of the rotation are separated into those resulting from motion in an external gravity field (namely, the tides) and those in an earth spinning in isolation, appearing as polar wander. Each category is affected by quantities of new data and changing observation techniques.

1.1 Tidal Distortion

The first effect is simple in concept but made complex by the complexity of the the Earth. Under rotation in the gravity field of the Moon and Sun, attractional forces affect the world ocean very strongly (figure 1.1). The interior is affected less and is here the subject of investigation. The time lag in the passage of the bodily tidal waves, a measure of dissipation within the Earth's interior, is hard to separate, because of oceanic interaction. In view of the acceptance of plate tectonics, the Earth can no longer be treated as an inert or passive body. Volumes of the upper mantle are maintained in unstable equilibrium by convection. The research position is tackled as follows.

1.1.1 Astronomical Situation

Recent years have been marked by increasingly successful efforts to monitor the spin-down of the Earth and the recession of the Moon. From these, the value of

6   Tectonic Consequences of the Earth's Rotation

Fig. 1.1. Response of the Earth to tidal forces. An observer at a point on the shore perceives that the marine tide, confined by the ocean boundaries, is oscillatory, flowing and ebbing twice daily. In contrast, bodily tidal bulges (constituting the earth-tides), pass the point as one-way waves. These progress unimpeded around the Earth, in a direction opposite its rotation. In figure, the offset of the Moon signifies the tidal phase lag, delimiting the dissipation.

the total tidal energy dissipation has been calculated, marine plus "solid Earth." To date, in modeling the solid-Earth fraction the Earth has been assumed to be virtually in eccentric synchronous orbit. The resultant static and entirely oscillatory tidal bulge does not in reality emulate the action of the tidal bulges, which progress as waves.

1.1.2 Planetary Viewpoint

To an interplanetary traveler it would be apparent that the Earth's tides are unique. Due to the comparative size of its satellite, the Earth has been described by astronomers (Kuiper 1955) as a double planet. Unlike Io, known to be the site of tidal input in its tectonics, the Earth is in asynchronous orbit about its orbital barycenter. Mercury is in almost synchronous solar orbit. Whereas the solar tide in Venus is much larger than in the Earth, the frequency of wave passage is less than one hundredth that on the Earth. Mars experiences only a minor solar tide. Its satellites are minute.

It has become apparent that the Earth is affected throughout by endogenous vortical flow, thermal convection being intrinsic in forming its surface features. In contrast to a regime of steady terrestrial heat escape, the regime on Venus is now seen to be one of long intervals of heat accumulation and tectonic quiescence. These are separated by episodes of catastrophic mantle overturn, causing resurfacing of the entire planet. Though not associated with mobile plate motion as on the Earth, volcanism on Venus is latitude-dependent, maximum in low latitudes.

He or she (if the planetary traveler is so constituted) might suspect that vorticity input by the Earth's unique wave tides is responsible for a principal part of the surface distortion, dubbed "geotectonics" by its inhabitants. The traveler might be forgiven for supposing that plate tectonics is a secondary effect, and that without the existence of wave tides the environment of an earth thus devoid of subduction would be Venusian, and sterile (Bostrom, 1998a,b).

A first attempt has therefore been made in chapters 3 and 4 to examine consequences of rotational strain (vorticity) induction. To separate earlier investigations, it has been necessary (chapter 3) to recount models of dissipation in an earth experiencing spheroidal and oscillatory deformation, as in a second-degree eigenvibration. In these, the critical variables are loss in self-reversing shear and its range of frequency-dependence. On this basis assuming the absence of a secular term in the deformation, to elucidate frequency-dependence it has been necessary to turn to the concept of an absorption-band, and long-standing investigations of the energy loss in terms of the Gibbs free energy of the relaxation process.

Energy loss must take place in the Earth as a result of such processes, but the loss in self-canceling spheroidal deformation must apparently be minor in comparison with that due to the secular distortion, producing cumulative vortical flow.

Modeling "approximations" of the solid-Earth loss, excluding the distortion, has hitherto made it necessary to attribute almost all of the reduction in the Earth's rotation to loss in the seas. Shallow-sea losses demonstrated by Taylor (1920) are potentially large enough to account for the dissipation, but still so poorly controlled that the process is exceedingly hard to quantify. To elucidate a complicated situation, it has here been necessary to chase down promising and rational lines of research which have proved unproductive.

1.1.3 Observation and Measurement

Gravimetric data based independently on satellite orbital elements, satellite-borne radar, and surface gravity observations now restrict mass distributions which are physically possible.

In consequence, since 1970 the Earth has been seen to be over-flattened relative to its rotation rate, a condition only glimpsed at the time of appearance of "The Rotation of the Earth." The restriction and implications as to its cause are allocated to chapter 6, with reference also to the inferences of Goldreich and Toomre (1969) as to displacement of the principal axes of moment. Implications as to the viscosity of the lower mantle (a factor at first believed to prevent convection), are tentatively referred to as a tidally induced excess of flow, to be expected in lower latitudes.

Orbital gravity information supplemented by ground truth has been employed by Rapp (1988) to project an image of the gravity gradient at an elevation of 160 km. Unlike satellite radar altimetry, his projection provides uniform and "seamless" coverage of continent and ocean alike. Based on comparison with geologically "known" mantle plumes (producing surface "hotspots"), Rapp's data suggest the existence of an extended number of these entities. His projection seems also to describe a preferred orientation of asthenospheric features in the tropical Atlantic and Pacific (chapter 5).

At the time of writing, the extended data-set obtained by the dual Topex-Poseidon radar satellite, measuring geocentric sea-surface height, is still being examined. Through assigning a quadratic dissipation factor, the apparent marine dissipation can be reconciled with the computed solid-tide contribution based on spheroidal oscillatory deformation combined with seismologic values of the elastic quality factor, Q, or other models.

Since initiation of the International Geophysical Year (1951), global observation of the solid tides has been owed to undertakings of the Observatoire Royal, Brussels, Belgium. Unexpected observed values have caused speculation as to the reduction factors that should be applied; namely, reference model and correction for oceanic effects. Independent measurement techniques, of which the potential is at present only glimpsed, suggest (chapter 7) that this subject will shortly be less one-dimensional. Observations employing the U.S. Global Positioning System (GPS), of local geocentric-distance changes, and Very Long Baseline Interferometry (VLBI) may provide Love numbers pertinent to long profiles (hence large earth samples). On the basis of published Topex analyses, it may perhaps be concluded that, without independent data obtained by the Brussels group, models employing fictitious tidal action would have been accepted without question.

Also noted is promotion by the Centre International des Marées Terrestres (CMT) in Brussels of the only surface instrument, the tiltmeter, capable of directly detecting vorticity-induction. The bulge stress formulated in standard works (Melchior 1983) makes no claim to be other than static. The dissipation that would result from periodic sign reversal, producing oscillatory spheroidal deformation, is unrelated to that resulting from rotation of the Melchior stress field without reversal, as in actuality. The output of a tiltmeter, assuming self-canceling spheroidal oscillation as commonly modeled, vs. rotational distortion with cumulative displacement, is illustrated in figure 1.2. Electronic amplification almost removes restrictions as to signal acquisition. Unfortunately, placement of these instruments in deep boreholes is necessary, in order to overcome site effects that otherwise restrict their use.

1.1.4 Geological Earth

An attempt has been made (chapters 4 and 5) to progress from imposition of a static bulge on an inert earth, to passage of the tidal waves in an earth "pre-strained," as by endogenous convection.

Investigations of mineral structure in pressure/temperature space (Thompson 1992; Gasparik 1993) have shown that under upper-mantle conditions there exist tens of discrete phase states. Ongoing pressure change under convection causes material in regions of vertical flow to exist constantly in a condition of marginal stability. At depth the tidal pressure-change rate greatly exceeds that in the vertical flow, raising the possibility of modulation (chapter 4).

For many years investigations by the U.S. Geological Survey (Shaw 1970; Shaw, Kistler, and Evernden 1971) have pointed to the possibility that where localized as in in regions subject to flexure, tidal dissipation could contribute significant heat deposition. Combined with phase instability, the strongly heterogeneous nature of the the upper mantle points to an enhanced importance of this process. We might expect (chapter 4) that concentrations of dissipation occur in subduction regions,

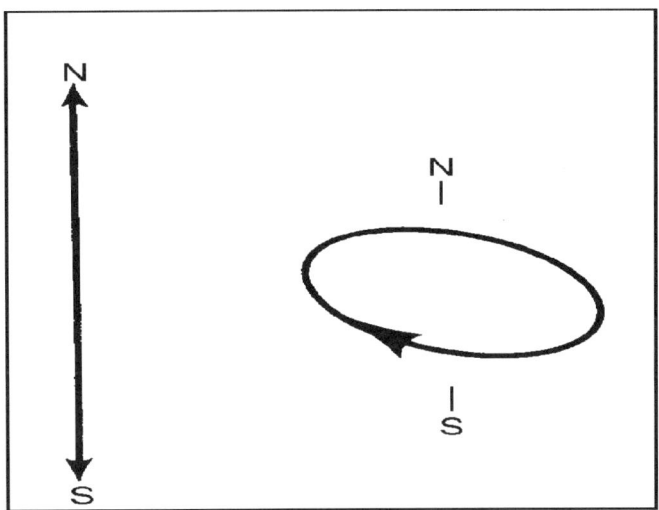

Fig. 1.2. Trace of vertical pendulum acting as tiltmeter at mid-latitude in Northern Hemisphere, after removal of site effects. *Left*: Installed in an earth in nonreal synchronous orbit, experiencing oscillatory spheroidal tidal deformation; as commonly modeled. *Right*: (drawn after Melchior 1983): In real Earth, having asynchronous orbit, rotating reference to the external gravity field. Trace is an ellipse, in which clockwise sense is function of the Earth's unvarying rotation direction; area and azimuth describe loss in induced vorticity.

as a result of the action of the tidal pressure pulse on material at the point of transition.

1.1.5 Synchroneity

A degree of synchroneity is apparent between tidal action and tectonic events, in the manner found by the USGS and others (Klein 1976; Dzurisin 1980; Weems and Perry 1989; Mauk and Johnson 1973; Shirokov 1983). Commonly termed "triggering," to be operative initiation entails unrecovered displacement favoring tidal dissipation; under highly nonlinear conditions, endogenous convection is prone to lead to tidal interaction. It is logical to suppose that to a greater or lesser extent, a preponderance of the advance in global tectonics must take place in tidal phase, rather than in antiphase. The investigations by Klein (1976), alone insofar as is known in taking vorticity into account, suggest that efforts by the USGS and others to define intervals of enhanced proneness to seismicity may be worth continuing. Apparently no investigations have been made as to a correlation, if any, between global seismicity, such as identified by Romanowicz (1993) and Du (1994, 1996), and long-term variations in tidal action.

10   Tectonic Consequences of the Earth's Rotation

Energy relations considered in chapter 4 make it apparent that no matter how high a correlation might turn out to exist between tides and earthquake incidence, expenditure in seismicity cannot represent the major part of the tidal dissipation. An inverse proposition may be correct, in that an unknown but substantial proportion of the energy released in seismicity may accumulate as secular elastic strain under the action of the wave tides. Information as to possible induction is already apparent, incidental to ongoing research. Besides "excess flattening," for instance, Gordon and Jurdy (1986) have demonstrated a persistent maximum of plate velocities in low latitudes.

1.2 Polar Wander

Rotation or tilting of the Earth about an axis such as x, in the plane of the Equator (figure 1.3), is experienced as true polar wander (TPW).

In an absolute setting, the Earth rotates in an inertial frame, affected only by axifugal forces and those resulting from conservation of momentum. Corroborating expectations of only slow external momentum transference, Williams (1993) cites evidence that the obliquity during Phanerozoic time has been comparable to the present.

As distinct from tectonic effects of tidal action, those possibly associated with

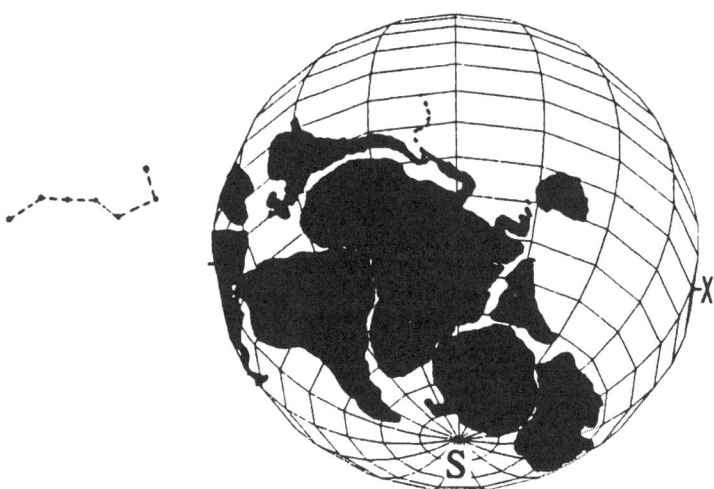

Fig. 1.3. Earth in rotation about an axis that is "almost space-fixed." Conservation requires that referred to an inertial frame the axis orientation can change only slowly, governed by slow transfer of angular momentum to external objects such as the Sun. Internally, mass shifts under geological processes cause reorientation of the axes of moment relative to the Earth, requiring geographic migration of the Earth's rotation axis and equatorial bulge.

polar wander stem from reconfiguration of the Earth under displacement of the bulge. The subject is hence exceedingly speculative. Whereas vorticity induction is susceptible to daily observation in the form of tiltmeter output, the effect, if it exists, of large-scale reconfiguration can only be assessed in terms of the disruption of Gondwana and limited information as to polar wander. As in much of geology, the "experiment" cannot be repeated.

The existence of TPW has been a subject of speculation since before the time of G. H. Darwin (1879), e.g., Garthe (1852). Subsequent to Darwin (1908) polar motion was estimated to be insignificant, prevented by "the strength of the Earth," i.e., immovability of the equatorial bulge. Immovability seemed to be corroborated by the discovery of the Earth's "excess flattening," interpreted (Munk and MacDonald 1960b; McKenzie 1968, 1972) as requiring viscosity in excess of 10^{26} P in the lower mantle, obviating the possibility of convection. Were geological time longer than available, the motion of Gold's beetles (Gold 1955) would effect polar wander.

In chapter 6, attention has been given to the demonstration by Goldreich and Toomre (1969) that, in the presence of convection, immobility of the Equatorial bulge is not an obstacle to TPW. As convection and bulge formation are gravity-driven, both are shaped by the same field, prescribed by conservation of momentum and placement of the moment axes. For more than a decade the effects identified by Goldreich and Toomre were postulated not to have practical significance, still on grounds that bulge strength would prevent it, or that isostasy would annul the effect of geological mass shifts (Dickman, 1979).

Data recovered by Besse and Courtillot (1991) point to the existence of post-Paleozoic TPW averaging several centimeters per year, but highly erratic in rate. Having in mind Goldreich and Toomre's analysis, Stevenson (1993a,b) points out that TPW seems surprisingly slow. Rapp's (1988) upward projection of the gravity gradient (chapter 5), may supplement data based on paleodeclination. If evidence is correct that the Indian Ocean triple-junction is the site of a hotspot or plume group stationary in the mantle, the uniquely rapid northward motion of the India plate may be composed of TPW added to plate motion.

In case of polar wander, a question arises as to how reconfiguration is effected, taking the form of migration of the equatorial bulge. The displacement volume required, especially if TPW is concentrated within episodes, is comparable with the total flow rate in the mantle measured in seafloor spreading and subduction. Viscosity distribution in the mantle is now seen (Peltier 1989; Forte, Dziewonski, and Woodward 1993) to be such that figure adjustment must disproportionately be effected in the asthenosphere.

The breakup of Gondwana and the formation of new oceans (chapter 6) may indicate that at the time of rapid TPW, figure adjustment calls for reorganization of the convection in the mantle. A gravity low of several thousand milligals, which would otherwise develop, is conspicuous by its absence, requiring that if TPW has taken place, "isostatic processes" on an unmatched scale must have accompanied it, initiating reorganization. The possibility seems to exist (chapter 6) that the outburst of volatile-rich volcanics in Mesozoic Gondwana was the result of destabilization, under the pressure decrement and areal extension.

1.3 Research Field

Perhaps more than in most fields, research as to the relation between the Earth's deformation and its rotation has been characterized by dominant personalities. This may be because it is hard to explore such disparate areas as astronomy and geology. The difficulty is illustrated in the contribution of four figures, each at the research frontier in his field.

G. H. Darwin's avocation was cosmogony, including the Earth's origins, unconstrained by geophysical data obtained subsequently. Jeffreys accepted Darwin's formulation of the bodily tides, with omissions only now apparent. Invoking constants pointing to Earth strength, Jeffreys opposed the recommendation of the geologist, Joly, that we investigate further the possibility of continental mobility. Of figures of the time, a fourth, Eddington, seems to have been most aware of the pitfalls awaiting anyone who ventures into territories each taking a life-time to explore—pitfalls now more numerous than ever.

The situation is that it may henceforth be possible to obtain some much-needed information, in particular as to the nature and magnitude of the tidal dissipation. At present, our information is limited to measurements obtained by the Observatoire Royal at Brussels. Difficult as these have been to obtain, requiring observation in uncultivated places, they provide our best estimate of internal energy deposition.

Geologically, energetics suggests that it may be necessary to observe tidal input in models of plate tectonics. That the major term in the distortion is secular and cumulative seems to require that we re-examine the mechanics of the dissipation, departing from the oscillatory models conjured by Love. The question perhaps must be faced (chapter 7), of whether in respect to long-term effects these constructs should be regarded as fictitious, in favor of comparing data with models incorporating the primary tidal action, distortion.

Considerations noted in chapter 3 suggest that the dissipation may be comparable with the viscous fraction in the convection. Is it possible that tidal input is responsible for the departure of the the Earth's primary surface features from those to be expected from random mantle convection? If so, it would be appropriate, whilst watching spectacular shoreline dissipation, for the observer to reflect that energy from the same source may accumulate in the lithosphere, contributing eventually to interplate tectonism.

A major challenge (chapter 7) is that, as the long-term tidal displacement is secular and devoid of a reference point, its observation presents formidable difficulty.

2

Historical Perception

Earth Rotation and Global Tectonics

2.1 Overview

In the past, analyses have made it seem unlikely that the Earth's rotation has affected global tectonics. Recent data suggest otherwise.

The situation is approached on the basis of some classical research. Awareness of the Earth's complexity has required transition from conceptual models by Newton, Halley, and Laplace; to those constructed by Kelvin and G. H. Darwin; to those used at present. Models were at first constrained only by astronomical data. Data now stem from seismologic mapping of the interior; from the paleomagnetic record; from ground-based gravimetry; and from synoptic observation of the Earth via satellite.

A fundamental factor has become the action of tides. One hundred years after Darwin, a problem still is that identified by Munk in 1968: "In 1920 it appeared that Jeffreys had solved the problem of tidal dissipation. We have gone backwards ever since." Munk speculated that the problem would be solved in five years. Unfortunately, we have progressed in the same direction. The dissipation is suggested by more abundant, but conflicting, evidence. To what extent is tidal energy expenditure, several times that (for instance) in global seismicity, attributable to marine processes, as against those taking place in the "solid" Earth? In respect to the marine portion, authoritative opinion varies: the shallow-seas tidal energy sink explains almost the whole of the dissipation, slowing of the Earth's rotation, and recession of the Moon. In opposition, the shallow-seas mechanism has status no better than "geophysical folklore."

Progression has been from tidal models of an elastic earth, to those admitting elastic defects seismologically defined. In investigating the "solid-Earth" portion of the dissipation, it is notable that transition has *not* been made from modeling the tidal Earth as passive, highly elastic, and inert, to taking into account its "plate tectonic" state, entailing permanent internal metastability.

14   Tectonic Consequences of the Earth's Rotation

Mathematically, following Darwin tidal models have assumed oscillatory spheroidal deformation, imposed by a geostationary, time-varying potential. In reality the tide-causing potential does not reverse, but rather rotates relative to the Earth. The principal tides exist permanently, as waves progressing in one direction around the planet. Under this regime the Earth is not subjected primarily to self-cancelling spheroidal deformation, but rather to cumulative distortion. Dissipation and displacement take place in the mode identified by Helmholtz in the 1850s, namely induced vorticity or rotational-flow. Love's justly celebrated numbers represent the response of the Earth to forces only *thought* to have been known. Because it is dimensionally the same as the vortical flow in the convection driving plate tectonics, cumulative distortion may conceptually couple to tectonic processes, accordant with observed values of the phase lag.

For clarity, the evolution of current models is first outlined.

2.2 Passive Earth

2.2.1 Mass and Figure of Rotation

Well-founded models of the Earth based on astronomy preceded geophysical data by several centuries. Halley (1697) drew attention to Newton's and Hooke's conclusion, based on the equinoctial precession, that the figure of the Earth is that "of a compressed sphæroid," in which the polar is 17 miles less than the Equatorial radius. The modern estimate of this quantity is 13 miles.

By reason of its precarious basis, Newton's estimate of the Earth's density (*"verisimile est quod copia materiæ totius in terra quasi quintuplo vel sextuplo major sit quam si tota ex aqua constaret"*—Principia, Liber III, Prp. X, Th. X) seems only fortuitously to bracket modern values. In some respects, Newton's assessment may be thought to have foreshadowed the concept of isostasy. His cantankerousness (see, e.g., Manuel 1968) was perversely a factor ensuring the dissemination of his views and stimulation of active research.

To many, Newton's attitude was cooperative rather than overbearing. His correspondents included the rural cleric and mathematician Colin Campbell, of whom he remarked (Turnbull 1960): "I see that were he amongst us he would make children of us all." MacLaurin (1746), elected, with Newton's support, Professor of Mathematics at Edinburgh at the age of 27, showed that the equilibrium figure of a homogeneous fluid earth is indeed an ellipsoid of revolution. His concept of level surfaces has endured as standard in problems of the potential.

The mass of the Earth, hence its mean density, became known more exactly as a result of the experiments of Cavendish (1798), employing known reference masses and the torsion balance designed by John Mitchell.

2.2.2 Tides

Halley (1697) pointed out that Newton attributed the ocean tides to the reduction in gravity on the Moonward vs. the anti-Moon side of the Earth. Newton and

Halley realized the complexities introduced by the geometry of the world ocean, but could not have foreseen that these would remain incompletely resolved in the twentieth century. Among other relations, in his *Mécanique Céleste*, "a monument of mathematical genius," Laplace (1795) showed that marine tides due to the motion of the Moon, including declination, consist of the superimposition of three species (figure 2.1).

The Laplace tides have since continually been used as reference. Similar, but smaller, solar tides are superimposed. Through development of his coefficients, specifying spherical harmonics, Laplace made available enumeration of components of arbitrarily higher order. In his treatment Laplace assumed that marine aqueous friction prevents the accumulation of resonances and unlimited growth of secular terms. An overview and critique of the Laplacian tides is provided by Lamb (1945).

Laplace perceived that the hour-of-day of historically recorded eclipses may register changes in the Earth's rate of rotation. The available data caused it to appear that the rotation has been approximately constant during the past 2500 yrs; and that, therefore, the Earth has not experienced cooling and contraction. He for-

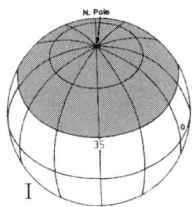

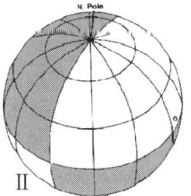

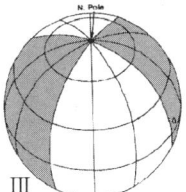

Fig. 2.1. Laplacian tides, in uniform ocean covering rigid earth. The shaded/unshaded portions in the figure represent areas where an equipotential surface is above and below its undisturbed level. At a point Species III forms twice daily, Species II daily, and Species I fortnightly for the Moon, half-yearly in respect to the Sun.

mulated the equilibrium figure of a body having a simple downward density increase and established that external to acting matter, the spatial second derivatives of a potential field V always sum to zero, i.e., that

$$\partial^2 V/\partial x^2 + \partial^2 V/\partial y^2 + \partial^2 V/\partial z^2 = 0, \quad \text{written } \nabla^2 V = 0.$$

The Laplace equation remains central to the concept of the potential field, essential, for example, in bounding density distributions defined by the orbital path of artificial satellites.

2.2.3 Tidal Energy Dissipation

Ekman (1991) notes that Halley (1697) had detected an anomalous acceleration of the Moon. This had been attributed to gravitational interaction between the Sun and planets.

In a paper identified by Becker (1898), Kant (1754) had pointed out that oceanic tidal friction must infallibly reduce the Earth's rotation. Working in the same area (Danzig) a century earlier, the astronomer Hevelius had obtained observations suggesting a difference between theoretical uniform time (UT) and time as measured by the Earth's rotation (Wünsch 1990). Becker (1898) recounts that in 1754, the Berlin Academy had offered a prize for a solution of the question: "Whether the rotation of the earth which produces the alternation of day and night, has undergone any change since the time of its origin? What is the cause and how can the fact be established?" The associated retardation in the lunar orbit was discovered by Laplace in 1824. In remarkable papers dealing with the role of rotation in cosmogony, Mayer (1863) computed that the length of the day is decreasing by 1/16 sec per 2500 yr under retardation by the ocean tides. Ninety years after Becker, we can accurately measure the reduction in the rotation, but are unable to establish the site of the friction, and its variation through geological time.

In the eighteenth century, the conservation of momentum was imprecisely defined (Cajori 1962), as still was the concept of "force" despite the relations formulated by Newton. The conservation of energy in its pre-subatomic form was identified by Mayer (1842) and the Danish natural philosopher Ludwig Colding (Elkana 1974) and formulated by Hermann von Helmholtz in 1847 (Helmholtz 1847; cf. Elkana 1974). Acknowledgment of the modestly presented claims of Mayer in favor of those by Joule was disputed by others in a debate of unparalleled acerbity, to be seen in 1863 issues of *The Philosophical Magazine*.

Prior to the unforeseeable revelation of radioactivity at the end of the century, it had to be supposed that the Earth is cooling. Its rate was rationally estimated by Kelvin (1863a), based solely on conductive transfer, using Fourier's formulation of heat conduction, indicating that its age is between 20 and 400 million years. Making reference to Mayer (1842), by then accessible in English translation, Ferrel (1864) confirmed an earlier conclusion as to the role of marine dissipation.

Analyses of the ocean tides, hence indirectly of the dissipation, based on elegant formulations but rudimentary data, were pursued by Kelvin (1875a,b), Darwin (1886, 1908), Hough (1897, 1898) and Lamb (1945). It was shown by Kelvin that Laplace's assumption that marine friction acts to produce an equilibrium state is incorrect, but that the impediment to tidal flow provided by the continents justifies Laplace's equilibrium assumption. Hough (1897, 1898) introduced an important modification, due to self-attraction of the deformed ocean. The analyses of Hough (1897, 1898) and Lamb (1945) established that tide height and phase at a point is complicated to an unknown extent by resonances as well as by seafloor friction.

The overall release of rotational energy represented by the deceleration is currently 3.37 TW, delimited by astronomical observation of day-length change and expansion of the lunar orbit (Christodoulidis et al. 1988; Marsh et al. 1990; Cartwright and Ray 1991). An abiding difficulty is to relate topical, "instantaneous" measurements, to rates over geologic time.

To account for the dissipation, Taylor (1920) calculated that dissipation concentrated in shallow seas may exceed that in the entire deep ocean. Based on oceanographic information of the time, Jeffreys (1921) suspected that shallow-water dissipation, as in the Bering Sea, might account for 80% of the world total. Records of the tides for major ports have been kept since before the Armada. Unfortunately, mariners favor sheltered harbors, making the data unrepresentative of the world ocean. Currently, the Topex-Poseidon project (chapter 4) has demonstrated that closely spaced and quasi-synoptic satellite data are necessary but insufficient to provide definitive data.

2.2.4 Solid Earth: Elastic Deformation

Until the late nineteenth century, models of the Earth ranged from a shell enclosing a fluid interior, supplying magma, to a spheroid of high rigidity. Energy dissipation and braking of the Earth's rotation were dependent on the marine tides.

Kelvin (1863b) reported an effect of tidal forces on the solid part of the Earth. By comparing the depth of long-period marine tides, reduced by yielding of the solid Earth under the tidal potential, with their equilibrium Laplacean value, Kelvin and later Darwin (1883) concluded that the solid Earth yields as if possessing the rigidity of steel. Elastic deformation reduces the marine tide to about 0.68 of its value were the Earth rigid. Kelvin concluded that it is improbable that an earth consisting of a shell less than 2000 miles (3200 km) thick could exhibit such rigidity. It was realized that the elastic deformation implies nothing as to energy dissipation.

2.2.5 Viscous Earth: Work of G. H. Darwin

In a series of papers, Sir George Darwin (1878, 1879a,b; 1908) examined the effect, not solely of lossless elastic yielding, but of imperfections in the elasticity, entailing energy dissipation. Darwin's involvement in the subject may have been stimulated by the interest in the marine tides of Whewell, Master of Trinity College Cambridge (Todhunter 1876), and by his cordial acquaintance with Kelvin. Whewell recommended a comprehensive survey of world tides including mapping cotidal lines—a project not yet complete.

Basing his estimates on Kelvin's formulation of the deformation of an elastic spheroid (Kelvin 1863b), Darwin substituted a time-dependent viscosity coefficient μ for a constant coefficient of rigidity, quantifying the tide in a homogeneous earth, of the viscosity of pitch; of glass; of iron; and of what has become known as a Maxwell solid. In a homogeneous earth, it was legitimate to model the lunar retarding torque as acting solely on the tidal bulge. The effect of inertia in respect to the decelerated interior was shown to be insignificant.

Darwin's formulation, (figure 2.2) further examined in section 2.4.2, corroborated (a) Kelvin's estimate as to the high rigidity of the Earth; (b) the view that a small imperfection of elasticity leads to only small change in depth of the ocean tide, but a significant phase change; (c) that integrated over time, a nonreversing force couple is created, akin to Kant's and Ferrel's marine retarding couple, tending to despin the Earth; and (d) that in a homogeneous, viscous earth, the couple tends slightly to displace westward the outer part relative to the inner.

It was shown that in such an earth, the retarding couple tends to westward-displace low-latitude regions faster than those at high latitudes as cos2[*latitude*].

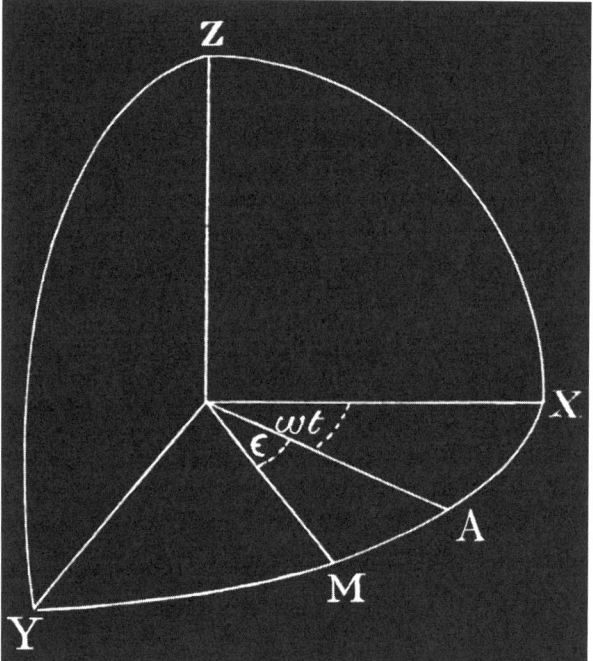

Fig. 2.2. Darwin's formulation of tidal despin and distortion. In a portion of the Earth, Z is the axis of rotation, XY the equatorial plane. M is the sublunar point, A the axis of the lunar bulge, ε the angle by which the bulge lags the overhead point, ω the Earth's angular velocity relative to the Moon. Darwin evaluated the retarding couple tending to rotate the bulge about the underlying viscous Earth, and the ensuing "screwing action." Reproduced from *Philosophical Transactions*.

In consequence, were the time sufficient, the retarding couple would result in a "screwing action." This would produce crustal "wrinkles," to be compared with mountain chains striking N-S at the equator, bearing to the east with increasing latitude. At the time, the Earth's age was limited to Kelvin's estimate of 10^8 yr, and the viscosity was that of cold rock or iron. Darwin (1908) lived to appreciate the profound change in estimates of the Earth's age, hence the time available for the action of the forces he had conjured, caused by the discovery of radioactivity.

As an example, he concluded that under a reasonable (but unverifiable) estimate of the Earth's viscosity, commencing at 45 Ma BP, a point at latitude 30° would by now be displaced 4.75′ eastward, one at latitude 60° 14.25′ eastward, relative to a point on the equator. In the "remote past," were the Earth and Moon to have been in closer proximity, because the screwing action varies as the sixth power of the Moon's distance, and the rotation rate of the Earth was then greater, the deformation might by now be prominent.

Energy input through dissipation, which otherwise would have extended Kelvin's cooling-based limit as to the Earth's age, turned out to be insignificant. Subsequent estimates of tidal action (Jeffreys 1924, 1929; Tomaschek 1957) point to the feebleness of the retarding torque in respect to an inert earth, seen to have a viscosity in excess of 10^{20} P.

Tidal displacement of the crust was proposed by a geologist, John Joly (1925, 1928), based on his postulation of mantle softening under radiogenic heat. Jeffreys (1926) could not countenance Joly's concept, on the supposition that "Tidal friction is then supposed to tow the crust around over the fluid layer." Jeffreys (1929, 322) pointed out that the required braking torque would annul the Earth's rotation in a year. Jeffreys (1929) reports persistent attempts at this juncture to relate tectonic phenomena to tidal forces. Eddington (1923) remarked that "The time has gone by when the physicist prescribed dictatorially what theories the geologist might be permitted to consider." Despite Eddington's wariness, exchange of views assumed a disposition that inhibited further discussion. As a result of his far-reaching investigations, Jeffreys (1954) noted that despite absence of a known cause, sundry solar-system satellites have been decelerated to the point of synchronous rotation, showing a constant face to their primaries.

Current estimates of the dissipation are based on Darwinian models (sctn. 2.4.2) of an oscillatory, geostationary tidal bulge, and passive Earth.

2.2.6 Displacement of Axes

Besides tidal action, Darwin (1879a) examined the influence of mass-redistribution on the geographic axis of rotation. At the time, bulk flow under convection was unperceived, and mass displacement was dependent on surface geological processes. It was concluded, as at present (see, for instance, Goldreich and Toomre 1969), that during any gradual deformation of the spheroid, the instantaneous axis of rotation will remain "sensibly coincident with the principal axis of figure." It was furthermore realized that, relative to an external frame represented by the ecliptic, axis orientation must be nearly invariant.

Due to the apparent high viscosity and limited mass displacement, Darwin concluded that in the extreme, migration of the Earth in respect to its axis of rotation can have amounted to no more than 15°. The extreme was based on estimates supplied by his father, Sir Charles Darwin, of continental uplift and Pacific basin subsidence in the light of atoll formation.

2.2.7 Internal Structure

The installation of synchronized seismograph stations made it feasible to trace raypaths, establishing elastic-wave velocity within shells forming the Earth. The period prior to military expansion of the global seismometer network saw the identification of the core (Gutenberg 1914) and density/depth regionalization of the mantle based on Jeffreys' seismological tables (Jeffreys 1924, 1939; Bullen 1936; Bolt 1991).

It is not possible to determine from first-arrival times whether the "20° discontinuity" (Jeffreys 1962) in reception of seismic signals from near sources represents a decreased downward velocity gradient or a reversal in velocity as maintained by Gutenberg (1948). Seismologic demonstration of the low-velocity zone at the top of the mantle, remarkably coincident with the asthenosphere of Barrell (1914) and Daly (1951), was achieved by Dorman, Ewing, and Oliver (1960), by matching the dispersion produced in velocity models with that seen in seismograms.

Regionally, a disparity between the oceanic and continental parts of the Earth was established as a result of the frequency-dependent velocity variation in long-path Love waves (Gutenberg and Richter 1951). It was realized that the seismologic evidence accorded with separation by the Austrian geologist Suess 60 years earlier of the Earth's surface into Atlantic and Pacific structures. At mid-century, much support, both geophysical and geological, could be found for the opinion that the ocean basins were of Archean origin (Daly 1951; Gutenberg and Richter 1951). The region of continental structure was considered by the latter authors to include the Atlantic and Indian ocean basins. Suboceanic epicenters, such as those clustering on the mid-ocean ridges, well identified by 1923 (Sieberg 1923), were ascribed to the release of "residual strain."

2.3 Internally Unstable Earth

Discovery of an internal energy source, radioactivity, has made it illogical to treat the planet as a passive body. Bulk flow (convection) under thermal density gradients causes geologically rapid mass displacement, entailing phase instability in minerals forming the upper mantle.

Until the end of the nineteenth century, orthodox opinion was confined to the concept of stationary continents and ocean basins. This view was undermined only gradually, following the revelation by Becquerel in 1896 of an intra-atomic energy source, "activité radiante spontanée ou radioactivité" (Becquerel 1903; Steensrup and Gerward 1996). The work of Becquerel and Curie (1904) was received with some reluctance by the research establishment, until then versed in classical thermodynamics. Kelvin (1903) hypothesized that the energy emitted by radium might be input by wave vector from an external source.

Curie's demonstration that radioactivity was rock-resident was followed by Rutherford's work at Montreal, described by Chadwick (1962), pointing to an earth having an energy source effective for billions rather than merely millions of years. Hitherto perceived as rapidly cooling, hence shrinking, the Earth has since been seen as fully tectonically active, and perhaps (Keondjian and Monin 1977) still to be heating. Internal sources (radioactivity and gravitation) are at present estimated to produce some ten times the energy released in the Earth's despin under tidal braking.

Until the mid-1900s, despite evidence as to the Earth's large internal energy production, prevailing opinion continued to admit only "fixist" tectonics. Suggestions by Wegener (1915), and earlier F. B. Taylor (1910a,b), as to the possibility of continental drift, were met by geophysical data as to the apparent strength of the oceanic crust (see, e.g., Jeffreys 1952).

Gutenberg's 1949 review (Gutenberg 1951), while less unyielding, conveyed the impression that continental displacement is unlikely in view of the ancient nature of the ocean basins. Gutenberg (1936) allowed for the possibility that the continents flowed apart in Archean times, leaving stretched continental residue between them. Models including continental motion (Taylor 1910a, 1928a; Ampferer 1925, 1941; Holmes 1928a, 1944), permitting processes seen today as beyond question, were received with incomprehension and unquantifiable objections. Those introduced to the geological sciences at this juncture will recall admonitions that whereas research set to examine continental drift would not prosper, it was safely possible to examine Gutenberg's *fliesstheorie*.

The possibility that internal convection causes crustal mobility was raised as early as 1839 (Hopkins 1839) and 1883 (Fisher 1889). Fisher's proposal was offered at a time when a lasting energy source was not evident. As pointed out by Sullivan (1974), Fisher's description of the volcanic mid-oceanic plateau and sinking of cooled crust beneath Japan bears notable resemblance to postulates of plate tectonics. In consequence of measurement of the Earth's heat efflux and estimates of the viscosity (Haskell 1936, Griggs 1939, Vening Meinesz 1948), it had become evident that the critical Rayleigh number in the mantle, $R_c \approx 1500$, mandating thermal convection, is exceeded by orders of magnitude.

Gutenberg perceived gravity as the immediate cause of seismicity (personal communication), and his finding that melting point is reached at a depth of 80 km (Gutenberg 1948) suggests the development of orogenic belts through convection (Pekeris 1935; Griggs 1939; Vening Meinesz 1958; see also Griggs 1972). Adoption of a viscosity following Lomnitz's law or Jeffreys' (1976) modification of it would cause disturbance of the hydrostatic state to be damped, causing convection, if started, to die out. Jeffreys based rational reservations as to convection upon those as to isostasy. In these it was concluded that "the hypothesis of viscous flow, always tending to produce exact isostasy, is false and contradicts both the geological and the geophysical evidence." The ridged topography of the ocean floor, now attributed to convection, was perceptible as resistance to isostasy; hence strength too great to permit convection. Sixty years before this, Darwin, (1908 x), in a paper still insufficiently recognized, had concluded that "The existence of an isostatic layer, at which the hydrostatic pressure is uniform, at no great depth below the earth's surface, is now well established."

At a time of mobilist orthodoxy, it is sobering to review the reluctance of 60 years ago to examine the possibility of continental displacement now "obvious." At a 1928 symposium of the American Association of Petroleum Geologists, whose members possessed unmatched exploration data, it was concluded that the question of continental drift well merited follow-up (van der Gracht 1928; Holmes 1928a,b). Despite this, in the succeeding decades geophysical formulations having a recondite basis were accorded greater weight than the record of historical geology. The factor permitting progress was the acknowledgment of convection in the mantle. In turn, this was dependent on acknowledgment of the paleomagnetic record.

2.3.1 Paleomagnetism

The existence of a mineralogic record of the magnetic field was indicated as early as 1930 by Koenigsberger (1930, 1936, 1938). In experiments having a remarkably modern ring, McNish and Johnson (1938 a,b) examined in detail the magnetization of a core taken by the Carnegie Institution from the Atlantic floor at 46° 03′ N. It was suggested that the observed magnetization represents "the fossil magnetization of the sediments, acquired by action of the Earth's field in the process of deposition" and that "if subsequent investigation supports this belief, this method will enable a study of the history of the Earth's magnetic field in past geologic age." "Information . . . may also be derived from the study which should furnish a means of dating sediments by their magnetic orientation."

Until the investigations of Koenigsberger (1930, 1936, 1938), Nagata (1941, 1952), and Néel (1951), it could be argued that the components comprising thermal remanent magnetisation (TRM) could not survive for geologic periods. Prior to the demonstration by Nagata (1952a,b) and Uyeda (1955) that reverse TRM is associated with identifiable petrology, it was uncertain whether field-reversed specimens recorded, as seemed unlikely, reversals of the main geomagnetic field. Subsequently it became necessary to acknowledge the occurrence of geomagnetic reversals.

LeGrand (1990) has recounted that, despite the pioneering data obtained by McNish and Johnson in the Atlantic, the uncertainties were judged so daunting as to lead to the abandonment in the 1950s of paleomagnetic research at the Carnegie Institution. Difficulties adduced included uncertainty as to the role played by magnetostriction and chemical changes. Uncertainty remained, furthermore, as to whether the geomagnetic field originated in the core or was an expression of rotation of the Earth as a whole as suggested by Blackett (1952; see also Sutherland 1903).

Subsurface measurements by Runcorn et al. (1951; Chapman 1948) indicated that the geomagnetic field increases downward, a relation unlikely in the event that Blackett's hypothesis was correct. Extensive paleomagnetic observations then carried out by Creer, Irving, and Runcorn (1957) established that, despite unknowns in the acquisition mechanism, paleomagnetic directions are coherent over large areas and preserved piecewise in the clasts of ancient aggregates. Apparently to preserve credibility, it was considered conservative by most investigators to invoke polar wander in preference to continental drift as an explanation of the paleomagnetic data. Using data accumulated in the late 1950s, Runcorn (1962a) pointed out not only that diverse and coherent paleomagnetic data pointed to a separation of the continents, but that a tenable explanation required some form of convection in the mantle.

The situation consequently arose, as described by McIlhinny (1993) in a commemorative address to the AGU, in which paleomagnetic data were seen by scientists on one side of the Atlantic as having established the existence of continental drift, but were little heeded by those on the other. Vening Meinesz (1958) had identified mantle convection as the only plausible explanation of the gravity anomalies discovered by him in what is now called Sundaland, and furthermore pointed to the role, presently seen as fundamental, of the transition region in the upper mantle. The seafloor spreading concept advanced by Dietz (1961) implicitly called for a process of convection in the upper mantle. A meeting of the Royal Society in 1965 was devoted to the subject of continental displacement and mantle convection. Wilson (1965) proposed the division of the Earth's surface by mobile belts into "several large rigid plates." General acceptance of convection, implicit in the plate tectonics concept, awaited the reinterpretation of a 1956 seafloor magnetic survey (Vine and Matthews 1963; Morley and Rochelle 1964) in the 1960s.

2.3.2 Metastability Under Convection

During the period in which paleomagnetics has elucidated the motion of continents, a combination of enhanced computer facilities and the expanded seismometer network has permitted identification of the settings in which flow is taking place.

Flow is active where cooled, dense lithosphere is foundering in the mantle and in regions of complementary upflow associated with the ocean-floor ridge system. The propagation of seismic waves has permitted velocity-mapping of the interior in increasing detail (e.g., Su, Woodward, and Dziewonski 1994).

Within this setting, mineralogic research (Thompson 1992) has shown that the material comprising the upper mantle exists in tens of discrete pressure/temperature phases. There operates a mineralogic hydration ⇔ dehydration cycle (Gasparik 1993; Nolet 1994), in which volatiles including water are drawn into the mantle in subduction zones and returned to the surface in volcanicity.

Under this regime, material under vertical displacement, hence pressure change, is subject to continued phase and state transition. Besides causing mass displacement, displacing the axes of inertia, convection causes volumes of the mantle to be continually in marginal equilibrium.

2.3.3 Polar Wander

Contrasting with the case of a passive earth, the existence of mantle convection has made likely true polar wander. Because secular polar wander would explain accumulating paleoclimatic data pointing to displacement of the continents, consideration was given early to the question of axis displacement, by Kreichgauer (1926), and to its effects by Kravetz (1927), Spitaler (1930), and others. Perceived permanent strength of the deep interior including the equatorial bulge, preventing polar wander, rendered such studies hypothetical. Permanent strength of at least 1.5×10^9 dyn/cm^2 down to the core, implying the existence of a minimum yield stress as distinct from a Newtonian viscosity, was entailed in considerations summarized by Jeffreys (1952), based on an assumed permanence of surface relief and gravity anomalies of wide extent.

Based on the damping rate of the Chandler wobble, Gold (1955) pointed out that feasible mass displacement in mid-latitudes would lead to geologically significant polar wander. The instantaneous rotation axis would move toward its new equilibrium position, causing slow figure adjustment, displacing further the axis of equilibrium.

Munk and Macdonald (1960a) regarded the recently shown excess flattening of figure, if real, to be "an almost certain measure of the Earth's finite strength to resist deformation from stresses, no matter for how long a time these stresses are applied." McKenzie (1966) concluded on the basis of the excess flattening that the viscosity of the lower mantle must be so great (6×10^{26} Stokes) as to prevent both lower-mantle convection and polar wander.

In a fundamental paper, Goldreich and Toomre (1969) have found that, given convection and the small axial-moment-differences now observed, migration of the earth body with respect to an almost stationary axis of rotation is inevitable, occasionally at rates exceeding plate motion. In a rotating convective earth, migration of the equatorial bulge is not inhibitory, but essentially forms part of the convection. Limited data (Courtillot and Besse 1987) suggest that true polar wander occurs, at an erratic rate. Tectonic aspects of the reconfiguration (Bostrom 1990) have scarcely been addressed (but see recently Kirschvink et al. 1997; Richards 1997).

2.4 Status: Classical Models

To a remarkable degree, investigations have followed the direction initiated by Darwin in the 1870s. Spherical harmonic specification of the tidal bulge, formulated by Laplace and employed to advantage by Kelvin, was adopted by Love (1909) and his approach followed with some fidelity by models such as those by Zschau (1978) and Wahr and Bergen (1986). The tidal deceleration of the Earth is tracked by the Bureau de L'Heure on a daily basis.

2.4.1 Instrumental Data

Data as to the solid tides have been obtained using satellite-mounted radar, surface gravimetry, and space-based geodesy.

2.4.1.1 Altimetric Satellite. At the time of writing, comprehensive coverage of the marine earth has been achieved by the Topex/Poseidon orbiting radar (Andersen, Woodworth, and Flather 1995). Separation of the seafloor part of the geocentric distance from the ocean-surface part has been accomplished by estimation (Ray 1996). Faute de mieux, the "solid" contribution is supplied by computed models. Current results are reviewed by Andersen, Woodworth, and Flather (1995).

2.4.1.2 Work of Brussels Earth-Tides Center. Global observation of the solid tides was made feasible by the development in the 1940s of the spring gravimeter. An account of early techniques has been given by Melchior (1993). Melchior recounts that the Shell Oil Company made a 15-day tidal gravity survey at 29 stations around the world in 1949. Commencing at the time of the International Geophysical Year, the Observatoire Royal at Brussels developed instrumentation having calibrated phase and amplitude response. Observation to the sub-microgal level is required. The response must be separated from oceanic effects extending to the center of continents and those due to the fluid core.

Brussels' Trans-World Tidal Gravity Profile was initiated in 1973 (Melchior 1983), using increasingly closely calibrated gravimeters. Melchior (1994a,b) has compiled a preliminary global data base of 211 stations having documented operating procedure and calibrated response.

Results appear as a residue after subtracting the effects noted plus the response of a reference model earth. A regional variation in the tide has become apparent, at first tentatively correlated with heat flow. Melchior (1989) has pointed to the uncertainty in the model to which observations must be referred. As in the case of the Topex tidal reductions, current models assume a high-Q earth and stationary, oscillatory bulge. In some regions, Melchior (1995a) detects a disparity between the phase lag in the semidiurnal and diurnal tides and notes the disparity in the imposed strain mode. Mention should also be made of the observations by Brussels of tidal tilts. Tiltmeters produce a signal that is not entirely oscillatory, capable of directly detecting cumulative distortion.

2.4.1.3 "Space-Based" Geodesy. At the time of writing, observations of the tidal distortion referred to an external, inertial frame have become feasible through the use of the Global Positioning System (GPS) and Very Long Baseline Interferometry (VLBI). Effects of the ionosphere, atmosphere, and world ocean present a challenge to signal separation. In respect to the dissipation, space-based techniques already provide independent comparison with gravimetric data. It is perhaps worth recalling that at the symposium summoned by the American Association of Petroleum Geologists in 1928 to cope with "embarrassing" evidence of continental drift, Wegener proposed "the possibility of checking its truth by repeated astronomical observations of latitude and longitude" (Wegener 1928). Regularly observed VLBI profiles employing stellar signal arrivals (Ryan et al. 1993; this volume, section 7.4.2) now show the trans-Atlantic distance to be increasing by some 17 mm/yr.

2.4.2 Tidal Models

The assumptions forming the basis of reference or datum models, as compared with the tidal distortion of the Earth, are summarized in table 2.1 and illustrated in figure 2.3.

Table 2.1. Models of the Semidiurnal Solid Tide: Basis and Assumptions

Model Type	(P)assive or (M)eta-stable Earth	Potential	Forcing stress	Period of tidal forces	Deformation, distortion	Dissipation mechanism	Cumulative displacement	Reference
1; bulge	P	Stationary relative to Earth	Reverses (oscillatory)	1/2 day	Oscillatory shear	Oscillatory pure shear; Q_{seis}	Nil	1;2;3;4;5;6;7
2; bulge	P	Same as 1	Same as 1	as in 1	As 1, + tangential (laminar) flow under the retarding torque	Same as 1	Insignificant, under retarding torque	Darwin's mechanism; 8,9,10
3; wave	M	Rotates relative to Earth	Nonreversing force-couples	Secular	Cumulative vortical flow	Vortical flow: viscosity & transitions	Unlimited, as for convection	11;12;13;14;15; 16;17

References: 1, Love 1909; **2,** Jeffreys 1924; **3,** Munk and MacDonald 1960, their eq. 11.7.1; **4,** Zschau 1978 62–65; **5,** Wahr and Bergen 1986; **6,** Dehant 1986; **7,** Dehant and Zschau 1989; **8, 9,** Darwin 1879, 1908; **10,** Tomaschek 1957; **11,** Danes 1973; **12,** Bostrom 1978a,b,c; 1981; **13,** Dubrovskiy 1985; **14,** Kosygin and Maslov 1986, 1989, 1990; **15,** Revuzhenko 1991; Revuzhenko et al. 1983; **16,** Bobryakov et al. 1983, 1991; **17,** Maslov and Noltimier 1993.

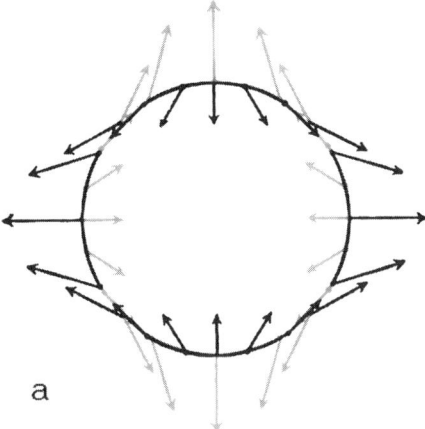

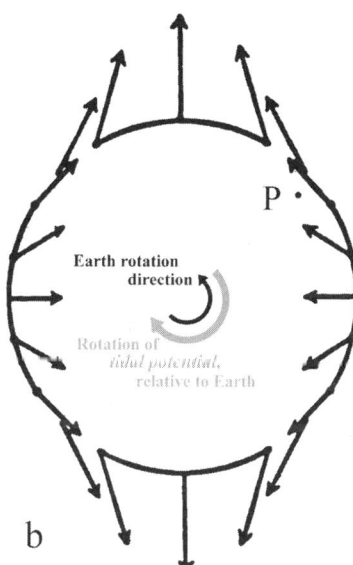

Fig. 2.3. Tidal models (c.f. table 2.1). (a) Types 1, 2: *Tide-causing external potential is stationary reference to the Earth* (*rotates with it*). Viewed from over North Pole. These models assume sinusoidal temporal variations in the magnitude of the potential (as by variation in the Earth-Moon distance). The force system oscillates directly between the black and gray states, passing through zero. Vectors after drawings by Sir George Darwin. (b) Type 3: *Actual situation; in these models, the Earth rotates relative to the tide-causing potential.* Stationing one's self at any point P fixed in the Earth, it will be observed that as the Earth rotates counterclockwise, the force system rotates clockwise, without reversal. "Bulges" permanently extant migrate east to west, existing as waves.

28 Tectonic Consequences of the Earth's Rotation

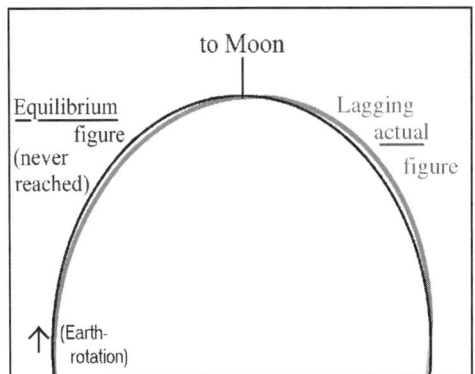
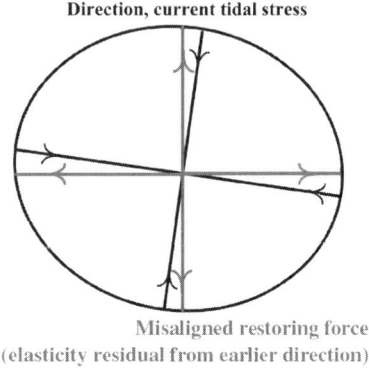

Fig. 2.3. Re-configuration of the tidal Earth, under rotating potential. (*Left*) Tidal figure continually adjusts, away from its lagging instantaneous form toward equilibrium in the joint field of the Earth and Moon. Its equilibrium figure (long axis aligned with the Moon) is never attained, because rotation is ongoing. In respect to internal stress, "actual" figure includes displaced oceanic masses. The principal solid tides, M_2 and S_2, represent waves perpetually in existence, moving around the Earth. (*Right*) Tidal stress-ellipsoid seen in section. It perpetually rotates relative to the Earth, never reversing in direction. The **black** forcing stress leads the **gray** residual opposing force. The latter lags due imperfect elasticity, of any kind. In the fashion that a geostationary potential creates body *forces*, the rotating potential brings into existence body *couples*, equally pervasive. As the phase lag is minute relative to the period of the wave, at any point the sense of the couples is unchanged whether the stress is increasing or decreasing (bulge forming or relaxing).

Being a result of the continuous re-configuration, the couples are referable to internal axes. The unrecovered distortion and dissipation create the phase-lag, hence the external torque accelerating the Moon and retarding the Earth. The retarding torque produces tangential stress 10^{-8} of the stress imposed by wave-passage, and insignificant displacement.

The geometry of the strain ellipsoid varies in oscillatory fashion, but its defining vectors rotate without reversal relative to particle lines. Couples evolve at all times, depths and localities. Displacement is thus cumulative and in the vortical mode identified by Helmholtz, independent of translation and extension. Having a separate existence, the system fosters only circulation devoid of translation.

Estimable models by Darwin, Love, and successors, employing a sinusoidally varying geostationary potential, replicate neither the principal displacement nor characteristic passage of the major tides around the Earth; but only geographically-stationary bulging, oscillatory displacement, and secondary dissipation. In them the black and gray force vectors lengthen and shorten without rotation, generating no secular couples.

Colloquially, a surface observer does not see the Moon eliciting and releasing a stationary bulge; instead, a wave accompanies the Moon around the Earth, passing the observer. Notably, the Moon does not act so as to "drag bulges," or any object, around the Earth.

Current models (Zschau 1978, 1986a,b; Zschau and Wang 1986; Wahr and Bergen 1986; Dehant et al. 1991) follow the classical formulation of the mathematician A. E. H. Love (1909). This axiomatically assumes an earth stationary with respect to the external, tide-causing potential. Under the influence of tidal forces thus defined (figure 2.3a), the model earth develops bulges, likewise stationary, on the Moon and anti-Moon sides. The potential and hence the deformation are reversed half-daily, the latter approximating the form of a spheroidal eigenvibration.

A Pole-centered projection (figure 2.3b) makes clear that in actuality the forcing system rotates without reversal, defining Type 3 (wave) models. The tidal stress ellipsoid, figure 2.3c, whose static form was formulated by Takeuchi (1951), Melchior (1983, his eqtns 5.23) and others undergoes complicated latitude- and depth-dependent evolution; subjected to Earth rotation, at all times and at all depths progression of the ellipsoid is accomplished via rotation, in one direction only. In the figure the black axial vectors perpetually lead the gray vectors denoting the restoring force. Adjustment can conveniently be visualized as continual Earth re-configuration towards equilibrium. Re-configuration (figures 2.3c, 2.4) induces cumulative rotational (non-potential) flow or circulation, distinct sui generis from oscillatory deformation under a geostationary potential.

Danes (1973) has modeled a peristaltic pump, formed in the low-viscosity channel bounded by the lithosphere and more rigid inner Earth. Vorticity-induction by wave tides, and kinship to flow under buoyancy, were identified by the writer (Bostrom 1978a,b,c; 1981a). Revuzhenko, Bobryakov, and Shemyakin (1983), and Dubrovskiy (1985) have taken up the action of wave tides. Bobyakov, Revuzhenko, and Shemyakin (1991) have examined distortable physical models subjected to wave-passage. Seeking to account for geotectonic phenomena, Kosygin and Maslov (1986, 1989, 1990) and Maslov and Noltimier (1993) have considered the action of Danes's pump, wave tides introduced by the writer, and computational models (Revuzhenko 1991).

Reconstructions of the orbital history of Kuiper's Earth-Moon double planet (Kuiper 1955; Mignard 1981; Touma and Wisdom 1994) have generally been based upon Kelvin-Darwin-Love (oscillatory) models of the solid tidal dissipation. Discussion has centered upon the relationship of the dissipation (phase lag) to a finite tidal frequency. The assumption of an unreal synchronous rotation of the Earth with respect to the system barycenter excludes any effects of zero-frequency vortical dissipation.

To trace this historically, in pioneering researches as to tidal deformation of a viscous spheroid, Sir George Darwin (1879a,b; 1908) based his approach on the formulation of Lord Kelvin (Sir William Thomson). Kelvin modeled an elastic earth. Darwin (1908, vol. 1) was greatly influenced by him: "Early in my scientific career it was my good fortune to be brought into close personal relationship with Lord Kelvin. Many visits to Glasow and to Largs have brought me to look up to him as my master, and I cannot find words to express how much I owe to his friendship and to his inspiration."

In comprehensive records, Darwin clearly defines his assumptions in text as well as algebraically (figure 2.5). His formulations were set forth prior to acceptance of the summation convention, abbreviating many-dimensional expressions,

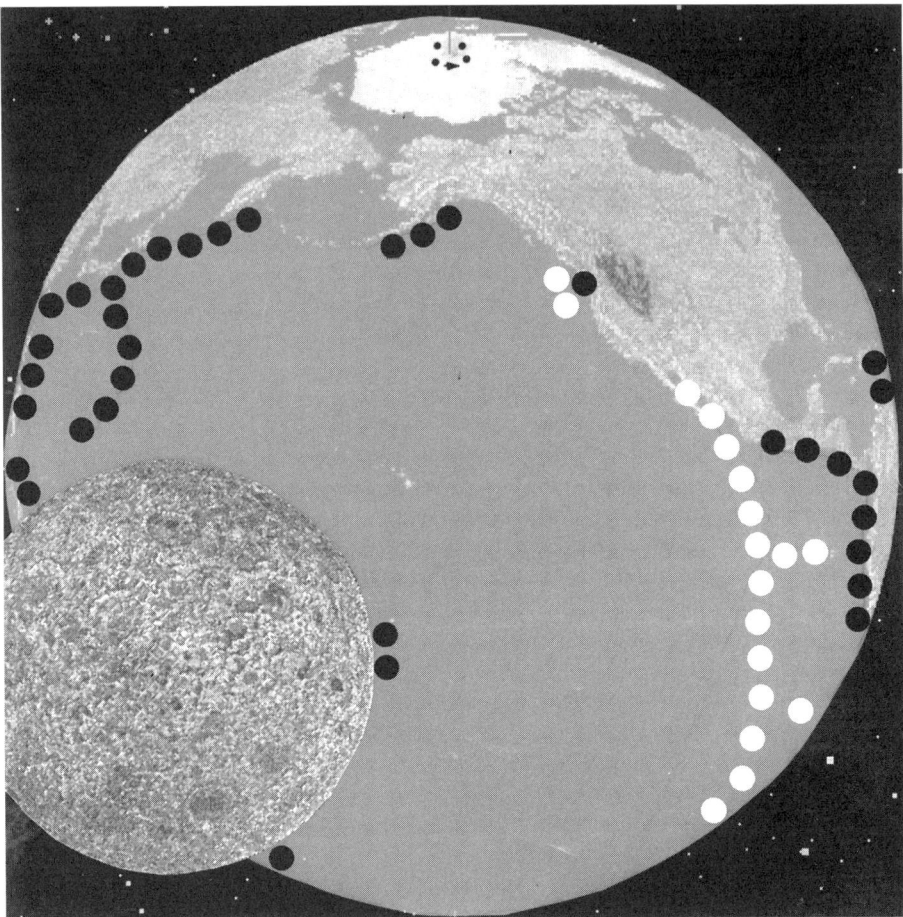

Fig. 2.4. Tidal models; type 3 ("geological Earth"): bulges always in existence, moving east to west around the Earth as waves. The lunar tidal wave of a few tens of centimeters passes a surface point 7.3×10^8 times/million years. Under imperfect elasticity distortion is vortical (namely, elemental circulation), and cumulative. The Earth is affected independently by flow in this mode under thermal convection. Due to it, volumes of the upper mantle are permanently at the point of phase and state transition. *Black circles*: Region of continuous pressure increase under subduction, containing material ready to invert to more stable phase under positive leg of tidal pressure pulse. *White circles:* Regions of upflow, sensitive to negative pulse-leg. (Orthographic projection from a point slightly outside the lunar orbit; 1700Z hr, 28 August 1994. Lunar "far" side: NASA image.)

> this inquiry joins itself on to that of my previous paper.
>
> In that paper it was shown that, if the influence of the disturbing body be expressed in the form of a potential, and if that potential be expressed as a series of solid harmonic functions of points within the disturbed spheroid, each multiplied by a simple time-harmonic, then each such harmonic term raises a tide in the disturbed spheroid, which is the same as though all the other terms were non-existent. This is true, whether the spheroid be fluid, elastic, viscous, or elastico-viscous. Further, the free surface of the spheroid, as tidally distorted by any term, is expressible by a surface harmonic of the same type as that of the generating term; and where there is a frictional resistance to the tidal motion, the phase of the corresponding simple time harmonic is retarded. The height of each tide, and the retardation of phase (or the lag) are functions of the frequency of the tide, and of the constants expressive of the physical constitution of the spheroid.
>
> Each such term in the expression for the form of the tidally distorted spheroid may be conveniently referred to as a simple tide.

Fig. 2.5. Darwin's textual specification (1879a, 448; 1908, 37), of his investigations of the tidal deformation of viscous spheroids. His "previous paper" deals with the bodily tides of viscous and semielastic spheroids, based on Kelvin's investigation of the elastic case. "Frictional resistance" hence phase lag are generated in the course of imposing time variation of intensity, of an external potential stationary with reference to the Earth. The models of Love (1909, 1927) are similarly bounded.

in full laborious to read. Kelvin had been interested principally in the question of the Earth's rigidity in response to a known force. He formulated an external potential (1875a,b), akin to that conceived by Laplace, in terms of which one may model the deformation to be expected in the form of superimposed spherical harmonics (Laplace's "coefficients"). This is legitimate on the special supposition of a potential stationary reference to the Earth, and generally in the case of elasticity.

Thus, extending Kelvin's approach (Kelvin and Tait 1895) Darwin modeled dissipation as a result of introducing a time term in the forces opposing the derivative of the disturbing potential. Parenthetically, he employed a viscosity having the form of Maxwell's recently formulated "modulus of the time of relaxation of rigidity," representing the time in which the initial stress has declined to e^{-1} of its initial value. Darwin's text and algebra make it clear that throughout his research he employed a potential fixed in orientation relative to the Earth, as responsible for the principal tidal deformation. In fashion of which the significance has not been conspicuous, the assumption of stationarity makes it axiomatic that his model limits the primary deformation and dissipation solely to the action of a function of the potential $V(x,y,z)$ of form $\nabla^2 V = 0$.

Shortly before this, Helmholtz (1858) had identified motion ("vortex motion") extraneous to a velocity-potential, writing it as the difference in shearing in orthogonal directions (u, v, velocity): $\omega_z = (du/dy - dv/dx) \neq 0$. The importance of Helmholtz's concept was recognized at the time. A translation by Tait, reviewed by Helmholtz himself, was published in *The Philosophical Magazine and Journal of Science* (Helmholtz 1858). Kelvin took much interest in vortex motion (Kelvin 1867), but would see no reason to incorporate it in the deformation of a purely elastic earth. Kelvin and Helmholtz were well acquainted, taking kindred pleasure in observing the velocity of gravity waves in water (Kelvin 1871): "About three weeks later, being becalmed in the Sound of Mull, I had an excellent opportunity, with the assistance of Professor Helmholtz, and my brother from Belfast, of determining by observation the minimum wave-velocity with some approach to accuracy ... The fishing-line was hung at a distance of two or three feet from the vessel's side ... The speed was determined by throwing into the sea pieces of paper previously wetted, and observing their transit across parallel planes." In a letter, they reported to *Nature* the "extreme closeness of this result to the theoretical estimate (23 centimetres per second)." In formulating vortex motion, Helmholtz originally conceived of an inviscid fluid, and at this time Kelvin's interests were focused on the possible existence of a "luminiferous ether."

In Darwin's time, literature concerning the elastic tides was already extensive (Darwin 1908, 238). As noted, his algebra and text specify that he confined simulation of tidal deformation to the effect of a geostationary time-varying spherical harmonic. Under "a frictional resistance to the tidal motion, the phase of the corresponding simple time harmonic is retarded." Misalignment of the bulge ensures creation of a retarding torque and minor secondary dissipation, under the tangential force.

Darwin searched specifically for nonreversing tidal forces other than the retarding torque: "As however, it was just possible that this general method of regarding the subject overlooked some residual tendency to secular distortion, I have given the subject a more careful consideration. From this it appears that there is no other tendency to distortion besides that arising out of tidal friction, which has just been discussed." In terms of his model he conducted an efficient "careful consideration," by means of expunging time-periodic terms in his formulation of the stress, preserving any (his couple $\mathcal{N}$, representing the retarding torque) tending to produce cumulative displacement. However, it is apparent that in deriving dissipation solely from a suppositional oscillatory stationary potential, Darwin's solution excluded Helmholtz's integrals of the second class, that describe the intrinsic, nonreversing primary distortion, and primary dissipation.

Writing in 1923, Eddington concluded that whether Darwin's attribution of the lag and retarding torque (to interior viscous dissipation) or Jeffreys' (to dissipation in shallow seas) was correct, the retarding torque acts as a surface-brake, tending in laminar flow to displace the outer layers of the Earth westward over the inner. As well realized (Munk 1972), in both cases the surface tangential torque is of a magnitude $\approx 10^{-4}$ dyn/cm^2, which can result in only insignificant displacement and dissipation. Its feebleness and the apparent thoroughness of Darwin's investigations were the rationale in Jeffreys' opposition (Jeffreys 1926, 1929) to Joly's proposal (Joly 1925, 1928), that we should look further into a joint effect, of tides and fusion under radioactive heat accumulation.

The surface torque contrasts with the stress created in bulge (wave) passage,

≈ 10^{+4}dyn/cm^2 (Takeuchi 1951; Bostrom 1976; table 3.2). Dissipation under the action of the surface-brake must be an insignificant fraction of the solid-Earth total. As a result of his observations of galaxy, nebula and solar-system formation, Eddington was much aware of the central role of rotation in cosmogony. Subsequent to the time of Helmholtz, it has been shown (Noether 1918) that angular momentum and kinetic energy are conserved independently of their translational counterparts, alone induced under a stationary potential.

2.4.2.1 Surface Motion. An extract from Helmholtz's 1858 paper is reproduced as figure 2.6. Vortical motion under the couples earlier described (figure 2.3c) is

> of magnetic matter outside or at the surface of the fluid must be taken so as to satisfy the conditions at the surface. Each magnetic mass can also, as we know, be replaced by electric currents. Thus, instead of using for the values of u, v, w the potential-function P of an external mass κ, we get quite as general a solution if we give ξ, η, and ζ outside of, or at the bounding surface of, the fluid any values such that only closed current-filaments exist; and then the integration in (5 a) must be extended to all space in which ξ, η, and ζ are different from zero.
>
> § 4.
>
> In hydrodynamic integrals of the first class it is sufficient, as I have shown above, to know the motion of the surface. By this the whole motion in the interior is determined. In integrals of the second class, on the other hand, the motion of the vortex-filaments in the interior of the fluid must be found with reference to their mutual action, and with attention to the conditions at the surface, by which the problem becomes much more complicated. Even this problem can be solved in certain simple cases—namely, when rotation of the fluid elements takes place only in known surfaces or lines, and the form of these surfaces or lines remains unchanged during the motion.
>
> The properties of surfaces bounded by an indefinitely thin sheet of rotating elements can be easily deduced from (5 c). If

Fig. 2.6. Explication by Helmholtz (in Tait's "certified" translation of his 1857 paper) of the nature of his "integrals of the second class," specifying vortex motion. In this he points out that internal motion conforming with these is indeterminate purely from surface motion. His integrals of the first class, determinable in terms of surface motion, describe irrotational flow under the spatial derivative of a potential. The preceding text refers to his analogy between induction and flow of electrical current (see Lamb 1945), and induction of material vortical flow. Helmholtz carefully pointed out that earlier, Euler had identified cases of fluid motion in which no velocity-potential exists.

devoid of a velocity potential, placing it within his category "integrals of the second class." It appears that, in general, rotation of a potential with respect to a planet of imperfect elasticity must create couples resulting in "Helmholtzian" vortical motion (rotation of elements relative to the planet itself; under self-organization, culminating in circulation). With benefit of hindsight, it is apparent that it is axiomatically impossible to model tidal distortion of the Earth using as input a stationary potential, devoid of a rotational dimension.

To date, despite Helmholtz's explication (cf. also Stiller 1971), it has been assumed that tidal displacement can be determined solely from the Earth's surface motion. Immediately we admit rotation with respect to the tidal potential, it becomes evident that the motion in the interior cannot be determined from the surface motion; that is, in terms of the characteristic numbers h and k of Love (1909, 1927). Like Darwin, Love (1911, section 59) modeled tidal deformation solely in terms of a stationary potential. This falls into Helmholtz's domain "integrals of the first class"; there exists a velocity potential of which the Laplacian sum of the

Fig. 2.7. (Left) Hermann von Helmholtz, 1821–1894. Primarily a physician, Helmholtz was responsible for nonmedical achievements of the first rank, in present connection formulating the conservation of energy (1847), and rotational flow (1858). Kelvin (Koenigsberger 1906) pointed to Helmholtz's identification of the electron (Kragh 1993), rejected by Faraday and Maxwell, as opening the way to "magnificent harvests of new and astonishing truth ... splendidly illustrated by Becquerel's discovery of radioactivity," and forecasting the electromagnetic waves of Maxwell and Hertz. Portrait from Koenigsberger. (Right) Lord Kelvin (Sir William Thomson), 1824–1907. Kelvin constructed models of an elastic tidal earth, with a view to determining its rigidity ("that of steel"). To do so, he expertly mobilized Laplace's formulation of the potential, constructed of spherical harmonics. His portrait (MIT, 1987) is reproduced by kind permission of the Massachusetts Institute of Technology.

Fig. 2.8. (Left) Sir George Darwin, 1845–1912. Darwin extended Kelvin's tidal formulations to the case of viscous spheroids and the Earth-Moon interaction. Restrictions on his Earth models, imposed by age-estimates by his friend and mentor, Kelvin, were removed by Becquerel's revelation of "activité radiante spontanée." His portrait is reproduced by courtesy of the Royal Society. (Right) Professor John Joly, 1857–1933. Joly may have been first (1903) to perceive the impact on the Earth's age of Becquerel's revelation of radioactivity, in providing escape from age limits imposed by Kelvin's estimates of cooling rates, and in suggesting the possibility of fusion and flow (Royal Society archive).

second derivatives is zero. Love's mechanism cannot induce cumulative rotational flow. As Gibbon might put it, restricting ourselves to Love models we seek in vain the passage around the Earth of the primary tidal bulge.

For these reasons, observed complex values of the Love numbers provide a correct measure of the dissipation, but not of the responsible mechanism. Love's oscillatory mechanism has compelled us to employ a high-frequency (semi-diurnal) elastic quality factor in modeling, rather than one of indefinitely longer period, applicable to steady stress. Failing evidence to the contrary, it must be supposed that dissipation in vortical mode, as this represents the primary tidal distortion, constitutes its major part.

It seems logical to conclude that Kelvin, Darwin, Love, and successors subsumed purely oscillatory models of the bodily tide from the Laplacian origins of this subject, the tides of the world ocean. Nevertheless, unlike the ocean tides, the bodily wave tides are not blocked at the continental margin, but proceed unimpeded around the Earth. Failing recognition of the distinction, the paradoxes in tidal theory noted by Michelson (1974) cannot be resolved. Insomuch as cumulative distortion of the entire mantle induces zero-frequency displacement, it seems unsafe to suppose (e.g., Christodoulidis et al. 1988) that the bodily tides have perturbation frequencies identical to those of the ocean tides. Remarkably, Kelvin (1849) described a means of modeling the motion of a fluid under deformation of the bound-

Fig. 2.9. (Left) Professor Augustus Edward Hough Love, 1863–1940. Love extended the mathematical work of Darwin, constructing models of the bodily tide and identifying what have become known as "Love numbers," used in describing the response of the Earth to forces thought to be known. At Oxford, remembered as a formidable practitioner of croquet. His portrait is reproduced from obituary notices of Fellows of the Royal Society. (Right) Sir Arthur Eddington, 1882–1944, "the most distinguished astrophysicist of his time" (Chandrasekhar 1982). His interests extended from relativity to terrestrial tidal forces. Chandrasekhar draws attention to Eddington's remark *in the year 1920*, stemming from his work "The Internal Constitution of the Stars" and identification of the energy source of the Sun: "it seems to bring a little nearer to fulfillment our dream of controlling this latent power for the well-being of the human race—or for its suicide." From a drawing (1933) by Augustus John, reproduced by courtesy of the Master and Fellows of Trinity College Cambridge.

ing surface that would be applicable to motion within the Earth—provided the semidiurnal tides were modeled correctly as waves, rather than as a stationary spheroidal deformation.

Formative contributors in the eclectic field of astronomy, mathematics, the tides, geophysics, rheology and geology are shown in figures 2.7 to 2.10.

Of the personalities illustrated, the interests of Eddington most nearly encompassed the field. Eddington seemed to suspect that excluded from the algebra of Darwin and Jeffreys, there lurked a term essential in specifying the bodily tides. In hindsight, it would appear that omitted antisymmetric terms may be responsible for much of the dissipation. Although governing a primary energy flux, as little attention currently is being given to Eddington's concerns as formerly was devoted to paleomagnetism and mantle convection (LeGrand 1990; McElhinny 1993). Research as to the secular distortion seems at the moment to be confined to the more

Fig. 2.10. (Left) Sir Harold Jeffreys, 1891–1989. His country's foremost geophysicist. His data as to the permanent strength of the Earth brought him into apparently inevitable opposition to Joly's hypotheses (1928), which would now be termed "mobilist." His portrait kindly lent me by Lady Jeffreys. (Right) Professor Stanley Keith Runcorn, 1922–1995. Perhaps more than any of his contemporaries, Runcorn perceived the central importance of establishing the existence or otherwise of convection in the Earth's mantle; he identified, furthermore, means of observing the resulting surface-displacement: palæomagnetism.

remote parts of eastern Siberia (Kosygin and Maslov, 1986, 1989, 1990; Revuzhenko 1991; Maslov and Noltimier, 1993).

2.5 Present Stage

2.5.1 Instrumental

Due to remarkable effort on the part of instrument designers and the Observatoire at Brussels, the problem of detection of the periodic tides has largely been overcome. The observation frontier is now presented by lack of a reference-model or datum; and no less important, in compensating for the effect of the world ocean.

The correction for oceanic effects (attraction and loading) is so large as to mask gravimetric values as far as the center of the continents. At present no cost-attainable means is in sight for mapping this factor (chapter 7). The oceanic problem is so formidable because it confronts an *interaction* (plate 1, frontispiece); in the heterogeneous Earth we have little idea of the contribution of the bodily tides to the marine tides—and vice versa.

A perceived limitation in global gravimetric observations is that these have yielded values so large and regionally variable as to have attracted scepticism. Values have been referred, apparently universally, to Darwin-Love models which omit the primary tidal action. The electronic tiltmeter (chapter 7) offers hope of instrumental observation of vorticity induction, if emplaced in deep, expensive bore-holes.

2.5.2 Geological

Orthodox tidal models (table 2.1, type 1) impose a geographically stationary bulge on an inert earth. Their action bears little relation to the passage of waves inducing vorticity.

In respect to sustained external forces, the Earth is strengthless; internal material is perpetually unstable and at "failure" point, under convection representing vortical flow already in existence. Those of us interested in tectonics have scarcely progressed since the time of Kelvin and Darwin. The Earth may have "the strength of steel" with respect to short-term, periodic forces. In the case of forces having infinite duration, the description is without meaning.

Despite Eddington's exculpation, the astronomer and geologist may be forgiven for supposing that in tidal models we have disregarded not only the Earth's orbital elements and the principal displacement mode; the most fundamental characteristic of our planet is instability, continually created by the "activité radiante spontanée ou radioactivité" of Becquerel and Curie—now perceived to dominate tectonics.

The multifarious processes comprising convection must take place in the presence of unremitting, unidirectional forces of the kind identified by Helmholtz. It will be found that under the effect of stress diffusion, these may be comparable with the minute torques which, integrated throughout large cells, drive buoyancy convection. How significant may this be?

In what follows, contemporary models, based on an inert earth, are first reviewed (chapter 3). An attempt is subsequently made (chapters 4 and 5) to identify the action of the bodily tides in an earth more nearly "geological."

3

Tidal Action in a Uniform, Inert Earth

Static Bulge, Mobile Bulge

3.1 Overview

The solid Earth is elongated some tens of centimeters in the direction of the Moon by the Moon's attraction and the counterbalancing orbital force. The elongation is evident as tidal "bulges" on the Moon and anti-Moon side of the Earth.

We discuss tidal models of two types. In the first, (section 3.5), deformation takes the form of an oscillating bulge. This is geographically stationary and the Earth is entirely passive, in the sense that it is devoid of internal dynamic processes. In the second type (section 3.6), a similarly passive earth rotates in the gravity field of the Moon, causing the "bulge" to take the form of a wave, passing continually from east to west. Its action, intrinsically asymmetric, is different in kind from that in the first type, imposing a cumulative distortion. In each type, bulges produced by the Sun are superimposed.

Dissipation due to internal friction is signaled as a lag in development of the bulge. The misalignment of the bulge and the Moon forms a couple that slows the Earth and expands the lunar orbit. The couple varies as the inverse sixth power of the lunar distance. The dissipation in the Earth including the oceans is known astronomically, amounting to some 3 TW. As a scale, the energy released in global seismicity is about one-tenth this figure, and the escape of internal heat some ten times greater.

The tides in the world ocean, and of this the shallow seas, are thought to account for most of the dissipation. These interact with those in the solid Earth. To apportion the dissipation it is necessary to quantify either, but preferably both. Unlike the solid tide, the marine tides are solely oscillatory, their motion blocked by the continents. Lengthy but scattered seaport records have been supplemented by almost synoptic observation of the sea height, using radar altimetry. Separation of the seafloor (solid-earth) part of the motion is at an early stage.

The difficulty may be summarized in terms of the uncertainty in the elastic quality factor Q. Estimates range from about 13–270, the dissipation varying inversely. Low-Q estimates are supported by difficult gravimetric observation and to some extent by early VLBI data. High estimates, indicating smaller tidal action, are yielded by models assuming a static bulge and oscillatory deformation. Models of the first type exclude a zero-frequency term. In the following, for intelligibility it has been necessary to recount the assumption in various dissipation mechanisms.

Observationally, with the exception of tiltmeters, instruments such as gravimeters sensitive to the solid tides produce solely scalar, oscillatory output. Being insensitive to vortical motion, dissipation is signaled as the phase lag without vectoral attribution. With respect to models of the second type the anelastic displacement contains a cumulative part, as a result of the bulge or wave always passing in one direction. The distortion represents vortical tectonic "flow." In this chapter the vorticity is referred to earth-fixed axes.

Until the present, dissipation under the solid tide has been modeled as noted on the basis of a uniform (radially symmetric) Earth, drawn in figure 3.1. Nevertheless seismology has made it evident that the interior is mechanically inhomogeneous. More fundamentally, the Earth is affected by endogenous convection, itself representing flow in vortical mode. Thus, in reality, tidal distortion is imposed on an earth in which volumes are perpetually at the point of yield and minerals at the point of phase and state change. An overview of tidal action taking some account of these factors is postponed until chapter 4.

3.2 Tidal Deformation

Under the external gravity potential, the world ocean may be expected to exhibit tides of the three species identified by Laplace, shown in figure 2.1. As a first approximation Laplace took it that under the influence of dissipation the ocean assumes its equilibrium form. Proudman (1960) concluded that, in respect to tides of very long period (19 y), this condition prevails; that in respect to the semiannual and annual tides, equilibrium is probably reached; but that tides of shorter periods are unlikely to follow the equilibrium law.

Love (1911) described the deformed solid Earth in terms of the radius change δr and the observable disturbance in gravity. If V_o is the potential including that of the rotation before disturbance, V_t is the tidal potential, and V_d the potential of the displaced mass, the potential of the whole is

$$V = V_o + V_t + V_d - \delta r.g.$$

From this, $\delta r = h(V_t/g)$, and $V_d = k.V_t$, where h and k are constants of proportionality referring to the displacement and the potential. As described by Melchior (1983), Love's numbers h and k have been determined via techniques employing tiltmeters, gravity observations, and extensometers. These yield linear combinations of h and k. A kindred constant, Shida's number l, describes lateral displacement. Since Love's time, his numbers have been viewed as the result of imposition of a geostationary disturbance, as distinct from passage of a wave.

Tidal Action in a Uniform, Inert Earth   41

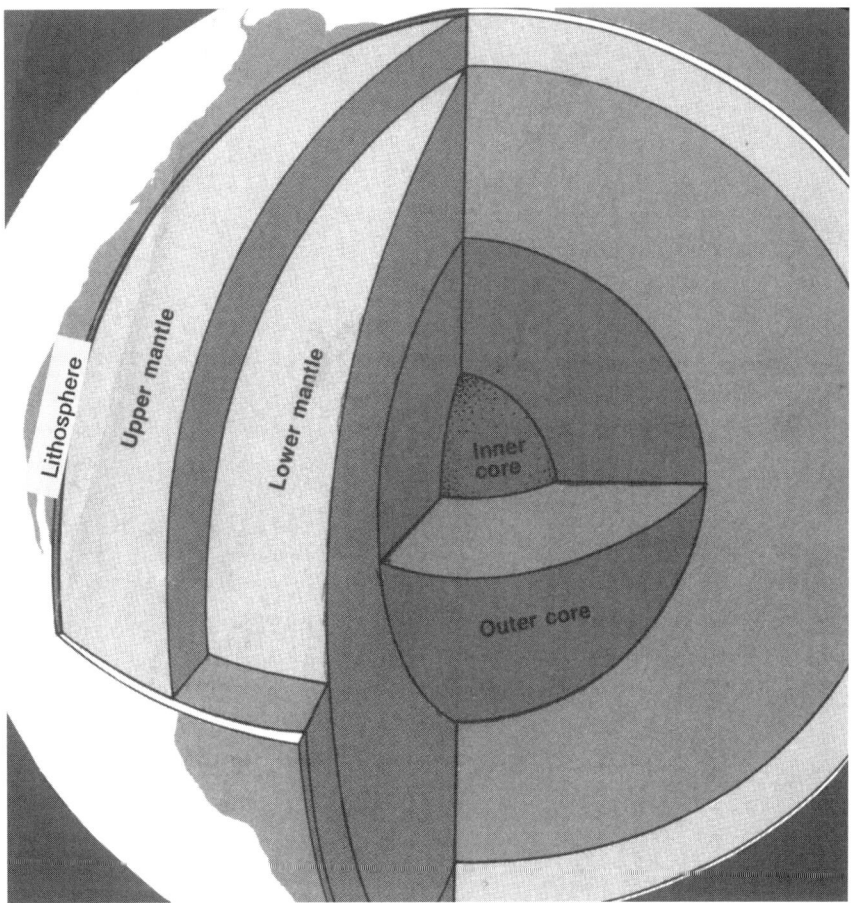

Fig. 3.1. Uniform (radially symmetrical) passive Earth, basis of orthodox solid-Earth tidal models. Average thickness of lithosphere is 75 km. Base of underlying upper-mantle shown at 600 km. At scale, the world ocean is film that is not visible.

Specimen values of the Love numbers h and k, based principally on latitude tides as being least subject to oceanic effects, are $h = 0.58$ to 0.64; $k = 0.23$ to 0.32 (Bonatz 1978). Love numbers of degree higher than 2 reflect yield in the shallower part of the Earth. The lunisolar range in δr at the Equator is 78 cms, and geographically variable. The range in gravity is 0.24 mGls.

In 1950, Takeuchi (1951) computed the static tidal deformation of the Earth based on seismologically bounded, radially symmetric density and elasticity. Melchior (1983) has developed expressions for the stress and strain within an earth constrained by seismologic values of the elastic constants and density. Wahr (1981) has computed the body tides of an elliptical, elastic, oceanless earth. More recently, Wahr and Bergen (1986) have examined the effect of anelasticity.

Earth cannot be perfectly elastic. Due to relaxational processes, the value of

Love numbers must vary with the period of the tidal species. Observations are additionally affected by attraction and loading effects of the marine tides and by resonances. In investigations important in the present regard, Zschau (1978) and Dehant (1987a,b, 1995) have expressed the imperfection in elasticity in terms of time-dependent Love numbers.

3.3 Total Tidal Dissipation

Astronomic observation of satellite orbits, the expansion of the lunar orbit, and the deceleration of the Earth (Munk and MacDonald 1960a,b; Mulholland 1980) delimit the total tidal energy dissipation, taking place "somewhere within the Earth," including the oceans. In detail, difficulty is met in distinguishing ephemeral events from trends longer than the observation series. Long-term, acceleration resulting from phenomena such as change in the Earth's inertial moment must be separated from that due to tidal braking.

With respect to internal processes, satellite observation of the geocentric height of the ocean surface (Cartwright and Ray 1991; Schrama and Ray 1994) has made earlier dissipation estimates obsolete. Schrama and Ray (1994, table 5) estimate that the dissipation in the tides ($M_2 + S_2 + O_1 + K_1$) is

$$3.3 \text{ TW} = 3.3 \times 10^{12} \text{ W} = 3.3 \times 10^{19} \text{ erg/s}$$

of which an unmeasured part is attributable to the solid Earth. The rate tallies within limits of accuracy with astronomic estimates based on the motion of satellites, including the Moon (Mulholland 1980; Lambeck 1988).

Long-term observations of the total dissipation have taken the form of daylength changes recovered from historically-recorded astronomical data (Munk and MacDonald 1960; Stephenson 1978; McCarthy and Babcock 1986; Lambeck 1988; Pang and Yau 1994). A major correction is required (Pang and Yau 1994) to account for the decrease in flattening, thence in inertial moment, therefore in despin, subsequent to Quaternary deglaciation. The record has been extended in the form of growth-lines preserved in organisms dating back 400 Ma, to the Paleozoic (Wells 1963; Scrutton 1978; Lambeck 1979) and earlier times as patterns in tidal rhythmites (Williams 1989, 1997). The length-of-day (l.o.d.) has been found to increase at the rate of about 2 ms per century (Lambeck 1988). Within accuracy limits, l.o.d. changes accord with the dissipation represented in lunar laser-ranging data and the record of altimetric satellites.

From the point of view of tectonics, an objective must be to separate the marine from the solid-Earth portion of the dissipation.

3.4 Dissipation: Marine Component

The world ocean amounts, by volume, to a film or coating on the surface of the planet (figure 3.1) but is held responsible for most of the dissipation. The geocentric sea-surface height has been mapped over almost the entire marine part of the

Earth by altimetric radar satellite (Schrama and Ray 1994). The geocentric sea-surface height contains the combined aqueous plus seafloor tide.

Long-term tide-gauge records, which measure ocean tide minus seafloor tide, are available at seaport localities, mostly unrepresentative of the open ocean. Locally, a combination of tide-gauge and altimetric data (Mitchum 1994; Molines et al. 1994) can assist in separation of the seafloor and marine portion. Seafloor observations using pressure-recorders (Ray et al. 1994) may in time effect the separation, but are dependent upon formidable logistical effort.

Based on the altimetric data, Cartwright and Ray (1991) have established that the energy input in respect to the seafloor plus aqueous tide is strongly concentrated in low latitudes (figure 3.2). The dissipation factor $Q_{tidal} \cong 13$ with respect to the oceanic two-thirds of the Earth, is corroborated within the confines of geoid model GEM-T2 (Marsh et al. 1990). The latter refers to the oceanic plus continental Earth and is based on observed orbital parameters, independent of altimetric data.

Correlation is not evident between the low-latitude work-input and the scattered shallow-sea and shelf areas credited with the bulk of the dissipation. In this connection, Cartwright and Ray (1989) note that "the really disturbing fact is that the few direct evaluations of what is thought to be the principal dissipation mechanism, namely, friction on the beds of shallow seas, have always fallen short of the supposed [orbitally derived] global dissipation rate."

The situation is taken up, in light of later developments, in section 3.8, and in chapter 7.

3.5 Solid-Earth Dissipation: Models that Assume Bulge Is Geostationary

On-land gravimetric observation of the solid tides contains information as to the internal dissipation. Unfortunately, in terms of measurement the observation also

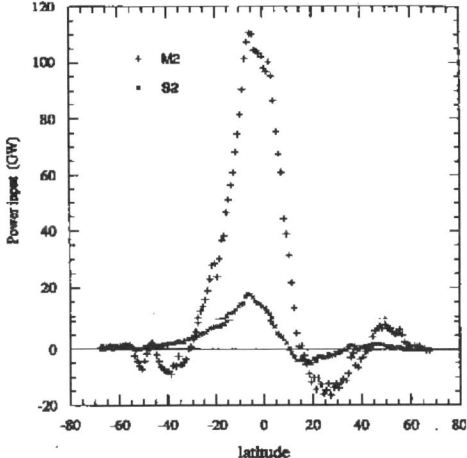

Fig. 3.2. Power input (rate of working) in the oceanic part of the Earth under the principal semi-diurnal tides, M_2, S_2, reproduced from Cartwright and Ray (1991). Summed in zones (for each latitude). Input is measured in terms of geocentric height of ocean surface, read by radar satellite. The observed quantities incorporate the seafloor tide, and therefore must be corrected by estimation for motion of the seafloor (Cartwright and Ray 1991 tables 3,4).

contains a major but unknown contribution by locally variable oceanic attraction and deformation. The effect extends to the interior of the continents. Estimation of the "solid" dissipation has accordingly been limited by absence of data as to the ocean tides. As noted, the situation is advancing, in that ubiquitous oceanic data, although containing an unseparated component of seafloor motion, are becoming available through the use of altimetric satellites.

After best possible correction for oceanic effects, gravimetric observations of the solid tide at a point are commonly presented as a departure from theoretical models of long standing. (Jeffreys 1924; Munk and MacDonald 1960a; MacDonald 1964; Kaula 1964; Zschau 1979; Dehant et al. 1991).

Jeffreys (1929, sect. 14.21) modeled the tidal deformation in response to the semidiurnally modulated external potential. With respect to a dissipation mechanism, he noted (sect. 14.422) that under elasticoviscosity, the distortional strain consists of two parts:

$$\mu E = P + 1/t_1 \int P dt$$

where P is a tangential stress and E the corresponding distortional strain; $1/t_1$ is a constant with the dimensions of a time.

Under long-term stress the first term becomes dominant. In confining the period to half a day this term was excluded. The associated phase lag determined in terms of seismologic variables and the damping of the free nutation is then of the order $2° \times 10^{-5}$, and the "solid" dissipation is insignificant.

Munk and MacDonald (1960) formulated the work done by a body force, F_i derivable from a potential U as a surface integral

$$dW/dt = \int \rho u_i n_i U \, dS$$

(where ρ is density, u velocity, and n an outward normal to the surface.)

Use of an observed value of the potential Love number and specimen values of the phase lag, hence dissipation (Munk and MacDonald, 1960a [eqs. 11.7.2, 11.7.3]), at semidiurnal frequency, yields values of the solid-tide dissipation (table 3.1).

Munk and MacDonald note that the first value of Q quoted, although based on the observed dissipation, is implausibly low if internal dissipation is estimated theoretically, in terms of a frequency band extending from seismic to Chandler-wobble frequencies ($30 < Q < 200$).

Kaula (1964) has examined solid tidal friction in terms of a Q constant throughout the Earth, in such a fashion that it is possible to vary Q with amplitude

Table 3.1. Phase Lag and Dissipation Corresponding to Values of Q

Q = 9	2 ϕ = 3°.2	$-\langle dE/dt \rangle = 2.7 \times 10^{19}$ erg/sec
40	0°.7	0.6×10^{19}
100	0°.3	0.25×10^{19}

and frequency. The stress is generated by the action of a geostationary potential varying at semidiurnal frequency. Kaula tested a value of $Q^{-1} \cong 0.075$ which yields a dissipation of 1.75×10^{19} erg/s. He points out that a smaller Q was considered by MacDonald (1964) on the grounds that an unexpectedly large tidal dissipation may be the result of a magnitude-dependency in respect to the deformation.

Working from an internal standpoint, Lagus and Anderson (1968), figure 3.3, computed the potential energy implanted by the semidiurnal tides in the form of a static spheroidal deformation, as a function of depth and the seismologically observed elastic constants. The dissipation processes to be investigated must then determine the fraction that is cyclically released in heat.

Zschau (1978) has identified shortcomings in earlier formulations of the dissipation (Munk and MacDonald 1960; Kaula 1964) and has computed the phase delay to be expected in the physical, radial bulge as distinct from the gravimetric "bulge." The phase shift was earlier computed using the difference in the phase of the stress and the strain in a restricted mantle element. Zschau points out that in the case of an extended massive body such as Earth, the potential of the displaced matter must itself be taken into account. The ratio between the phases of the radial and the gravimetric bulge may be as much as 10, rendering the latter very small.

Evaluations of the dissipation by Zschau employ a static spheroidal model of the tidal deformation topologically similar to that of Lagus and Anderson (1968), restricting dissipation to that in oscillatory shear and assuming incompressibility.

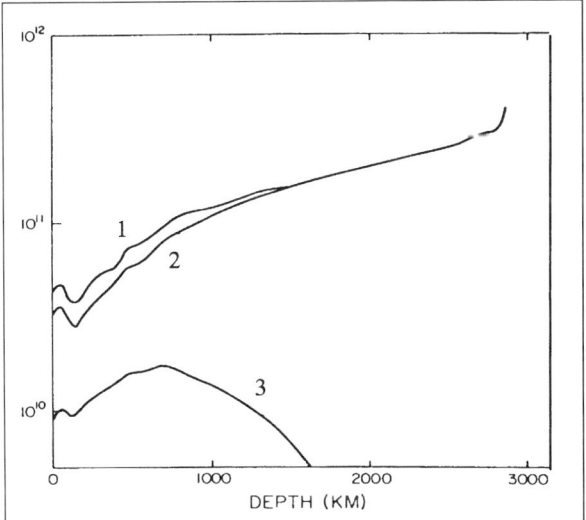

Fig. 3.3. Distribution of elastic energy with depth with respect to the semidiurnal tides as a spheroidal deformation, after Lagus and Anderson (1968). In erg/km^3 (vertical scale): 1 is the sum of energy associated with 2 (rigidity modulus) and 3, Lamé's constant; the mode is likened to the seismologically observed free oscillation $_0S_2$.

46   Tectonic Consequences of the Earth's Rotation

Dissipation was based experimentally on a homogeneous Q, and on a depth-dependent Q that was seismologically derived. Under these conditions, the lag to be expected in the gravimetric tidal bulge is of the order 0.001°. This quantity is immeasurable using available techniques.

To extend the restricted period range (up to 1 hour) of seismologically observed Q, Zschau (1983, 1985, 1986a,b) employed the concept of the absorption band combined with observations of damping of the Chandler wobble. Q pertinent to the damping of the Chandler wobble is bounded by observations by Ooe (1978), Okubo (1982), and others. These are based on the observed period increase (reference the theoretical, "elastic" period) which is due to anelasticity. The absorption band, figure 3.4, is the periodicity range of stresses within which Q is thought to be predictable. Munk and MacDonald (1960) have brought forward reasons to suppose that in respect to this deformation mode, dissipation varies little over a wide frequency-range.

Dissipation can be approximated empirically, using a function of the style $Q = c\,\overline{\omega}^{\alpha}$, in which $\overline{\omega}$ is frequency and a is a chosen exponent. Examining a range of possible values of α, Zschau (1986a,b) estimates that:

$$1.2 \times 10^{18} \text{ erg/s} = 0.12 \text{ TW}$$

is the most probable solid-Earth dissipation. This quantity is associated with a global $Q_{solid} = 270$, and most probable physical (as against gravimetric) phase lag

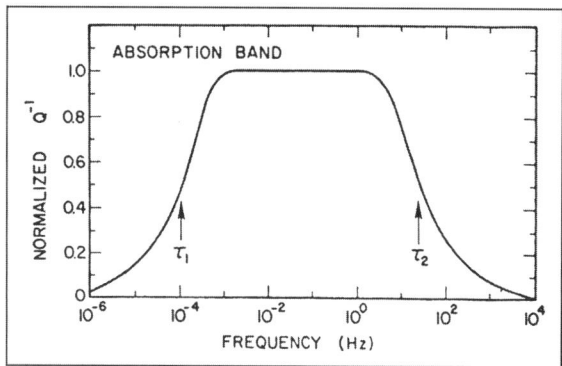

Fig. 3.4. Concept of absorption band (Liu, Anderson, and Kanamori 1976), displaying normalized Q^{-1} against arbitrary frequency. The graph consists of superimposed closely-spaced components of absorption vs. frequency, beyond which absorption decreases steeply. τ_1 and τ_2 are half-amplitude points. Absorption is seismologically observed to be very little frequency-dependent between periods of 1 s and one hr. Reproduced from Lundquist and Cormier (1980).

0.21°. Zschau again draws attention to the order-of-magnitude difference between physical and gravimetric phase lag.

Absorption has been formulated (Liu, Anderson, and Kanamori 1976; Anderson and Minster 1979; Minster and Anderson 1980) in terms of the dislocation microstructure of mantle materials under stresses of various duration. Liu, Anderson, and Kanamori (1976) introduced the concept of an absorption band of the type shown in figure 3.4. As noted, the flat portion of the spectrum is formed from the superposition of components peaking at closely spaced frequencies.

To extend further the predictable absorption band, employing still a spheroidal deformation mode, Zschau and Wang (1986) have sought estimation of dissipation through use of an expression

$$G = c - \phi/\rho_o$$

in which G is the Gibbs free energy of the relaxation process, ϕ is a seismic parameter, ρ_o is a density, and c a constant of the relaxation process (Anderson and Given 1982). On the basis that c is known, Zschau and Wang extend dissipation estimates to a period of 1 billion years, thereby taking account of such viscous processes as post-glacial isostatic recovery. The phase delay then expected in the semidiurnal gravimetric tide remains less than one hundredth of a degree (see also Dehant and Zschau 1989), too small to be observed.

Zschau (1983) has pointed out that in view of viscosity ranging down to 10^{18} P in the upper asthenosphere (Vetter and Meissner, 1977; see also Hales and Bloch 1969; Haxby and Weissel 1986; Hager 1991; Richards 1991; Lliboutry 1991; Bailey 1998), that region may be subject additionally to dissipative "Maxwellian flow." The latter is constituted of oscillatory (self-cancelling) displacement.

Wahr (1981) has computed the tidal deformation of a rotating, elastic, oceanless Earth, a computation extended by Dehant (1985, 1992, 1993; see also Wahr and Bergen 1986) to the case of an earth imperfectly elastic. In each case, the disturbing force (Wahr's 1981 eq. 2.2) is a lunisolar tidal force having the form

$$f(x,t) = \nabla V(x,t)$$

derived from the time variation of a potential stationary with respect to the Earth.

The tidal gravimetric factor δ (Dehant 1993) is the Earth transfer function relating the gravimetric body-tide signal and the vertical component of the gradient of the external tidal potential at an observation station. The gravimetric tide and its lag is a measure of the dissipation, carrying no information as to its mechanism.

3.5.1 Identification of Tidal Q; Origin of the Phase Lag

Reviewing, the dissipation, hence phase-lag, forthcoming from theoretical models has been based on spheroidal deformation within a finite frequency range, in most cases centered on a nearly semidiurnal frequency. Having then to be based on seismologic values of Q, the dissipation is minute.

Models thus constructed have in common that they assume variation of a potential stationary in orientation relative to the Earth.

48   Tectonic Consequences of the Earth's Rotation

In actuality, the tidal action does not stem from variation of a force representing the spatial derivative of a stationary potential, but instead, from forces created by its rotation (cf. section 2.4.2). The deformation, correctly a distortion as it represents vortical motion, now includes a zero-frequency component, in respect to which the dissipation bears no relation to non-rotational oscillatory systems.

In what follows an attempt has been made to gain insight into the nature of the distortion to be expected, *still within a "passive" or inert Earth*.

3.6 Distortion Under Rotation of the Potential

3.6.1 Orbital Mode

Tidal deformation of the Earth, reflected in the output of a sensor such as the tiltmeter (figure 3.5), is determined by the Earth's orbital characteristics.

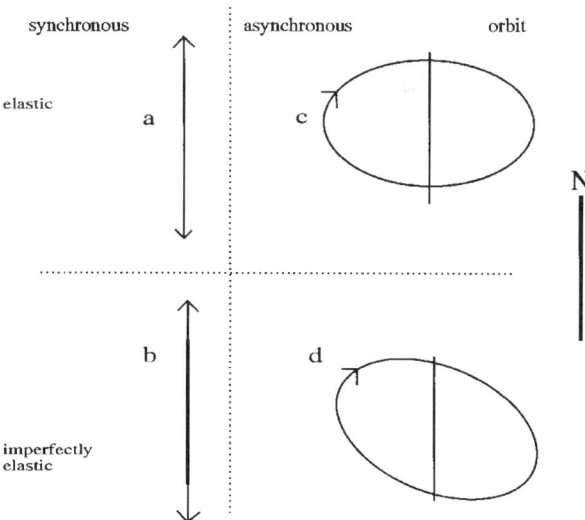

Fig. 3.5. Trace written by vertical pendulum at mid-latitude, Northern Hemisphere: (a) On elastic body in synchronous orbit (always one hemisphere facing its primary); north-south straight line is traced by pendulum, due to increase and decrease in stationary tidal bulge. (b) Still in synchronous orbit (orthodox tidal models), trace written by body that is imperfectly elastic; trace is N-S line slightly extended, and with phase-lag. (c) Body that is elastic and in rotation (asynchronous orbit) relative to external field: infinitely thin *ellipse* (no phase lag). (d) (Actuality) Earth-body, in rotation and imperfectly elastic; phase lag/lead in E-W and N-S components causes axes of traced ellipse to be tilted clockwise in azimuth.

In synchronous orbit, supposing purely elastic strain, in mid-latitudes the tiltmeter trace would take the form of an oscillatory north-south line. Introducing frictional losses, still synchronous, due to phase lag the N/S line would be slightly extended (figure 3.5,b). This is the trace to be expected under oscillatory tidal models, simulating $_0S_2$ eigenvibration, *supra*. In effect orthodox models assume that Earth is in synchronous orbit with varying distance to a disturbing body, resulting in the periodic formation of a static bulge. The dissipation mode then resembles that in Io.

In contrast, astronomy and everyday experience demonstrate that the Earth is in asynchronous orbit, rotating relative to the gravity field of the Moon and Sun. Although not described in the signal of gravimeters and single-component extensometers (which is purely oscillatory), the deformation mode is represented in the tiltmeter trace. The trace to be expected and which is well observed (e.g., Melchior 1983) is written in one direction only; clockwise in the Northern Hemisphere, counterclockwise in the Southern in respect to the major tides, M_2 and S_2. Under perfect elasticity the trace would be an ellipse with axes aligned N-S and E-W. The existence of dissipation causes the axes of the ellipsoid to be tilted in azimuth (figure 3.5d). This is the trace which is to be expected and is observed in experiments displaying the trace of pendulums (e.g., Melchior, 1983). The phase relations in the record of multidimensional extensometers (Bostrom and Vali 1968) contain similar information.

3.6.2 Nonreversing Force Couples

Within Earth the tidal stress axes rotate without reversal in one direction. The stress consists of a symmetric part S of the stress tensor, responsible for oscillatory deformation, plus an antisymmetric component A (figures 3.6a,b). In the Equator plane the antisymmetric part A is:

$$A = \begin{bmatrix} 0 & A_{r\theta} & A_{r\varphi} \\ -A_{\theta r} & 0 & A_{\theta\varphi} \\ -A_{\varphi r} & -A_{\varphi\theta} & 0 \end{bmatrix} \quad (-A_{\varphi r} = A_{r\varphi})$$

(Bostrom 1981) in which r is the radial direction, φ is tangential east-west, and θ is north-south. $A_{r\varphi}$ and $A_{\varphi r}$ are non-cancelling, but increase and decrease half-daily without reversing (figure 3.6c). A component of A dubbed $T_{\alpha\alpha}$ is evaluated in table 3.2. A component of the associated displacement resembles the toroidal component of plate motion, postulated by Runcorn (1957) and later observed.

The genesis of A is that the tidal stress ellipsoid, rotating without reversal, is continually misaligned with the restoring forces earlier excited, represented by the elasticity after discounting the loss in viscosity (figure 2.3c). A is not balanced, in the sense that angular momentum and vortical motion is induced in the material of

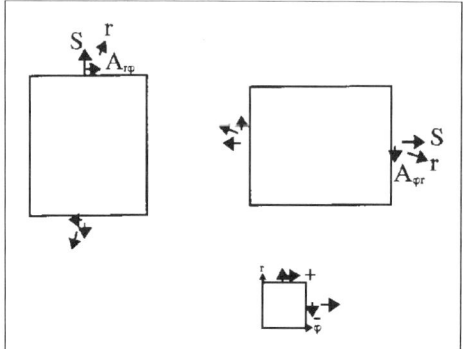

Fig. 3.6. (a) Symmetric S and antisymmetric part A of the tidal stress tensor. The regime is: twice-daily, S_{rr} and $S_{\varphi\varphi}$ alternately stretch and compress particle-lines, producing and reversing the tidal bulge, and causing dissipation in oscillatory, self-canceling shear. $A_{r\varphi}$ and $A_{\varphi r}$ increase and decrease as the point approaches and passes the Moon, anti-Moon, and quadrature meridians, but never reverse. Under imperfect elasticity, shear in this component is always in the same sense, resulting in cumulative vortical displacement. Localized coordinates, referred to the principal axes. φ, east, r, vertical, [θ], the equatorial plane.

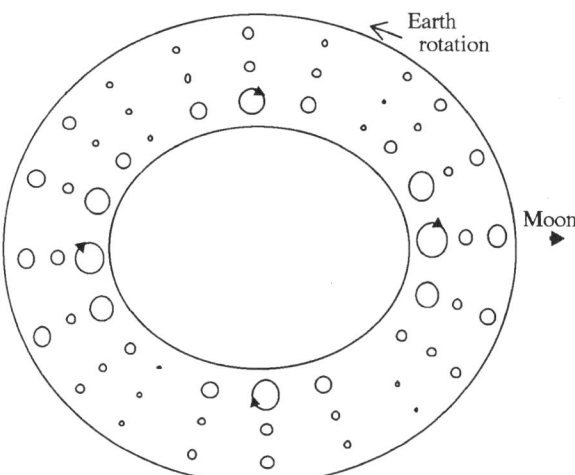

Fig. 3.6. (b) Torque at depth in equator plane due to rotation of the tidal potential, hence of semi-diurnal bulge. Computed by imposing rotation upon Takeuchi's evaluation (Takeuchi 1951) of the stress imposed by a *stationary* bulge (Bostrom 1981). Torque varies in magnitude as radius of symbols. Rotation sense is opposite to that of Earth rotation.

Tidal Action in a Uniform, Inert Earth 51

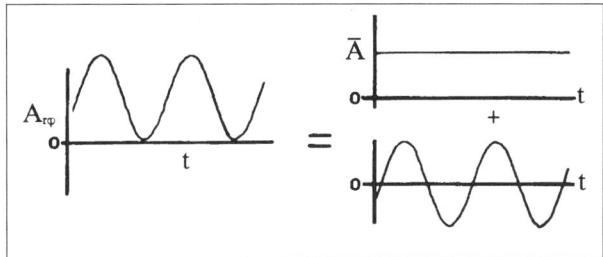

Fig. 3.6. (c) Oscillatory and secular parts of antisymmetric tensor-component A. The torque at a point varies in magnitude but never reverses, consisting of the sum of an oscillatory and secular component.

the mantle. The Earth's rotation is reduced by the portion of the retarding torque excited by the phase lag which is contributed by dissipation in this mode. In considering the action of the oscillatory, symmetrical part S of the stress ellipsoid, I have argued in chapter 4 that an additional dissipation fraction is contributed in bulk-viscosity losses, due to non-reversible phase transitions in regions of secular vertical flow.

3.6.3 Cumulative Distortion

The horizontal stress $T_{\alpha\alpha}$ imposed by a geostationary tidal bulge is illustrated in figure 3.7. In magnitude,

$$T_{\alpha\alpha} = \mu(0.0440 - 0.221 \cos 2\alpha)\{(3MrG)/gR^3\}$$

(following the formulation by Hoskins [1920]; see also Morgan, Stoner, and Dicke [1961]), in which: M is the Moon's mass; r and R the Earth's radius and Earth-Moon distance respectively; μ is the shear modulus; P, an arbitrary point in the shallow mantle, subtends α with respect to the Earth-Moon line. The Moon is here shown in the equator plane.

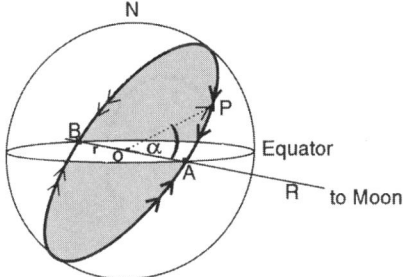

Fig. 3.7. Stress due to static tidal bulge. The horizontal stress $T_{\alpha\alpha}$ varies geographically as $\cos 2\alpha$. α is the angle subtended at Earth's center by the sub-Moon point A and a point such as P in the shallow mantle. Moon is shown in the plane of the Equator, P at latitude 30°N. The stress is directed as pressure or tension along great circles, such as BPA.

Tomaschek (1957, fig. 1) displays the variation in T with lunar declination. Table 3.2 lists values of T at points at various angular distances from the sublunar and antipodal points, i.e., at distances from the maximum of the semi-diurnal bulge.

The tidal bulge is not in fact geostationary. Introducing rotation of the Earth relative to the gravity field of the Moon, it will be clear that T rotates, as points such as P pass by the lunar/antilunar meridian. For the same reason the tangential stress quantified in Kaula (1964, table 4) does not reverse in the fashion applicable to an oscillatory formulation of the dissipation, but instead continuously rotates, introducing a secular term.

During rotation the three-dimensional stress ellipsoid continually leads the ellipsoid describing the elastic restoring force (cf. figure 2.3c). Perhaps more conveniently visualized, the strain ellipsoid likewise rotates without reversal (figure 3.8a). The vortex lines describing the orientation of the unrecovered displacement, dipping south in northern latitudes, become horizontal, aligned N-S, at the equator. The pendulum trace there is an E-W line.

It is incorrect to compute the dissipation based on the energy difference between end states, as in static eigenvibrations or in an earth in synchronous orbit; continuum mechanics requires piecewise integration over a continuous path incorporating all intermediate strains. Rationally, it is apparent that forces acting over a long (curved) path rather than the shortest distance represent dissipation proportional to the long, actual path.

This is most clearly evident in the review by Gross (1953) of Boltzmann's treatment of "elastic after-effects," i.e., viscoelasticity, at the molecular level, cited also by Liu, Anderson, and Kanamori (1976) in the form of the equation

$$\varepsilon(t) = \int_{-\infty}^{t} (d\sigma/dt) \cdot \Psi(t - \tau) d\tau.$$

The strain ε results from the sequential imposition of stress $\sigma(\tau)$ throughout its total history up until the instant t. Ψ is a function simulating creep. Under nonreversing rotation of σ relative to the medium, as distinct from sinusoidal time-variation in its magnitude, the residual strain is cumulative, in mode vortical or

Table 3.2. Horizontal Stress T (figure 3.7) as Function of Angle α

$\alpha°$	$T_{\alpha\alpha}$ dyn/cm^2
0	40,000
11	36,955
23	28,284
34	15,307
45	0
56	−15,307
68	−28,284
79	−36,955
90	−40,000

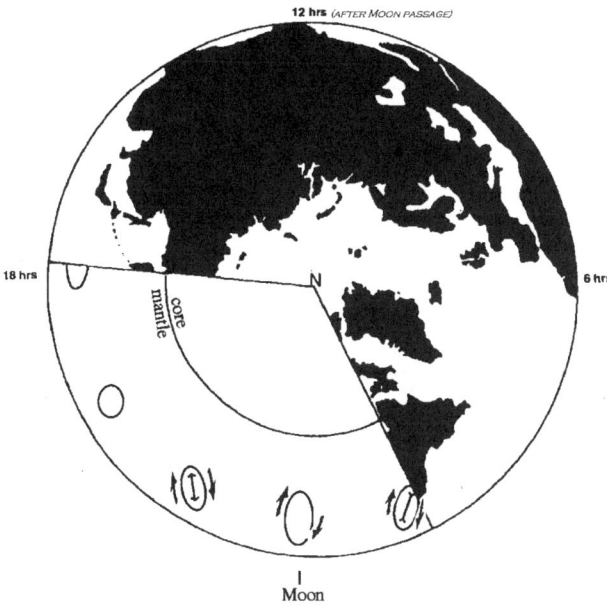

Fig. 3.8. (a) Orientation of the strain ellipsoid relative to particle lines in the shallow mantle. At a point the axes of the the stress ellipsoid, hence the strain ellipsoid, rotate as Earth rotates relative to the Moon (cf. figure 2.3c). ↔ and →—← indicate that the eccentricity of the strain ellipsoid increases and decreases (oscillates) under evolution of the stress ellipsoid. In contrast, its rotation relative to particle lines does not reverse but continues in same direction past quadrature points. The rotation direction is opposite to that of Earth. Under imperfect elasticity the rotation imparts cumulative vorticity, as in figures 3.8 (b,c), adjoining.

toroidal flow. The distortion under imperfect elasticity is illustrated in figure 3.8b, in an extended medium amounting to vorticity.

The distortion is compared in figure 3.9 with that resulting from the externally applied retarding couple, the surface-acting brake of Darwin (1908), Jeffreys (1921), and Eddington (1923). The dissipative processes responsible for tidal lag were attributed by Darwin (1879a,b), absent rheological information, to oscillatory deformation under a stationary potential plus surface-acting viscous friction under the tangential retarding couple excited by the dissipation. Darwin modeled the couple exerted on density-bounded layers by shallower elemental prisms. He and his successors found no other nonreversing stress. Tomaschek (1957) employed Darwin's formulation of the dissipation mechanism, employing viscosity derived from the damping of the Chandler wobble and post-glacial recovery rates. The part of the phase lag attributable to the solid Earth is then less than 1 arcsec. Failing identification of an alternative mechanism, the preponderance of the dissipation has since been relegated to the marine tides. The marine/solid-earth coupling mech-

54 Tectonic Consequences of the Earth's Rotation

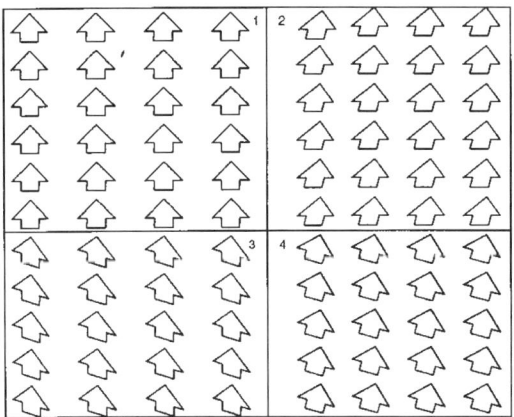

Fig. 3.8. (b) Distortion under stress system rotating relative to particle lines in a continuous medium. The mode is styled by Means (1976; see also Sommerfeld 1950) progressive non-coaxial simple shear. The vector sum of the shears imposed during revolution of the stress axes amounts to a rotation $\overline{\omega}_z = (du/dy - dv/dx) \neq 0$, where u and v are displacement, devoid of translation. Volume elements oriented as in 1 at first experience simple shear as in stage 2. Under continuing rotation of the stress axis, the residual strain element in 3 is added to that in 2. The sum as in 4 represents distortion.

In an elastic medium, the strain is recovered and residual distortion is nil. In the imperfectly elastic mantle, a fraction $1/Q$ of the rotation in 4 is residual and cumulative. In this mode, each element rotates relative to its neighbors, also relative to earth-fixed axes, for instance the axes of figure. Under imperfect elasticity, nonreversing rotation imparts cumulative vorticity, as in adjoining figure 3.8(c).

anism, effecting braking of the rotation, has been a subject of discussion (Schwiderski 1985a,b; Melchior 1989). Seafloor friction in the open ocean being minute, it has been suggested that braking is effected via the excess in pressure on the east flank of the continents.

With respect to the solid-earth portion of the dissipation, the comparative magnitude of the stresses suggests that its contribution to the retarding torque is due principally to the dissipation that results from re-configuration (internal) torques.

3.6.4 Displacement Under Tides Advancing as Waves

At all points in the interior, the tidal strain ellipsoid rotates without reversal. With respect to the anelastic, cumulative distortion, the vortex lines dip south in the

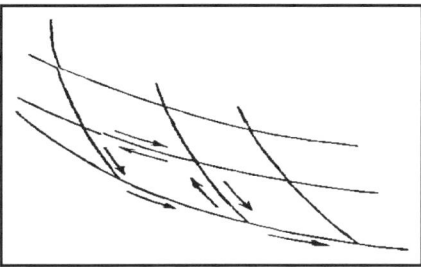

Fig. 3.8. (c) Illustration redrawn from Lamb (1945), demonstrating vortex motion or circulatory flow as formulated by Kelvin (1869), immediately following its identification by Helmholtz (1858). Overall circulation is here sinistral, but between individual elements shear is dextral and dissipatory. In Lamb's notation, with axis of vorticity normal to page, circulation is I(ABCD . . . A) = $(dw/dy - dv/dz)$ $\delta y\, \delta z$, i.e. integral taken around a closed curve defined by chosen points.

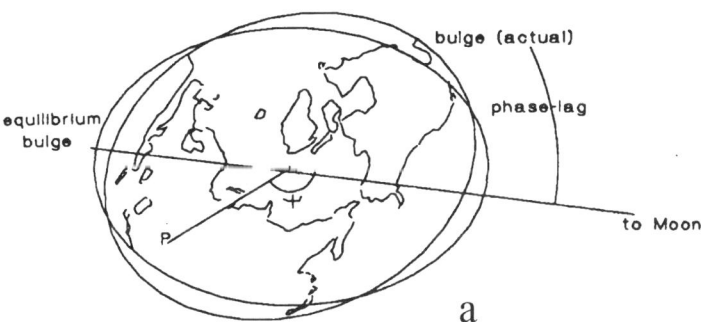

a

Fig. 3.9. Distortion as a result of the retarding couple, Eddington's surface brake, in comparison with that incurred during progression of the tidal bulge. (a) The external retarding couple arises as the difference between the attractions acting on material in hemispheres defined by the Earth-Moon line. Material in the hemisphere nearer the Moon (in the figure, the Americas hemisphere) is on the average attracted more strongly. Additionally, the material more strongly attracted is located further from Earth's rotation center, increasing the couple. At a point P in the Earth, during its revolution the lunar forces reverse. Material is displaced alternately east to west and west to east. Although reversing, integrated about the Earth the slightly different couples sum to form the torque acting to reduce the rotation and accelerate the Moon. Averaged over the surface during one revolution, the tangential force is of the order 10^{-4} to 10^{-5} dyn/cm^2.

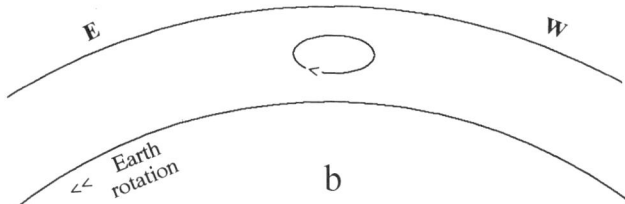

Fig. 3.9. (b) Secular internal torques and sense of the resulting displacement (a circulation). Passage of the 50 cm wave produces nonreversing stresses of the order 10^4 dyn/cm^2 (table 3.2) in the upper mantle. The tidal bulge progresses west at 15°/hour by means of continual adjustment towards an equilibrium figure never reached, which would be co-axial with the Moon. Anelastic losses cause a lag in the re-configuration, believed to be between 0.1° and 0.5°, producing internal torques which are a function of the tangent of the lag-angle. These are nonreversing. The sense (sign) of the distortion is to cause eastward displacement of deep material relative to shallow. The torques generated by re-configuration are internal and do not directly decelerate the Earth; the resulting dissipation causes a phase lag which does so.

north hemisphere, sketched in figure 3.10. It will be observed that their southward dip decreases with latitude, becoming zero at the equator. Vortex lines are there aligned N-S and the pendulum trace becomes an E-W line. The dip sense is reversed in the Southern Hemisphere.

Tectonically, the sense of the induced strain is to cause displacement in the mode of figures 3.11a,b. This mode is not instrumentally perceived, as by gravimeters and altimeters, except indirectly as a phase lag due to the dissipation. Due to the existence of a lithosphere resting on a much less viscous asthenosphere, stress

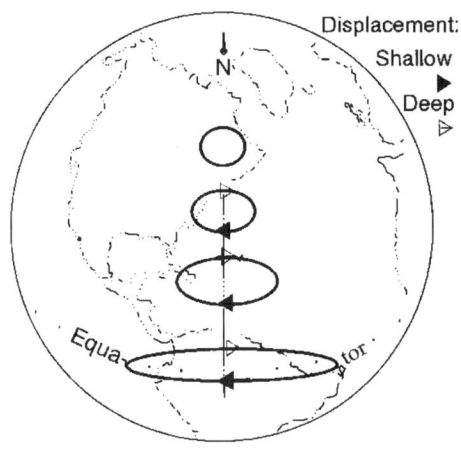

Fig. 3.10. Vorticity induction: secular displacement direction (deep/shallow) and vortex line. Moon in equator plane. Not to scale; the effect is maximized as $\cos^2$ of latitude. The magnitude of the induction is a function of Q_A, an unknown vorticity-effective elastic-quality factor. Its characteristic frequency (zero) lies entirely outside the seismic/Chandler band. It seems apparent that Q_A must reach its minimum in the upper mantle. In addition, the induction must vary regionally (chapter 4).

Tidal Action in a Uniform, Inert Earth 57

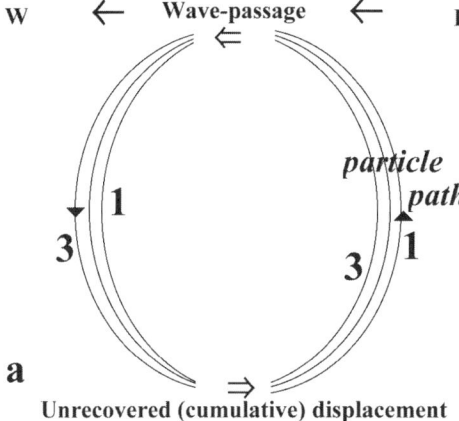

Fig. 3.11. (a) Particle motion (describing "creeping ellipse"), during three passages of a wave tide, under imperfect elasticity. In case of perfect elasticity, ellipses would be superimposed (no cumulative displacement).

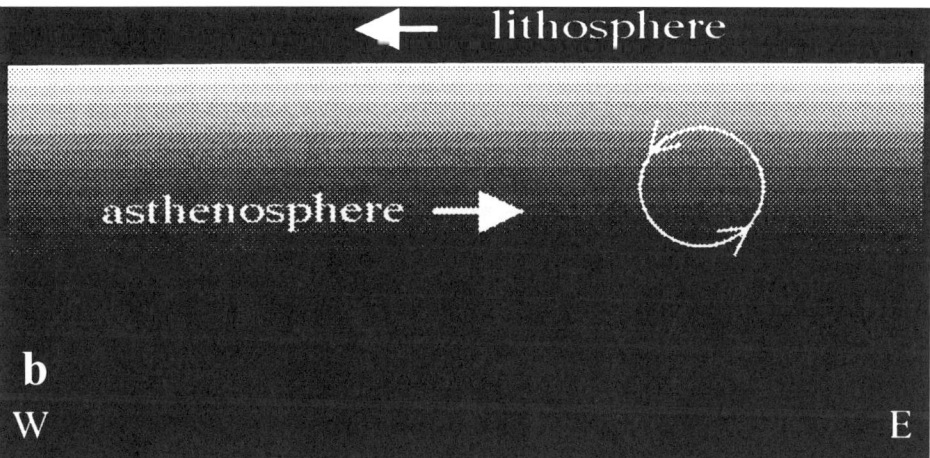

Fig. 3.11. (b) Vorticity induction and displacement direction of upper mantle relative to the lithosphere. Induction is effected in each volume element of an imperfectly elastic asthenosphere, acting so as to displace shallow and deep material simultaneously in opposite directions. In low latitudes, the axes of vorticity, normal to inscribed circle, are horizontal and aligned N-S into the paper.

diffusion (Elsasser 1969; Bostrom 1981) favors lateral summing of the stress and its incorporation in tectonic processes (chapter 4).

Testing specimen values (figure 3.11b), in low latitudes if Q in a layer 100 kms thick in the upper asthenosphere is 200, and strain under M_2 is 2×10^{-8}, westward displacement of the top of the layer (base of the lithosphere) vs. deep material is 10^{-3} cm/semidiurnal tidal cycle. Being cumulative, this amounts to 0.73 cm/yr; 7.3 km/Ma. Q pertinent to the antisymmetric part of the stress tensor is in actuality unknown, but due to the layered geometry may be less than the tested value (e.g., see Hales and Bloch 1969; Anderson and Sammis 1969). Based on well-constrained values of the elastic constants (Takeuchi 1951), in low latitudes the strain is somewhat greater than the value tested. Due to the colatitude increase in the vorticity, displacement is not that of the lithosphere as a whole. Low-latitude regions tend to lead those in higher latitudes, tending to a clockwise rotation at surface of segments in the Northern Hemisphere and vice versa.

The displacement may be compared with net-lithosphere-rotation found in plate motion. Runcorn (1973) pointed out that motion in a lithosphere shell can theoretically be separated into toroidal and poloidal vector components of velocity, v_p and v_τ. These describe respectively motion normal to a radius and in rotation about it, and only v_τ motion can displace the axes of moment.

Separation into toroidal and poloidal motion was eventually effectuated, by O'Connell, Gable, and Hager (1991) and Ricard, Doglioni, and Sabadini (1991); see also Minster et al. (1974). Accepting standard plate-motion components, the separation displays net rotation of the lithosphere concluded to be 1–5 cm/yr. The v_τ component, summing to a net rotation of the lithosphere relative to the hotspot population in the underlying mantle, was attributed to geographic viscosity variations having the geometry of lithosphere plates.

A no-net-torque condition may be imposed, the Earth to be unaffected by external torques. The predominantly westward sense of the rotation is then attributable to chance. On the other hand, if displacement under the zero-frequency component of M_2 and S_2 is taken into account, the sense is not fortuitous.

The reason for the perhaps surprisingly large value of the displacement is that immediately tidal action is recognized to be that of a wave, producing asymmetrical rather than purely symmetrical spheroidal deformation, even a small defect in elasticity results in displacement cumulative through time—abundant in geology. Entering the realm of speculation, the possibility is examined that lithosphere displacement results in the evolution first of a boundary layer, then of addition to the convective circulation in the mantle (sect. 4.3.2).

Using data representing worldwide petroleum exploration by Exxon, Nelson and Temple (1972; see also Roeder and Nelson 1971) concluded that aspects of global tectonics require the existence of flow entitled by them "mainstream mantle convection," causing preferential eastward displacement of deep layers relative to those nearer the surface. Doglioni, Moretti, and Roure (1991) have identified similar motion affecting the Mediterranean region. The phenomenon was apparently not further explored by Exxon, for the reason that observation conflicted with theory. The only recognized nonreversing tidal force was at the time the minute lunar retarding torque. It would appear that, in the long term, rotation of the strain ellipsoid contributes vorticity devoid of translation. Lithosphere-asthenosphere displacement is relative. The associated dissipation entails a phase lag, conjuring the secondary retarding torque. The causality order is such that the minute tangential shear ($\leq 10^{-4}$ dyn/cm^2) imposed by the latter is responsible for only insignificant displacement.

Jeffreys (1926) early identified the extreme weakness of the retarding torque, in dismissing a postulate by Joly (1925). Recognizing as early as 1903 the potential of radiogenic heat, Joly (1903, 1909, 1925, 1928) conceptualized the existence of an asthenosphere, intermittent mantle convection, and continental drift. At the time of introduction of the plate tectonics paradigm it was postulated (Bostrom 1973) that the retarding torque may bias the direction of extant convection. In respect to plate tectonics Jordan (1974) concluded with Jeffreys that the retarding torque is too weak to "drive the lithosphere." The enormously larger stress associated with passage of the tidal wave (Bostrom 1976) has made it illogical to appeal to the retarding torque, either to bias the convection or to "drive the lithosphere."

3.6.5 Dissipation in the Secular Component

Investigations by Guegen and Mercier (1973), Weertman (1975), Anderson and Minster (1979), Anderson and Minster (1981), and others indicate that the rheology of mantle minerals is controlled by microstructural processes of dislocation-climb and -glide. Steady-state creep is limited by slow self-diffusion, whereas loss in transient small stresses (as in seismic waves) is limited by comparatively fast diffusional processes and the creation of point defects.

In a searching review, Minster and Anderson (1981) draw attention to the critical roles played by yield stress, reference frequency, strain duration, and temperature. Following Anderson and Minster (1979), in the power-law expression

$$Q(\overline{\omega}) = c \times 10^3 \overline{\omega}^\alpha$$

(where $\overline{\omega}$ is frequency and c a constant applicable to the material of the mantle), the exponent α can be chosen so as to specify empirically the variation in the elastic quality factor, Q, with frequency. Zschau points out that over a frequency range (absorption band) covering seismic to the Chandler wobble, a single value of α can be found that satisfies observed values of the absorption.

Assuming the same functional dependence, no matter the value allotted to $\alpha > 0$, the attenuation increases without limit in the direction of greatly extended periods. Numerically (figure 3.12), commonly observed α values between 0.16 and 0.25 produce Q in the range 60 to 282 at a 14-month period (Chandler wobble). These accord with seismic/Chandler-based whole-earth values of Q. However, in terms of the same arbitrary function, for very long period and secular strain Q would drop to values below 40. Employing a kindred formulation

$$G = c \cdot \Psi/\rho_o$$

(where G is the Gibbs free energy pertinent to the relaxation, Ψ a seismic parameter, ρ density and c a constant of the material), Zschau and Wang (1986) recognize that for geological time periods the value of Q drops drastically. At an undeter-

60   Tectonic Consequences of the Earth's Rotation

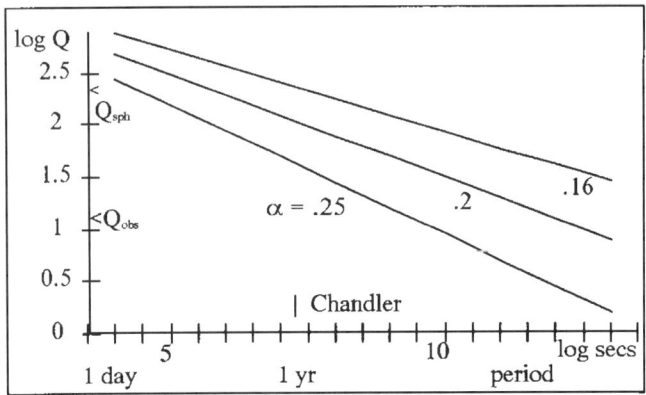

Fig. 3.12. Q relative to stress of increasing period, as dependence on an exponent, α. On y-axis (see text): $<Q_{sph}$ marks theoretical value of Q in respect to the semi-diurnal tides, assuming spheroidal oscillatory strain and seismologic attenuation. $<Q_{obs}$ marks value of Q observed, as gravimetric phase lag.

mined low-frequency limit, Q may be defined by Maxwellian flow, specified in Minster and Anderson (1979, Eq. 5 and 11). It then matches the viscosity observed in such processes as post-glacial isostatic recovery.

In the course of alternative investigations as to dissipation, O'Connell and Budiansky (1977), Borch and H.W. Green (1987), and D. H. Green, Cooper, and Zhang (1990) have examined the behavior of Q with temperature variation, and as attenuation in mantle containing partial melt, making plain its dependence on such factors as volatiles content. In the case of partial melt (D. H. Green, Cooper and Zhang 1990) the dissipation increases without limit (Q ⇒ 0) as the period is extended.

With reference to the tidal phase lag, values attributed to Q as a result of ongoing investigations are marked in figure 3.12 (y-axis). Because of the presence of the Moon, as pointed out by Zschau (1978), the gravimetrically observed phase lag is much reduced relative to that of the solid-Earth bulge. Based on the formulation,

$$Q^{-1}_{seismol/Chandler} = \tan \phi \cdot (-0.0508)$$

his most-probable value Q = 270 (Zschau 1986a,b) is associated with a phase displacement 0.011°, too small to be observable. Q = 270 is theoretically derived, based on oscillatory deformation within the absorption band covering seismic frequencies and the Chandler wobble.

The greatly different "observed" value of Q marked in the figure, Q ≈ 12.1, corresponds to the gravimetric phase lag 0.24° suggested by global measurements of tidal gravity (Melchior 1989, 1995b). Melchior notes that the phase displacement is extremely difficult to separate from noise (see figure 3.13 and later discussion).

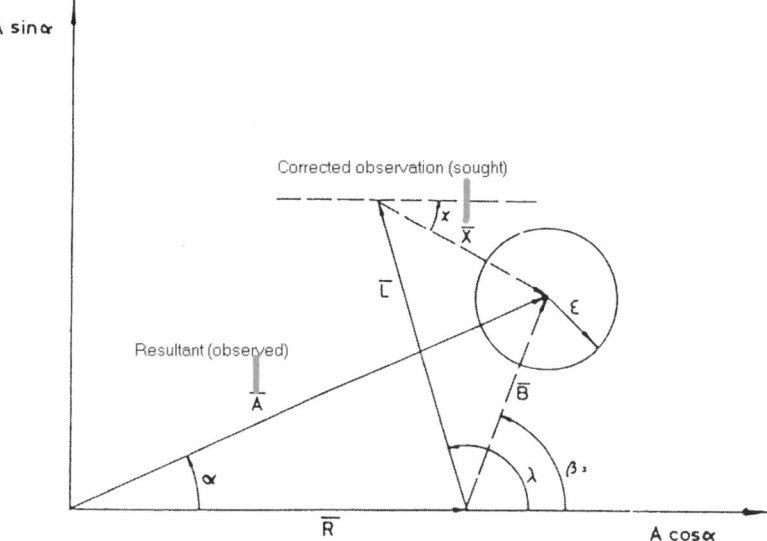

Fig. 3.13. Vector components comprising observed tidal gravity. (R, ρ) is the computed reference standard, having zero or minute phase angle based on low-dissipation (seismologic Q) model earth. $\mathbf{A}(A,\alpha) + \mathbf{B}(B,\beta)$ is observed gravimetric amplitude and phase. $\mathbf{A}(A,\alpha) + \mathbf{L}(L,\lambda)$ is resultant to be expected after estimated contribution $\mathbf{L}(L,\lambda)$ by marine tide; $\mathbf{X}(X,\chi)$ is residual sought; ε is estimated margin of error. After Melchior (1989).

3.7 Interaction, Bodily and Marine Tides

Estimation of the dissipation is much complicated by two-way interaction between the seafloor and marine tides (Groten and Brennecke 1973).

Using elastic values of the solid-Earth deformation, Hendershott (1972) concluded that tidal motion of the seafloor (his seafloor "heaving"), by its effect on the overlying ocean, is responsible for about one-third of the energy input to the oceans. This conclusion is modified by Schwiderski (1980, 1985a,b). Hendershott's derived value of marine tidal Q (Q = 34) is conspicuously larger than that (Q ≈ 13) deduced lately by Cartwright and Ray (1991), who used satellite altimetric data and assumed high, seismologic values of "solid" Q. Zahel (1978) has used an alternate parameterization.

A value of solid-Earth tidal Q only fractionally smaller than that assumed on a seismologic (oscillatory) basis would lead to increased "heaving," with enhanced conversion of tidal energy in the overlying ocean into baroclinic modes and dissipation, in the fashion described by Sjoberg and Stigebrand (1992). In this respect, the amplitude of the solid tide in the Equatorial Pacific reaches tens of centimeters. The phase difference between the geocentric sea-surface tides, now measured by Topex/Poseidon satellite altimetry (Schrama and Ray 1994, pl. I b), and Schwiderski's model, assuming an orthodox high-elasticity Q (Schwiderski 1980; Schrama and Ray's (1994) Pl. I d) is conspicuously large; see discussion, section 7.4.5.2.

62 Tectonic Consequences of the Earth's Rotation

More recent estimates of the marine dissipation have been constructed, which tend to reduce the difference between satellite-derived observation of the total dissipation, and estimates of the part attributable to the solid Earth. Current estimates of the marine portion have been reviewed by Andersen (1995). The major input factors, water velocity and the dissipation function, are not independently derivable. Using Seasat data at a time of meteorologic disturbances in the China and Yellow Sea region, Moon, Tang, and Choi (1991) found that a commonly employed value, 0.0025, of the quadratic bottom friction coefficient best fits local data, but the uncertainty accounts for more than the spread in overall dissipation estimates.

Ray (1996) has separated a part of the TOPEX signal that is attributable to the formation of internal baroclinic tides by the Hawaii ridge. Dissipation via this mechanism leaves proportionately less dissipation attributable to the shallow seas. Morozov (1995) attributes a possible 1.1 TW to internal tides. Munk (1997) has pointed out that added to shallow-seas estimates, this would sum to more than the firmly established astronomic total for the entire Earth.

Such disparities tend to reinforce the reservations of Munk and MacDonald (1960a) as to the role of the shallow seas; in seeking the major mechanism of marine dissipation "The only possibility that has occurred to us is that the bulk of the dissipation is associated with internal (or baroclinic) tides." In several respects the situation wryly described by Melchior (1983) prevails: "It is ... surprising that the many calculations performed since 1923 on different data have always delivered the needed amount [of dissipation] even when the astronomical conclusions were considerably changing."

Water loading of the seafloor causes solid-Earth deformation. To date, the effect has been estimated, as a linear, rather than measured, function of water depth (but see Ray et al. [1994]). Deployment of seafloor manometers (Spencer and Vassie 1985; Cartwright, Spencer, and Vassie 1987), permitting separation of geocentric sea-surface height from seafloor motion, is at an early stage.

Estimation is made by convolution of the water-depth load at a point with a globally uniform, radially symmetric Green's function devised by Longman (1963) and Farrell (1972); see also Francis and Dehant (1987) and Francis and Mazzega (1990). The computation load is minimized by formulation in terms of high-degree spherical harmonics (Ray and Sanchez 1989).

Dissipation in loading thus computed is insignificant when based upon Zschau's seismologic estimation of the effective tidal Q (Dehant and Zschau 1989). Zschau (1978) suggests that at critical distances (80–100 km) from the load center, the presence of the asthenosphere may enhance dissipation. It has not yet been feasible to take account of regional variations in Q, suggested by regional earth-tide phase lags (Melchior 1994a,b). Open questions as to tidal loading are taken up in chapter 7.

3.8 Partition, Marine/Solid-Earth Dissipation

Employing the terminology of Platzman (1984), it will be evident from the foregoing that if D is the dissipation in tidal components, with subs identifying respectively: t, the total dissipation, delimited by satellite orbital parameters and laser lunar-distance observation; m, the marine dissipation; l, the dissipation in seafloor loading; and s, the dissipation in the solid tide excluding loading

then there is obtained directly from a satellite ocean-surface height only the combined quantity:

$$D_t = D_m + D_l + D_s.$$

G. H. Darwin (1879b) formulated the dissipation D_s to be expected in an oceanless, viscous, passive earth. There existed at the time an entire absence of data as to the Earth's rheology. His eq. 16 incorporates a periodic term describing dissipation under oscillatory (semidiurnal) stress. As noted earlier, in sequence the resultant phase lag is made responsible for the lunar couple, the surface torque, and secondary minor dissipation. Hypothetically, the retarding torque would eventually result in the formation at the equator of north-bearing "wrinkles," bearing eastward in higher latitudes.

Subsequently, seismology and observation of the slowness of isostatic processes, requiring high viscosity (Tomaschek 1957), resulted in a shortfall in dissipation thus formulated. The investigations of Taylor (1920) in the Irish Sea, extended worldwide by Jeffreys (1921), suggested that a major part of the dissipation is likely to be surficial and marine (D_m), effected in shallow seas and adjacent to shore.

Estimation of D_m, in particular in terms of its locus, has proved difficult. Munk and MacDonald (1960a,b) concluded that the role of dissipation in the Bering Sea, preponderant in Jeffreys' evaluation, is likely to be only 10% of his value. A subsequent estimate (Miller 1966) provided about one half the global value necessary to explain the decrease in rotation.

In a comprehensive review, Platzman (1984) categorizes estimates of dissipation D_m in terms of: friction models, which attempt to quantify dissipation as the product of water-velocity and a friction coefficient; and boundary models, estimating energy in terms of level changes. The model developed by Platzman expresses dissipation D_t in the entire Earth as a surface integral. Dissipation D_s in his bodily tide component $P(k^*)$ (eqs 24–26) is estimated in terms of the seismologic Love numbers of Zschau (1978), namely of an oscillatory strain.

The discrepancy picked up by Platzman, namely, a systematic tendency of both ocean-tide data and models not to account for the total dissipation D_t, is still conspicuous (Cartwright and Ray 1989). Lambeck (1988) points out that values are too uncertain for the bodily dissipation D_s to be determined on the basis of the difference $(D_t - D_m)$ between marine-friction estimates and satellite determination of dissipation in the whole Earth.

To an extent, Platzman's second (boundary or level-change) model of dissipation may now be quantified. Geocentric sea-surface height observations provided by the Topex/Poseidon project tally numerically with the astronomically derived dissipation for the entire Earth; although they encompass D_t, the sum of the solid-Earth and aqueous dissipation, only for the oceanic Earth.

An "empirical" ocean-tide model by Schwiderski (1980, 1985a,b) has been caused to fit data at tide-gauges while conforming elsewhere with hydrodynamic constraints based on seafloor topography. A paradox emergent in this and some similar solutions is that the couple integrated over marine phase-lead regions is so large as to accelerate the Earth more than phase-lag regions cause deceleration. As water/seafloor coupling is minute under low velocities in the open ocean, it has been proposed that the terrestrial braking torque must perforce be applied in the form of differential pressure on the eastward-facing slope of continents. To compensate for the large regions tending to accelerate the Earth, Schwiderski (1985) suggests that a bodily phase lag of 0.5 degrees is required, attributable to unquantified distortion by the oceanic pressure. In response to subsequently obtained satellite data, models have been constructed adopting various assumptions as to aqueous dissipation and seafloor loading (e.g. Le Prevost et al. 1994; Egbert, Bennett, and Forman 1994; Andersen 1995), further discussed in chapter 7.

In respect to observation, a nonzero phase lag in the bodily tide has been reported by the Centre des Marées Terrestres in Brussels (Melchior 1989, 1995a,b). Melchior's separation of the components comprising an observation is shown in figure 3.13. In view of unexpected values, stations for which the oceanic correction may be imperfect were culled by him. Until recently, separation of the small $X(\chi)$ has been additionally affected by the difficulty in correcting long-period instruments for phase distortion. A correlation of earth-tide values with regional heat flow and lithosphere heterogeneities has been suggested (Molodenskii and Kramer 1980; Yanshin et al. 1986), and the suggestion criticized on theoretical grounds by Rydelek, Zurn, and Hinderer (1991). Nevertheless, the existence of a regionally variable amplitude and significant phase lag, apparently incompatible with a uniform earth, appears real (Melchior, 1994a; 1995a).

3.9 Summary: Tidal Action in Inert, Uniform Earth

Melchior points out that a logical basis for reducing the gross value of the incongruity in (X,χ), figure 3.13, would be to increase the value of component ρ in the computed reference model. A lag value of approximately 0.24° best fits global cotidal maps based on the Earth Tide Data Bank maintained at Brussels (Melchior 1995b). Melchior emphasizes that the RMS error is difficult to evaluate correctly.

Theoretical as against measured versions of the energy partition between solid-Earth and marine tides (the theoretical confined to the assumption of a passive Earth having seismologic Q), are shown in figure 3.14. The total dissipation D_t (M_2 + S_2) may be considered 3.0 TW (Cartwright and Ray 1991, table 4; see also Marsh et al. [1990]; Schrama and Ray [1994]). Cartwright and Ray's data set assumes values for dissipation in the nonoceanic part of the Earth and high elasticity (high Q_{solid}).

A several-fold disparity is conspicuous between estimated values of the solid-Earth dissipation based on seismologically-defined Q, and "measured" values, incorporating real processes. Solely as a scale, the energy expenditure in seismicity (Kanamori 1978) is a fraction of the smallest estimate of tidal energy input, that

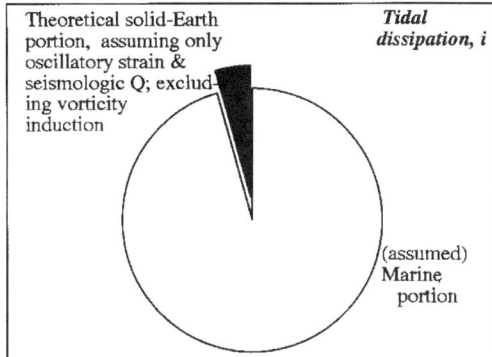

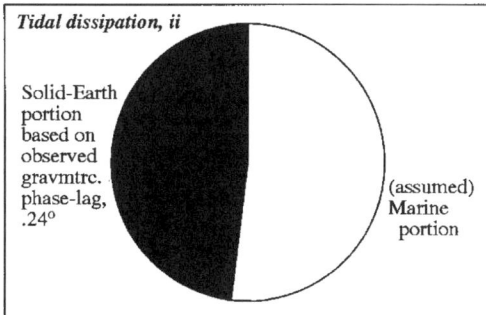

Fig. 3.14. Apportionment of tidal dissipation, marine as against that in the solid Earth, displaying several-fold discrepancy between theory and observation. In both figures, marine dissipation is allowed to fill the portion not accounted for in estimates of dissipation in the solid Earth. In each case, the total, conforming with observation of orbital elements and satellite altimetric data, is approximately 3.0 TW. (Top) Solid-Earth contribution as a proportion of the total, restricting dissipation to oscillatory strain (Zschau 1986a,b), employing dissipation factor Q within seismologic/Chandler absorption band. (Bottom) Solid-Earth contribution indicated by observed gravimetric tidal phase lag 0.24° (Melchior 1995b). The phase-displacement requires $Q_{solid} \approx 12$.

assuming solely oscillatory deformation (Zschau 1986). An unmeasured tectonic energy flux is almost certain to be associated with aseismic or "silent" tectonic processes (chapter 4), entailing processes of glide and flow.

In attempting to reconcile theory and observation, dissipation based on seismologic/Chandler values of Q is estimated (Zschau 1986) to amount to about 4.4% of the total. Total ($M_2 + S_2$) is estimated by Schrama and Ray (1994, table 5) to amount to 2.81 TW (2.81 × 10^{19} erg/s). Using solely mathematical manipulation, if we allot the value 0.1644 to the exponent α (figure 3.12), on the proposition that the asthenosphere contains a proportion of partial melt, Q may be made to drop to Q < 50 for secular stress while preserving the value 270 preferred by Zschau (1986) in the seismologic/Chandler range. It appears unlikely, nevertheless, that oscillatory $1/Q_{seis/Chandler}$ can feasibly delimit attenuation in respect to secular, vortical flow.

The phase lag 0.24° emergent in global gravimetric measurements of M_2 and S_2, now amounting to some hundreds (Melchior 1995), is equivalent to an earth having notably small tide-effective $Q_{solid} \approx 12$. The dissipation in this case is several times' estimates assuming oscillatory strain. If the geologically based net-lithosphere-rotation reported by Ricard, Doglioni, and Sabadini (1991) and O'Connell, Gable, and Hager (1991) is real, the specimen $Q^{-1} = 0.005$ tested in section 3.6.4 (figure 3.11b), leading to displacement 7.3 km/Ma, is low. A higher value would permit simultaneous reconciliation of the "observed" geologic displacement rate, the observed gravimetric phase lag, and conditions (e.g., Anderson and Sammis 1970) to be expected in the upper asthenosphere.

A 0.5° (physical-bulge) phase lag, suggested by Schwiderski (1985a,b) on the basis of the shortfall in his hydrologic estimate of the marine dissipation, is more plausible given these considerations. Subsequent hydrologic estimates (Le Prevost et al. 1994; Andersen 1995) more closely match the total dissipation, but like Schwiderski's, contain arbitrary coefficients in respect to the mechanism. It might be supposed that the difference (0.61 TW) between Schwiderski's estimate of the hydrologic dissipation and the value of the total, yielded by Topex/Poseidon data (Schrama and Ray 1994), lies in Schwiderski's assumption of high elasticity; this possibility is pursued in section 7.4.5. As apparent earlier, a decrease in Q_{solid} would lead to increased seafloor heaving in the fashion identified by Hendershott (1972), generating dissipative baroclinic waves. The volume of the asthenosphere is so huge in comparison with that of the oceans (figure 3.1), that dissipation estimates are highly sensitive to the value of Q_{solid}.

Using newly available VLBI data, investigators at the University of Toronto and Harvard-Smithsonian Center for Astrophysics (Mitrovica et al. 1994) find the imaginary part of the Love number in respect to the diurnal tide O_1 to correspond to a phase lag $0.7° \pm 0.5°$. If applicable, the corresponding gravimetric phase lag in M_2, S_2 is about 0.27°. The frequency of O_1 is within the seismologic/Chandler absorption band. Oceanic noise does not obscure O_1 in the fashion in which it obscures M_2, but the mode is far from identical. Based on similar VLBI data, Herring and Dong (1994) arrive at an estimated value $0.604 + i.005 \pm 0.002$ of the

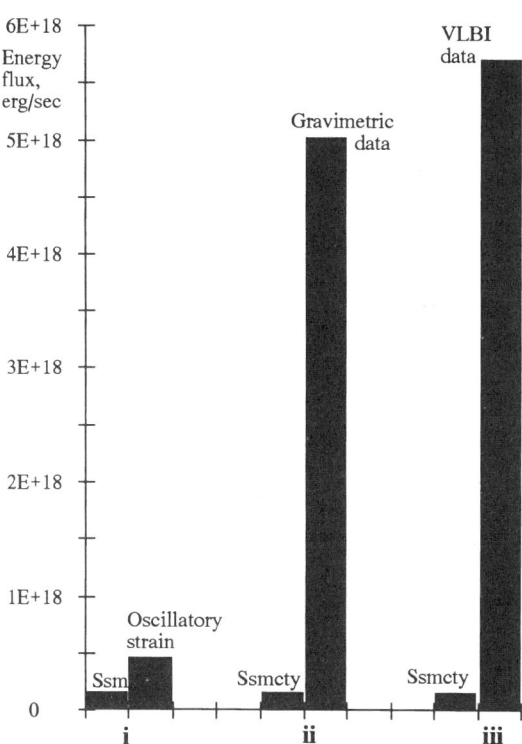

Fig. 3.15. Solid-Earth tidal energy input, estimated as compared with observed values based on the phase lag in M_2 and S_2. The estimate of the global expenditure in seismicity shown as comparison, 4.5×10^{24} erg/yr, is due to Kanamori (1978). Column pairs: **(i)** Global energy expenditure in seismicity as against tidal energy deposition in the solid Earth (Zschau 1978, 1986a,b), based on oscillatory (semidiurnal) strain and seismologic values of Q. **(ii)** Seismicity expenditure against dissipation as measured by global gravimetric phase lag 0.24° (Melchior 1995b). **(iii)** Seismicity in comparison to dissipation in M_2 and S_2 based on the VLBI observations of Herring and Dong (1994).

second-degree harmonic radial Love number h[2] (defining the physical bulge) in respect to M_2. Based on these data (figure 3.15, col. iii) the dissipation is greater than indicated by surface gravimetry, and accords with Schwiderski's (1986a,b) shortfall in the dissipation.

In respect to the disparity between observed and theoretical values of the phase lag, it should perhaps be recalled that altimetric satellite data are confined to the area occupied by the world ocean. Antithetically, surface observations of gravity are confined to dry land, in most places having continental lithosphere.

The disparity between estimated values of the solid-Earth dissipation and the observed phase lag has elicited comment for several decades. In their fundamental treatise, Munk and MacDonald (1960a) concluded that bodily tides can account for from 10 to 25% of the observed dissipation, and that the energy loss is a factor of 100 greater than found by Jeffreys. It may be thought that the situation scarcely has changed.

The foregoing assumes an inert Earth and parameters globally uniform. In the following chapter, the problem is addressed of tidal action in an earth more geologically plausible.

4

Tidal Action

Pre-Stress Under Convection

4.1 Overview

Tidal stress is imposed on an earth affected throughout by convection, thus perpetually prestressed. The convection encompasses mineral phase-changes, changes of state, and irreversible processes of differentiation. In consequence, a fraction of the interior is perpetually in an unstable state, such that incremental stress is prone to lead to "failure." In certain circumstances endogenous stored energy can be released as a result of extraneous disturbance. Release may occur as a result of material being stressed beyond a marginal limit, bringing about essentially an advance in the convection. In constructing "geological" models of the tidal Earth, it cannot be realistically considered passive.

Under rotation relative to the external gravity field the M_2 and S_2 "bulges" exist as a wave, traveling continuously around the Earth. As a result of the imposed stress axes rotating with respect to the Earth, imperfection in elasticity makes its appearance in the mode cumulative rotational flow (vorticity), identified by Helmholtz in 1857. As this is dimensionally identical to the flow in the endogenous buoyancy convection, and similarly prone to form cells, if the tide stress is significant the two must be interactive. In the current chapter an attempt has been made to evaluate the stress and tidal energy input, and relate these to potential geological processes.

With respect to the flow of material in convection, non-linearity and discontinuities are intrinsic. Within downflow (subduction) regions incremental pressure-increments cause non-reversing transition to high-pressure phases. Seismic energy is generated when slowly increasing strain progresses into nucleation and fracture, so that a stress increment releases accumulated elastic energy. In regions of upflow, state-change under depressurisation produces vulcanism, an effect prone to cascade as a result of the unloading.

The arrangement here followed is that there is first reviewed what is known of the convection in the mantle, including some recent numerical models. Opinion is divided as to whether the convection affects the mantle as a whole, or its upper and lower parts separately. The flow rate is typically a few centimeters per year. Much flow may be in the form of vertical, thin "plumes."

Rather than the uniform inert shell hypothesized in chapter 3, the lithosphere varies in thickness from several hundred kilometers beneath the older continents to zero in places beneath the world ocean. Lithosphere cools and founders, either as far as the lower-mantle interface or the core-mantle boundary. Thus, besides being prestrained, the Earth's interior is mechanically heterogeneous (figure 4.1). Under motion this results in regional stress concentrations, corresponding to order-of-magnitude rigidity variation.

Beneath the lithosphere Barrell's asthenosphere extends to about 400 km. Under old cratons the asthenosphere is minimally thick and may even be absent.

An attempt has been made to identify mechanisms of tidal energy deposition, representing dissipation paths. A candidate mechanism is the effect of the tidal

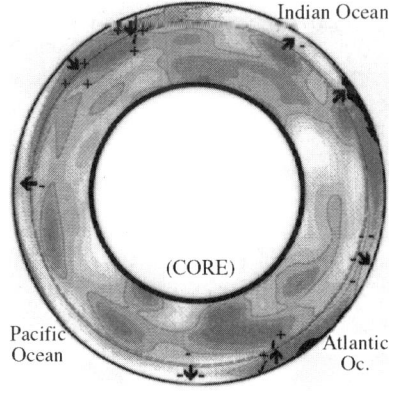

Fig. 4.1. A representation of the real or "geological" Earth, as distinct from passive and uniform models earlier considered (figure 3.1). Equatorial section. Regions of downflow and upflow are the site of steady increase (+) and decrease (−) in pressure respectively, and of discontinuous phase and state transitions. A fraction is continuously at the point of transition. Heterogeneity is indicated by lithosphere thickness variations (black). The lithosphere is vanishingly thin at the crest of the ocean ridges and reaches several hundred kilometers in thickness beneath the core of the older cratons. The structure of the convection in the mantle is suggested by variations in S-wave velocity from Su, Woodward, and Dziewonski (1994), indicated by background contours. C.I. 0.5% velocity-variation, darker faster. The oceans are systematically underlain by slow, presumably warm and low-Q, material. Two-layer convection is suggested by the velocity distribution of Su et al. Opinion is divided as to whether mantle plumes, an important category of upflow, originate at the core-mantle boundary; thin stems may escape detection as velocity anomalies. Similarly, uncertainty exists as to whether the mid-ocean spreading centers overlie upflow seated in the deep mantle, or upflow of shallower origin.

pressure pulse on material in the course of subduction. At depth the tidal pulse is many times the daily pressure increment in subduction. It is to be expected that in material already in a marginal state, transitions occur predominantly during the superimposed pressure increase, rather than during decrease. If substantial, the effect is of a degree of modulation of the convection.

An evaluation of viscous dissipation in vorticity-induction vs. that in convection suggests there may exist substantial tidal input, liable to be conspicuous in low latitudes. In respect to cotidal seismicity, dissipation in global seismicity is too small to account for much of the solid tidal dissipation, even if the correlation were strong. On the contrary, tidal energy accumulated as elastic strain could be responsible, as a component of plate motion, for a portion of global seismicity.

Field evidence for and against these processes is taken up in chapter 5.

4.2 Characteristics of the Convection

Recognition of plate tectonics and continental shift has entailed acknowledgment that the mantle is in convection. Its energy source is radioactivity and possibly, in part, gravitational compaction (Keondzhyan and Monin 1976; Sharpe and Peltier 1979; O'Connell and Hager 1980; Lubimova 1982). Observational constraints are provided by intramantle velocity anomalies, surface gravity, the magnetic record of seafloor spreading, and the geologic record, including isotope geochemistry.

Some thousands of ray paths derived from seismic events display the departure from travel times in a homogeneous Earth (figure 4.2). Tomography based on the inhomogeneities suggests density anomalies capable of producing the convection.

4.2.2 Dynamics

The flow to be expected on the basis of the quasi-density anomalies is a function of numerous, interactive variables. These include changes in viscosity with pressure and temperature and endo- and exothermic phase changes. Large-scale features of the convection have been reviewed by Olson, Silver, and Carlson (1990). Primarily, flow is driven by the foundering of cool, solidified segments of the surface layer (lithosphere slabs). A difficulty in forward-modeling convection based on the quasi-density distribution is that the flow tends to break into high-order components, including deep-seated thin plumes; such features are not resolved by present seismic data (figures 4.3a,b). Davies and colleagues (Davies 1984; Gurnis and Davies 1986; Hill et al. 1992) point to the probable importance of deep-seated thin plumes in the regime of convection.

Numerical models (Tackley et al. 1993) suggest that under mantle conditions (high Rayleigh number; internal heating, surface cooling; selected phase transitions; depth-dependent viscosity), convection takes the form of intermittently cascading downflows interspersed with geographically diffused upflow regions. Such structures correspond with geophysical indications (Creager and Jordan 1986) that downgoing slab material eventually descends into the lower mantle. Forte et al. (1993) point to the influence exerted on convection by aggregations of differentiate

Tidal Action 71

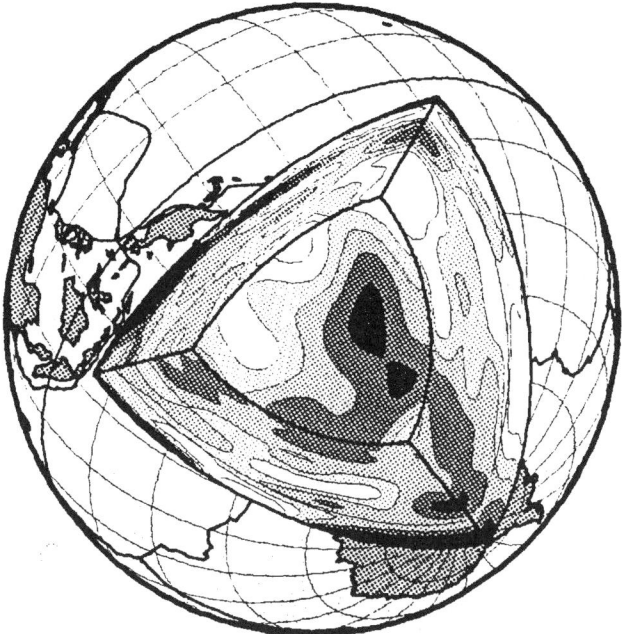

Fig. 4.2. Location of shear wave velocity anomalies within the mantle and at the core-mantle boundary, including regions up to degree 12. C.I. 0.5%; darker is faster. A thin line marks the 670-km velocity discontinuity. Despite density being in the denominator, slow regions are postulated to mark anomalously low density. Su et al. conclude that below the mid-ocean ridges, low-velocity regions persist into the lower mantle, rather than cease at shallower depth. Below about 2000 km the velocity regions are dominated by those of degrees 2 and 3. Reproduced from Su, Woodward, and Dziewonski (1994).

represented by the continents. Several models display the importance of endothermic and exothermic phase changes expected at the 400- and 670-km seismic discontinuities (see below).

4.2.3 Flow Rate

Velocity and flux constraints are imposed on the convection by seafloor spreading recorded as paleomagnetism. At surface, by this criterion extension varies in rate from 2 cm/yr to more than 20 cm/yr. At times in the past, these rates have been exceeded. Major rearrangements of plate motion are seen to have taken place, as at 147 and 120 Ma, (Nakanishi, Tuamaki, and Kobayashi 1992), for uncertain causes. It has been speculated on the one hand that the upwelling of "super-plumes" (Larsen 1991; Su, Woodward, and Dziewonski 1994) leads to plate breakup. On the other hand, Anderson (1994a,b) suggests that instead processes intrinsic to

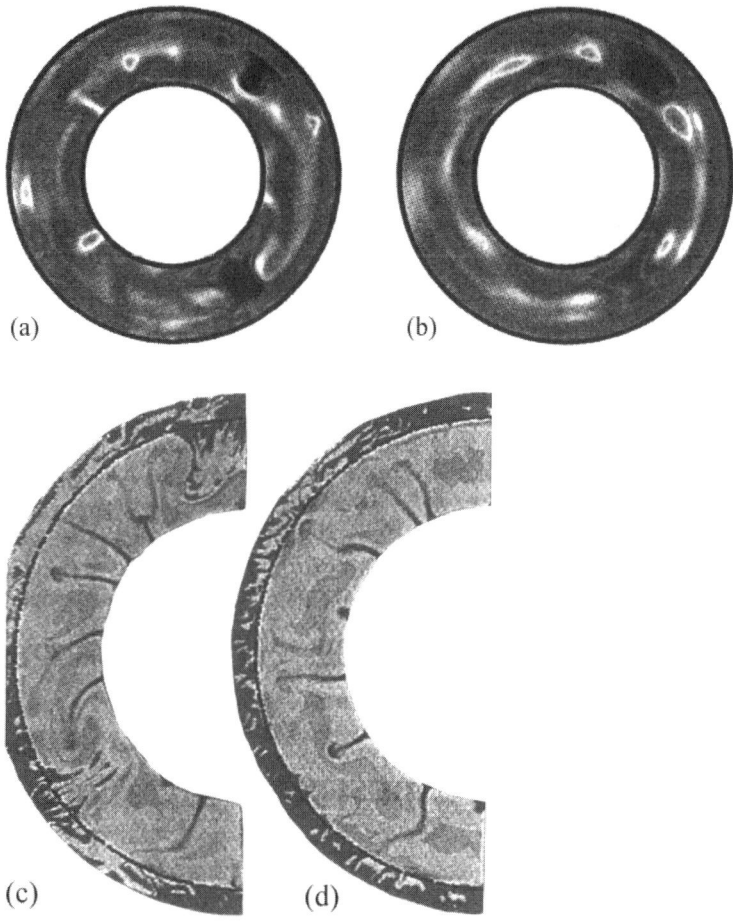

Fig. 4.3. Computed models of mantle convection affecting mantle as a whole, in comparison with regimes affecting separately the upper and lower mantle: (a) Convection structure (represented by temperature variation) assuming phase-change at 670 km, compared with (b), structure which would be visible assuming seismic mappability to degree, order 10,13. To this resolution, significant higher-order features such as columns are smeared or not identifiable. From Jordan et al. (1993). Convection structure now based on seismologic quasi-density distribution: (c) During epoch of regional whole-mantle system; (d) during epoch of two-layer system. From Woodward (1994).

plates themselves, in particular heat accumulation beneath thick continental lithosphere, must lead to their breakup and global reorganization. High-velocity (cool) anomalies in the lower mantle surround the Pacific Ocean (Olson, Silver, and Carlson 1990, fig. 2), suggesting both a systematic relation to subduction and its northward shift with time, relative to present coordinates.

4.2.4 Upper Mantle System

Uncertainty exists as to whether autonomous flow systems affect separately the upper and lower mantle (figures 4a,b,c,d). Evidence can be brought forward (Anderson 1987) that the convection is layered, separate systems ruling above and below velocity discontinuities at approximately 410 and 660 km (Nolet, Grand, and Kennett 1994, figs. 1,2,3).

Velocity-associated details of the asthenosphere and lithosphere have been brought into relief by Zhang and Tanimoto (1992). The upper mantle is laterally heterogeneous down to a modulus of hundreds of kilometers, perhaps to one yet smaller. In places, as in northeast Africa, low-velocity material appears deflected in its uprise by cold overlying lithosphere. Differences in the vertical extent of the

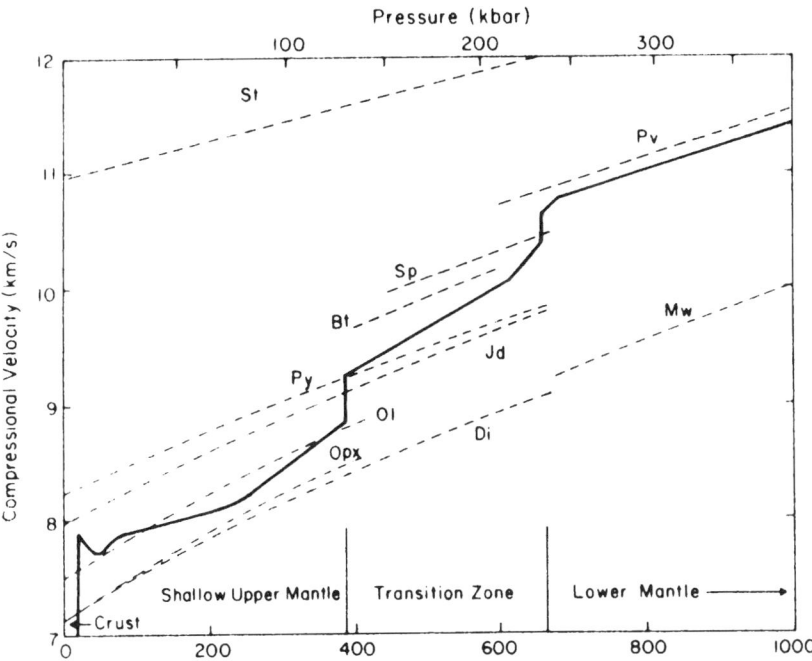

Fig. 4.4. P-wave sonic velocity in minerals compared with observed seismic velocity (thick line) in Gulf of California. Depth in km. Minerals: Bt, beta-phase; Di, diopside; Jd, jadeite; Mw, magnesiowustite; Ol, olivine; Opx, orthopyroxene; Pv, $MgSiO_3$, perovskite; Py, pyrope; Sp, Mg_2SiO_4 spinel. (Reproduced from Gasparik [1993].)

low-velocity region agree with earlier suggestions that spreading is shallow-rooted and can be attributed to decompression-melting (Richards, Duncan, and Courtillot [1989]; Renne and Basu [1991]; White and McKenzie [1995]), as a result (rather than a cause) of lithosphere extension. Nevertheless, Su, Woodward, and Dziewonski (1994) find evidence that below the mid-ocean rift system, low velocities extend to depths greater than the upper mantle.

Seismologically mapped velocity layering may be the expression of either thermal, compositional, or phase transitions. Velocity steps at 410 and 660 km may correlate (Olson, Silver, and Carlson 1990; Gasparik 1993) with phase transformations from olivine structure to spinel and from spinel to perovskite, (figure 4.4). Experiments (Ito and Takahashi 1989; Akaogi, Ito, and Navrotsky 1989; Katsura and Ito 1989) suggest that the velocity discontinuity at about 410 km is attributable to the $\alpha - \beta$, γ transitions in $(Mg_{0.9}FeO_{0.1})_2SiO_4$ at about 1440°C, and the 660 km discontinuity to $(Mg,Fe)SiO_3$-perovskite and $(Fe,Mg)O$ at about 1600°C (figure 4.5). Some flow models (figures 4.3c,d) suggest that the convection in the mantle is time-variable, in that intervals in which whole-mantle flow prevails alternate with those in which the regime is regionally two-layered.

4.2.5 Circulation of Volatiles

The volatiles content of the convection (Peacock 1990; Thompson 1992; Gasparik 1993; Philipott 1993; Nolet 1994) affects physical properties, to the extent that the hydration cycle may conceivably be responsible for the identity of "the upper mantle." Water is drawn into the mantle in subduction zones in the form of pressure-sensitive hydrous minerals (figure 4.6a,b).

Within the flow under convection a hydration-dehydration cycle exists (Nolet 1994), effecting transitions in not less than tens of magnesian silicate species (Thompson 1992). Within subduction regions (figure 4.6), released H_2O and CO_2 may trigger partial-melting reactions in overlying material. Most subducted water becomes resident in high-pressure hydrous minerals, eventually to be returned to the surface in volcanism, under conditions of decreasing pressure in upflow. Bell and Rossman (1992) believe that OH may be also resident in nominally anhydrous mineral phases, to all depths in the mantle; see also Irifune et al., 1998.

Complementary upflow regions are marked by the volcanic mid-ocean rift axes and mantle plumes. Peacock (1990) notes that in contrast to the regime prevalent in subduction zones, magmas derived from upwelling asthenosphere beneath ocean ridges contain only minor volatiles.

4.2.6 Rheology of the Upper Mantle

It has become evident that in the upper mantle empirical formulation of strain rates in terms of power law creep has become an inadequate descriptor. In modeling the rheology of the asthenosphere, Karato and Wu (1993) (figure 4.7) have separated the effects of diffusion- and dislocation-creep and have incorporated the effect of water saturation.

In diffusive creep, the strain rate increases linearly with stress. In contrast, in

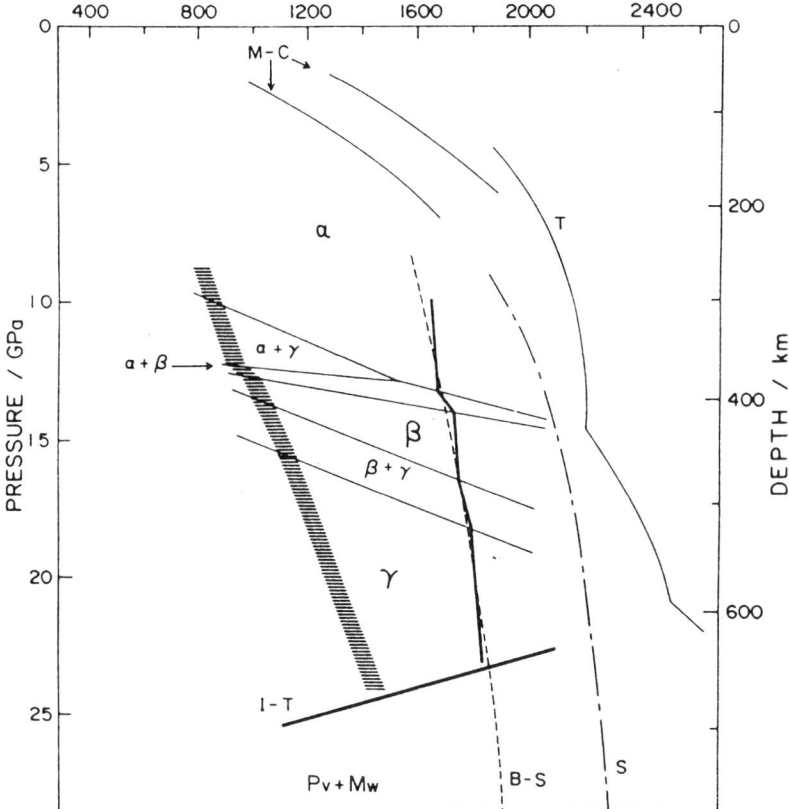

Fig. 4.5. Transitions expected at 410 and 660 km velocity discontinuities. Temperature (degrees Kelvin), pressure-depth and $\alpha - \beta$, γ transitions in system $(Mg_{0.9}\text{-}FeO_{0.1})_2SiO_4$. Normal mantle: thick shaded curve. Downgoing slab: hatched. I-T, boundary for spinel dissociation. M-C (Mercier and Carter 1975), pyroxene geotherms under oceanic (high-temperature) and continental (relatively low-temperature) regime. B-S and S (Brown and Shackland 1981; Stacey 1977), estimated geotherms. In comparison, T is melting curve of "dry" peridotite (Takahas 1986). Akaogi et al. note that the width (thickness) of the 400-km discontinuity appears to be about 18 km. From Akaogi, Ito, and Navrotsky (1989).

dislocation creep this relation is nonlinear. In the shallow upper asthenosphere the dominant mechanism is likely to be dislocation creep, with effective viscosity reaching a sharp minimum. Due to the effect of grain size, rapid increase in strain-rate occurs with stress increase, resulting in what these authors term rheological weakening and decoupling of the layers above and below a critical depth. Empirically, Wiens and Stein (1985) conclude on the basis of seismic data that a layer with viscosity $10^{19} - 10^{20}$ P results in decoupling of oceanic plates from the underlying mantle; see also Anderson and Sammis, 1970.

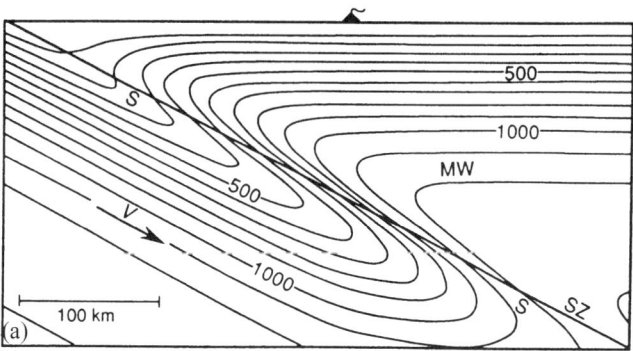

Fig. 4.6. (a) Subduction zone: computed thermal structure. Isotherms at 100° interval; note local gradient inversion. The distortion is due to the lag in thermal conduction as the cool slab, moving from the left, sinks into the warm underlying asthenosphere. Volatiles (V) are entrained in hydrous minerals descending below sub-arc mantle wedge (MW). SZ, shear zone on top of slab; S, oceanic crust forming top of slab. The effects are strongly dependent on the rate of subduction, the thickness of the slab, and its initial temperature. (From Peacock 1990).

4.2.6.1 Fracture. Difficulty is encountered in formulating the behavior of the visco-elastic but frangible lithosphere and crust. In particular, it is hard to construct convection models incorporating the stick-slip processes that characterize seismicity (Benioff 1951; Whitehead and Gans 1974).

In nature, beyond a characteristic deformation rate strain nucleates and a discontinuity (fracture) develops. If sudden, this releases seismic energy. Power-law rheology of the kind earlier proposed (chapter 3) cannot reproduce this behavior. Benioff (1951) postulated the existence of stick-slip fracture, under which resistance to displacement suddenly drops, releasing stored elastic energy and generating a seismic wave. In 1976, Whitehead and Gans formulated a function $g(u)$, representing the friction force characteristic of a given displacement velocity, u. Beyond a prescribed rate the friction sharply decreases.

The operation of $g(u)$ in the case of steadily advancing stress results in the periodic sudden release of strain in a seismic-like sequence (figure 4.8). Bercovici and colleagues (Bercovici 1993; Bercovici and Wessel 1994) have employed a similar function to model the behavior of viscosity-heterogeneities, representing lithosphere plates. The structure assists in explaining the unexpectedly large proportion of dissipation in the mode of toroidal vs. poloidal motion (O'Connell, Gable, and Hager 1991; Gable, O'Connell, and Travis 1991; Ribe 1992), generated by what has been assumed to be purely vertically acting buoyancy.

4.7.6.2 Upflow Regions. These are inherently unstable because viscosity decreases, with possible evolution of gaseous fraction, under a regime of maintained

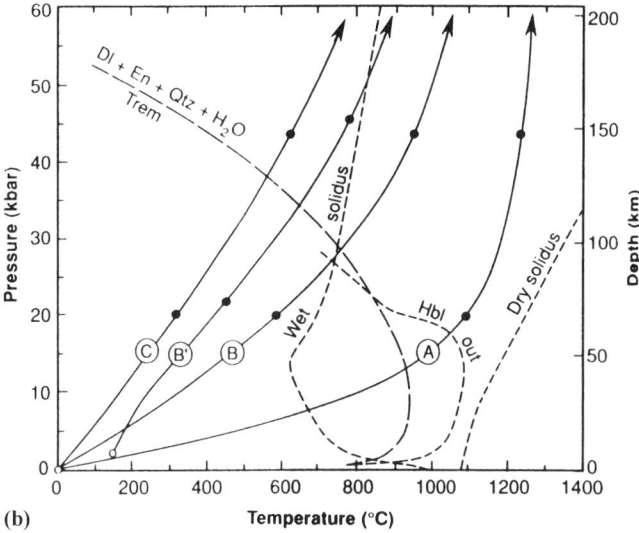

Fig. 4.6. (b) Subduction zone: effect of water content on melting-point and mineral phases for various slabs, with respect to the system [basalt + H_2O]. Paths represent those of: A, young (warm) oceanic lithosphere; B, top of crust; B', base of crust; C, cool, older slab, based on laboratory investigations. Dots are at million-year intervals. Note large difference between dry and wet solidus. After Peacock (1990).

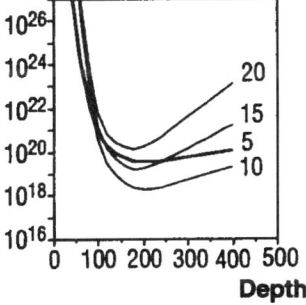

Fig. 4.7. Computed change in effective viscosity with depth, assuming diffusion creep (thick curve) and nonlinear dislocation creep (thin curves) in the upper mantle, in the case of relatively cool material entraining aqueous mineral phases. Characteristic: a sharp minimum (low-strength channel) with respect to dislocation creep; and crossover of the regimes, favouring localized deformation and formation of zone of mechanical decoupling. Quantities: assumed activation volume, in cc/mol. (Karato and Wu 1993).

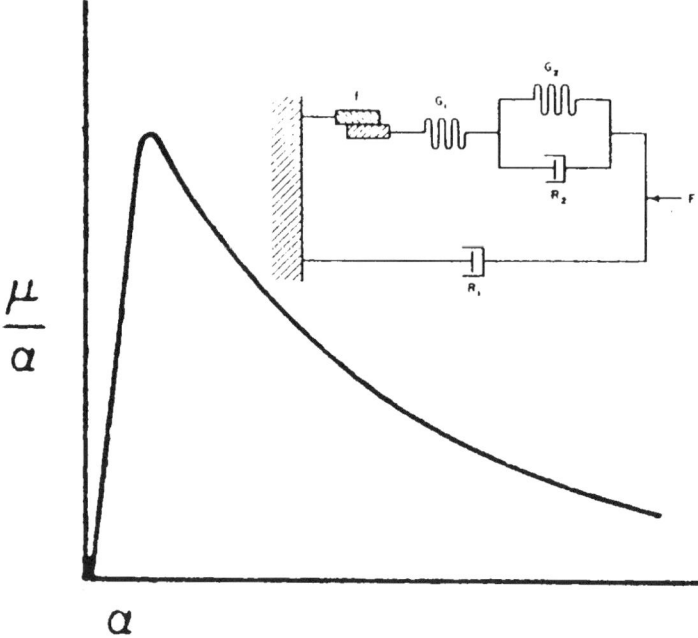

Fig. 4.8. Inset: Stick-slip element f, incorporated by Benioff (1951) in strain circuit diagram, to represent fault capable of sudden, seismogenic motion. With motion commencement, frictional resistance drops to a fraction of its static value. Graph represents drop of resistance to motion (y-axis) described by function g(u), written by Whitehead and Gans (1974) to emulate seismogenic motion.

temperature but decreasing pressure. Events are prone to be self-sustaining, due to cascading as a result of unloading of overburden.

White and McKenzie (1995) conclude that the voluminous volcanics associated with basalt floods and mantle plumes, figure 4.9, are the result of decompression melting, perhaps initiated by lithosphere thinning.

4.3 Dissipation Paths

With respect to containng a stress increment, an earth strained to the limit under convection is a leaky vessel. Birch has remarked that terrestrial energy degradation must be Protean in its complexity and elusiveness. Among other effects, incremental pressure can result in state change and voluminous differentiation. Some potential processes are explored in what follows. All encounter nonlinearities, in the form of state and phase transitions, fracture, and differentiation.

Tidal advance takes place due to continuous, nonreversing reorientation of lunar and solar equipotential surfaces relative to the Earth. The elemental distortion

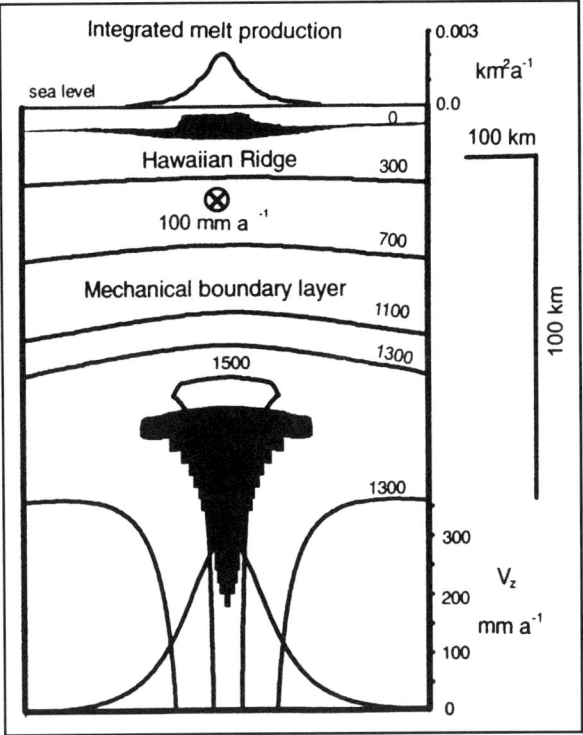

Fig. 4.9. Uprise of Hawaii plume, modeled by White and McKenzie, 1995. Black: the region of melt-generation. Isotherms: degrees Celsius. Vertical velocity component: mm per year. Vertical exaggeration, 4.6:1.

to be expected in an imperfectly elastic mantle, outlined earlier (section 2.4.2; figure 2.3c), in an extended medium takes the form of a vorticity. The resulting tidal "bulges" are not symmetric and static, but progress laterally. Bulges always in existence move around the Earth. Under elasticity other than perfect, the distortion imposes motion in the mode identified by Helmholtz (1858).

Describing the possible motion modes at a point, translation, deformation, and rotation, Helmholtz recognized that rotation cannot be generated by the "rectangular" action of a stationary potential. Having regard to a velocity potential ϕ, with respect to which the velocity is

$$u_i = d\phi/dx_i$$

it became evident that the relation

$$d^2\phi/dx^2 + d^2\phi/dy^2 + d^2\phi/dz^2 = 0 \qquad (4.1)$$

as under "rectangular action" does not prevail, and that

80 Tectonic Consequences of the Earth's Rotation

$$(du/dy - dv/dx) \neq 0. \tag{4.2}$$

In the hands of Helmholtz and successors (summarized, for example, in Serrin 1959), equation (4.2) has been shown not to be an algebraic curiosity, but to have physical meaning and to characterize motion common in fluid flow.

Models based on time-varying but geostationary astronomical equipotential surfaces, which axiomatically produce oscillatory spheroidal deformation, lack a dimension (rotation), and cannot describe the tidal Earth. Deformation in the static mode is sketched in figure 4.10.

The principal tides, M_2 and S_2, always present and mobile invariably in one direction, require realistic modeling. An attempt to display the distortion associated with these is incorporated in figure 4.11, for reference alongside heterogeneities in the Earth that are seismologically mapped. Large-scale tectonic processes, including fracture and differentiation encompassed in the convection, act in the long term as the motion of a fluid, in tectonic flow having similar dimensions.

The dissipation can correctly be represented (Zschau 1978; Platzman 1984) by a surface integral of the components of degree n of the time average

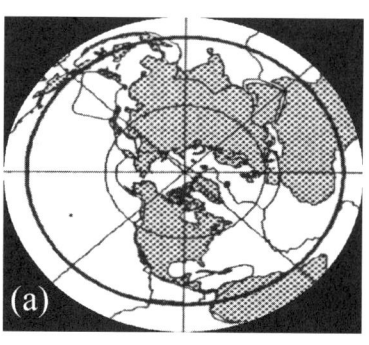 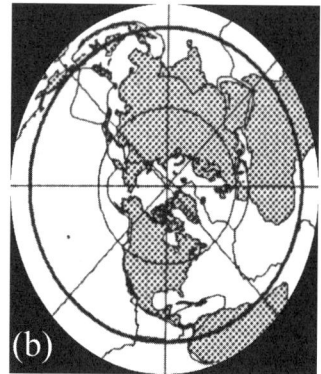

Fig. 4.10. Unreal earth: Bulge development oscillating between configurations (a) and (b) (cf. Lagus and Anderson 1968; Wahr and Bergen 1986; Zschau 1978; Dehant 1987a,b). The induced strain is similar to that in second-degree eigenvibration, and to body such as Io in synchronous orbit. As bulge is geographically stationary, principal axes of strain ellipsoids increase and diminish without rotation reference to particle lines, inducing self-cancelling shear. Minor tides of this form affect the Earth, caused by small lunar-distance variation.

Deformation, displacement, and dissipation conform with equation (4.1). Likening such a body to the Earth, Munk and MacDonald (1960) identify the tidal work done:

$$dW/dt = \int_v \rho u_i \partial \Omega / \partial x_i \, dV = \int_S \rho u_i n_i \Omega \, dS$$

in which: ρ is density, u_i is velocity. In the surface-integral n is normal to the surface. The force system is derived from a *geostationary* potential, Ω.

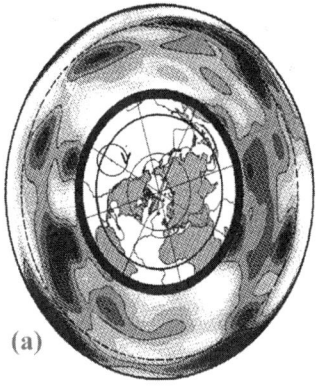 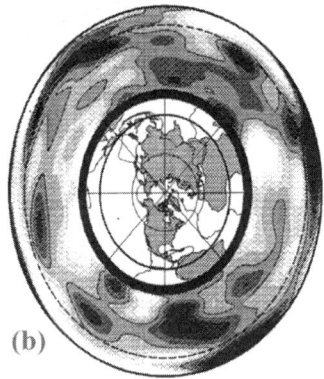

Fig. 4.II. Tidal Earth in actuality: the Earth rotates relative to the external gravity field, here static relative to page; "bulges" form mobile wave. Over time the elemental strain ellipsoids rotate relative to particle lines (no reversal), promoting only secular vortical flow (circulation). Dissipation, conforming to equation (4.2), is a function of strain path, not represented by difference in extrema.

In specimen interval, (b) succeeds (a) by 2 hours (30°). Inscribed ellipse, in Pacific in (a): sense of the induction. Equatorial section suggested by V_s anomaly, reproduced from Su, Woodward, and Dziewonski 1994. C.I., 0.5%; darker is "denser." Dash line: base of low-Q asthenosphere.

$$W = (4\pi GR)^{-1} \Sigma_n (2n+1) \int < \overline{\Omega}'_n \partial \Omega''_n / \partial t > dS \tag{4.3}$$

in which Ω' and Ω'' are the astronomical and secondary potential, R is the radius of an all-enveloping surface, S are surface elements. < and > designate a time average. Having astronomic information as to Ω', Ω'' can be measured.

The functional inclusion of Ω'' ensures that equation (4.3) incorporates dissipation due to both rotational and irrotational flow and such effects as bulk-viscosity. However, as Platzman (1984) emphasizes, observations based on (4.3) cannot separate dissipation taking place in the solid Earth from that in the oceans. The effects of solid-Earth and oceans are simultaneously integrated. By the same token, such observations cannot separate dissipation under the oscillatory, non-rotational portion of the tidal action in the solid Earth, liable to be minute, from that due to "Helmholtzian" vorticity induction (for convenience, TVI).

Representing as it does the fundamental tidal action, even prior to taking into account the non-linearities inseparable from mantle material in convection, it seems inevitable that the induction is responsible for the principal bodily dissipation. The question of reference models, against which to measure observations, is postponed until chapter 7.

Discounting atmospheric effects (Platzman 1984), three dissipation routes thus exist: that via the seas; that in potential (irrotational) flow within the solid Earth; and that via non-potential, rotational flow, due to vorticity-inducing distortion. Dissipation in all three media contributes to the phase lag in satellite-derived data, likewise that in the semidiurnal tides gravimetrically observed at surface.

The constitution and state of the upper mantle in particular are such as to render it susceptible to dissipation in induced vortical flow. Although tidal forces are greater at depth, the viscosity of the lower-mantle being orders of magnitude greater, dissipation there is likely to be smaller. The apparently significant extent to which tidal action may affect the D″ core-mantle boundary region (Bostrom 1998c), recently found to be susceptible, is not here pursued. An attempt is made to examine the extent to which vortical flow may bias, or couple to, the pre-existent, endogenous convection, having congruent flow mode.

The dissipation in an earth affected jointly by tides and buoyancy convection can be written (Kaula 1968)

$$\int_{\text{Earth}} (\sigma_{ik}^C v_{i,k}^C + \sigma_{ik}^T v_{i,k}^T + \sigma_{ik}^C v_{i,k}^T + \sigma_{ik}^T v_{i,k}^C)\, dV \qquad (4.4)$$
$$\quad\quad\; [1] \quad\quad\; [2] \quad\quad\; [3] \quad\quad\; [4]$$

in which σ_{ik} is the deviatoric stress tensor, $v_{i,k}$ is the velocity gradient tensor, and integration is throughout the Earth's interior. Superscripted T and C refer to the tidal action and convection respectively, the cross-terms [3] and [4] then denoting interaction. In [2] and [4] the summation must be extended to include antisymmetric terms in σ_{ik}^T omitted in "orthodox" irrotational models, and also to zero frequency. Then:

Term [1] specifies the work done by buoyancy in overcoming viscous friction.

[2] . . . . the work in tidal distortion. As earlier made evident, this term includes a zero-frequency component in σ_{ik}^T, contributing cumulative vorticity. Its effect is considered in section 4.3.2.

[4] contains elements in σ_{ik}^T acting on material maintained at the point of fracture, state change or phase transition by the convection. Their possible action is examined in section 4.4.

4.3.1 Tidal Vorticity Induction (TVI)

Under successors to Helmholtz, equations formulating vortical flow under thermal buoyancy have been developed (e.g., Rayleigh 1916; Landau and Lifshitz 1959; Batchelor 1967), prescribing the simultaneous conservation of mass, momentum, and energy in a viscous fluid.

The viscosity of the mantle varies with pressure and temperature. Due to non-reversing pressure transitions, bulk as well as shear viscosity are present. Analytic solutions of the convection initialized as a density distribution cannot be written. Figure 4.3 illustrates solutions reached by stepped numerical integration.

Figures 4.12a,b compare the rotation of the tidal strain ellipsoid, concomitant with that of the stress ellipsoid inducing vorticity, with the vorticity present in "observed" thermal convection. It may be noted (figure 4.12a) that the apparent oscillatory component of the deformation, represented by the growth and decline of the strain ellipsoid, entails rise and fall of the Earth's surface (change in geocentric distance), directly apparent in gravimetric data. However the alternation is achieved in no respect by stress reversal, but solely through rotation of the stress ellipsoid.

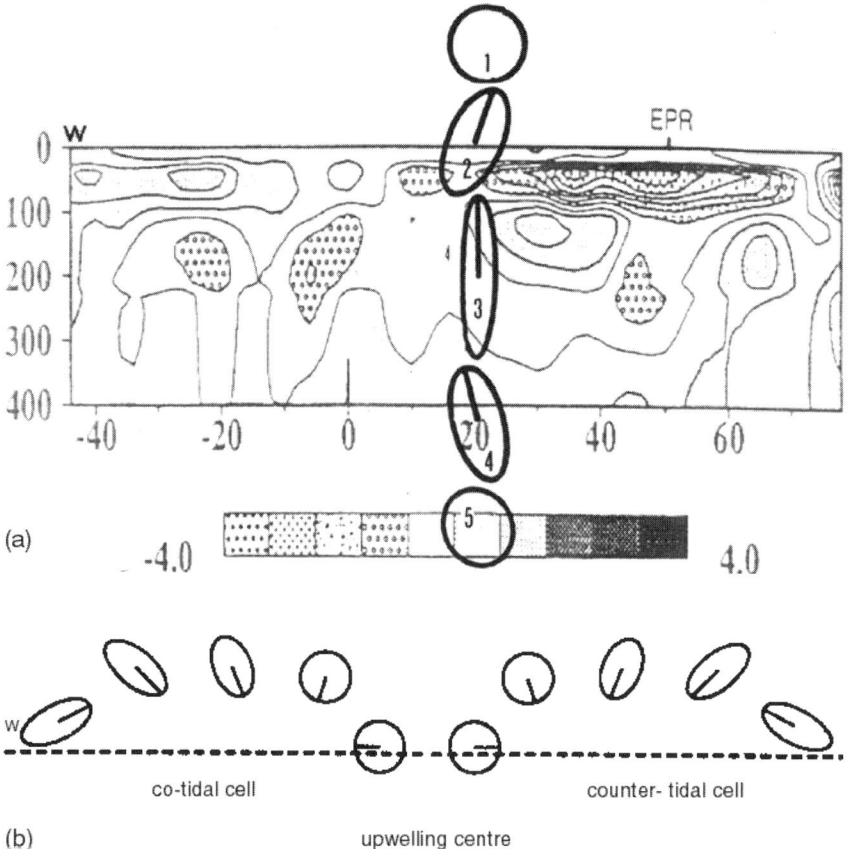

Fig. 4.12. Vorticity jointly under tidal induction (TVI) and thermal buoyancy: (a) Tidal strain ellipsoid in asthenosphere section. The background structure is west-east trans-Pacific profile D prepared by Morgan et al. (1995), representing anomaly (−4% to +4%) in vertical shear velocities. EPR: crest of East Pacific Rise with subjacent low-velocity material. Note vertical exaggeration. In any volume element during semidiurnal cycle, strain-figure acts sequentially as shown at 20° marker, at times 1–5: As locality approaches sub-lunar point, ellipsoid forms (time 2) and extends toward the Moon; during formation, particle line extends and rotates counterclockwise; as tidal bulge reaches maximum (time 3), line extends and rotates further; having passed sublunar point, (time 4), line contracts and extension reverses, but rotation continues in same direction. At time 5, extension has been removed by cancelation, but Q-dependent residual of the vortical motion (noncoaxial progressive simple shear) remains and is cycle/cycle cumulative. At quadrant positions, formation of the horizontal strain ellipsoid adds rotation in identical sense (cf. figures 3.8a,b). (b) Strain ellipsoids, rotation, and extension of particle lines in thermal convection, in upper part of convection cells in the Pacific basin (a, preceding), seen from south. In respect to the cell on the left, (sinistral circulation), tidal contribution enhances the convection; circulation in cell on the right, having opposite sense, is impeded. Vorticity in the grossly extended trans-Pacific limb of the EPR is cotidal, and in the diminishing east limb is countertidal. The cotidal west limb of the Mid-Atlantic Ridge, carrying the Americas, likewise is dominant.

84 Tectonic Consequences of the Earth's Rotation

In terms of gravimetric observations, the difference is indistinguishable. In respect to dissipation, integrated over the surface (Eqtn. (4.3)), the difference $\Omega' - \Omega''$ picks up dissipation whatever the mechanism.

Taking up from section 3.6.3, the outer Earth consists of lithosphere resting upon an asthenosphere six orders less viscous. With reference to term [2], equation (4.4), under non-reversing passage of the tidal bulge the stress $A_{r\varphi}$ exerts a tangential force on an element of the lithosphere with respect to the underlying asthenosphere (figure 3.11b). The force is part of a couple, sinistral in the figure, tending to shift the lithosphere west pari passu with eastward displacement of material at depth. The existence of A is not a product of the retarding torque. The causality (entropic) order is that rotation of the tidal stress ellipsoid results in vortical dissipation in an imperfectly elastic mantle, resulting in phase lag in the tidal bulge, hence a contribution to the retarding couple.

Professor Eduard Berg drew my attention to the action of stress diffusion, based on the work of Kolsky (1953; see also Gordon 1965; Elsasser 1969; Berg and Lutschak 1973; Bott and Dean 1973; Bostrom 1981). In the present context, with respect to a small secular stress such as A, the lithosphere acts as a stress transmitter. The situation is comparable to that of a floating ice sheet. In respect to an ice sheet wind-caused tangential stress minute per square centimeter is additive over many kilometers, summing at the shoreward end to the stress that crushes ships. An analogous aspect of stress transmission is seen in buoyancy convection. The torque generated in a volume element by the horizontal density-gradient is minute, within an extent of 1 cm being too small to overcome a viscous opposing force. Becoming coherent in cells, thus integrated over large distances, the torque results in convective flow overcoming large viscous forces.

With respect to the lithosphere, the role of stress-diffusion progressively becomes dominant. Following Elsasser (1969), upon application of $A_{r\varphi}$ to an elemental vertical slice forming a part of the lithosphere (figure 4.13), the stress acting elsewhere in the slab is:

$$A(\varphi) = E\, \partial u/\partial \varphi$$

where E is an elastic constant of the slab; u, the displacement.

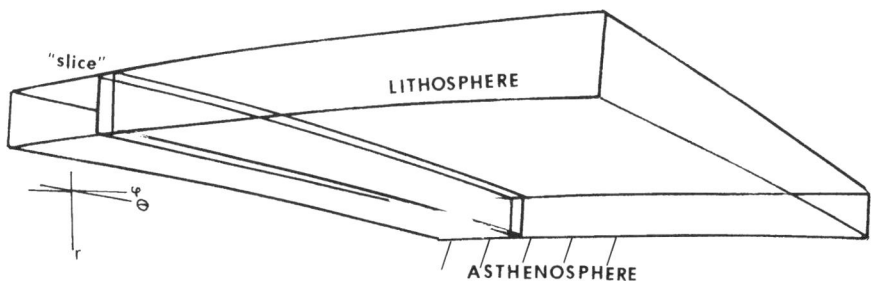

Fig. 4.13. Lithosphere, thickness h_1, and elemental slice seen from below, referred to Earth-fixed localized coordinates.

In the notation of Elsasser (1969), the diffusion distance is

$$\chi \approx \{(4h_1 h_2 Et)/\eta\}^{1/2}$$

where h_1 and h_2 are thickness of the lithosphere and asthenosphere; η, viscosity of the asthenosphere.

Estimates of viscosity in the upper asthenosphere vary by several orders of magnitude. Jeffreys (1976) found only limited evidence for the existence of subcrustal "finite viscosity." Pang, Yau, and Chou (1995), Peltier (1986), and Forte, Woodward, and Dziewonski (1994) find viscosity in the region of 10^{21} P. Based on postglacial uplift rates, Fjeldskaar (1994) estimates the viscosity of the upper asthenosphere to be less than 7×10^{19} Pa s. Hager (1991) finds that the global heat efflux, plate motion, and major geoidal feaures are best satisfied by an asthenosphere having a viscosity 2×10^{19} Pa s. Vetter and Meissner (1977) find viscosities an order less. Evidence has been brought forward (Anderson and Sammis 1970) that partially molten material tends to accumulate at the base of the lithosphere, acting as a lubricant with respect to horizontal motion.

At any point $A_{r\varphi}$ is minute, about 10^{-8} dyn/cm^2 (Bostrom 1981a, table 2). Values are derivable by differentiation employing the value of stress imposed by a static tidal bulge (Takeuchi 1951), based upon seismologic values of the elastic constants and density. For present purposes, Takeuchi's values have not changed.

Departing from Elasser's case, the stress is not of finite duration, and not applied solely to the end of a lithosphere plate. As time passes, each slab is acted on throughout by A plus the transmitted stress. The stress acting on a lithosphere element is then the small quantity $A_{r\varphi}$ *plus the horizontal sum of stresses acting on slices within diffusion distance.* The integration is represented by the convolution

$$A_{(\varphi\chi)} = (A_\varphi * A_\chi) \tag{4.5}$$

where A_φ, and A_χ are the primary stress and the stress transmitted by diffusion, respectively.

The displacement rate of the lithosphere is then

$$A_{(\varphi\chi)}(h_1 h_2)/\eta \cdot$$

The behavior of equation (4.5) and its consequences are illustrated in figure 4.14. It will be found that geologically speaking, the diffusion distance increases rapidly. Even for high-end estimates of the viscosity, within some millions of years transmitted stress extends to trans-oceanic distances. The trend is that the distance to which the stress propagates increases with increasing thickness of lithosphere and/or asthenosphere. The inferred velocity increases rapidly with decreasing viscosity, because this parameter affects both the diffusion distance (as $\eta^{1/2}$) and the resistance to motion. In consequence (fig. 4.14c), the displacement-rate reaches values too large to be plausible as compared to tectonic observations. On the basis of geotectonic analysis, O'Connell, Gable, and Hager (1991); Ricard, Doglioni, and Sabadini (1991); and Cadek and Ricard (1992) have inferred net westward displacement of the lithosphere reference to the underlying mantle of 1 – 3.6 cm/yr.

With reference to Elsasser (1969), Bott and Dean (1973), and earlier work by

86 Tectonic Consequences of the Earth's Rotation

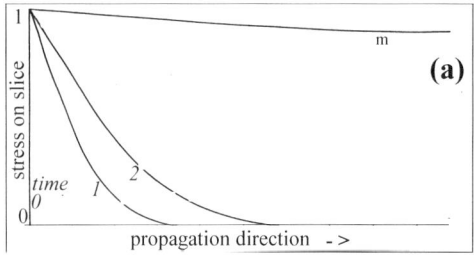

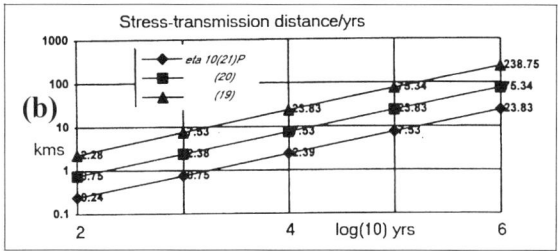

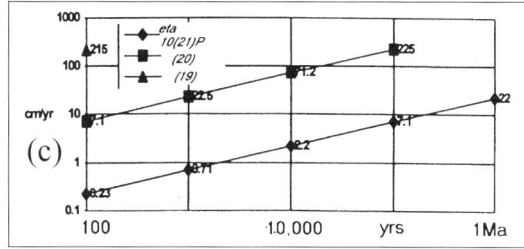

Fig. 4.14. Effects of stress diffusion. (a) Value of stress within lithosphere slab at times 1, 2, ..., m resulting from its continued application to elemental slice, commencing at time 0. Eventually (m), the stress throughout approaches that applied at one end. (b) Distance to which stress has been transmitted, reaching e^{-1} of its applied value, at times 100 up through 1,000,000 yr, for asthenosphere viscosity values 10^{19}, 10^{20}, 10^{21} P. (c) Theoretical displacement rates for test values of the viscosity.

Orowan (1958), it is apparent that in respect to a slab, stress-diffusion is self-limiting. Beyond an effective diffusion distance yield increasingly takes place by failure processes such as buckling and lithosphere thickening, and ultimately by departure from laminar flow.

In his treatise, Birkhoff (1950) reviewed unsolved problems as to boundary layers in hydrodynamics. Only a minority of the "paradoxes" he identified have become analytically soluble. A difficulty in computing numerical solutions, as alternative in this case, is that we possess few controls as to the rheology of the layer underlying the lithosphere; the points raised by Anderson and Sammis (1970) as to the viscosity of the top of the asthenosphere still are pertinent.

Several investigators including Chandrasekhar (1961), Gebhart (1962), Tozer

(1965), and Rice (1971, 1972), below, have examined the regime pertinent to a "lithosphere" resting on substrate having asthenosphere viscosity. In examining the occurrence of the Kelvin-Helmholtz instability at a horizontal interface between moving fluids Chandrasekhar (1961, section 102) shows that instability develops, initially as "crinkles" at the interface, *for even the smallest difference in velocity*. Helmholtz attributed this to the effect of a sharp, discontinuous interface. Chandrasekhar generalizes, showing that K-H instability develops likewise in the case of a continuous transition. In what follows I speculate that contiguous to an asthenosphere already under convection, the continuous shearing and developing "crinkles" devolve into vortical flow, adding components to the vortical flow under buoyancy.

4.3.2 Contribution to Convection

It may be illuminating first to examine the transition from conductive stability to that in cellular convective flow, upon reaching a critical Rayleigh number. The equations of Rayleigh (1916) demonstrate that greater thermal transmission efficiency is present at points beyond his critical number, but do not clarify the physical transition process. Within a heated fluid reaching critical point, an infinitesimal horizontal buoyancy-gradient force is at first present between one volume element and the next; convective stability is attained via the development of flow in (self-)organised cells. The minute buoyancy force is then additive throughout cells by stress transmission, reflected by the high order of the term representing the spatial dimension, such as layer thickness, in the Rayleigh number. Imposed upon such a system, vortical stress under the wave tides is additive (spatially integrated) over the dimensions of large cells already extant; the process is potentially the more effective because the forcing stress is ab initio highly coherent, being in phase throughout the mantle. There appears a tendency to foster circulation in a non-random, cotidal direction, see section 5.5.4.

If significant, a tidal contribution must appear in the form of a bias in the buoyancy flow, accompanied by dissipation and phase-displacement. Viscous dissipation under the induction as against that under the thermal convection is considered as a function of the dissipation number.

The equation of flow is written (Chandrasekhar 1961):

$$\rho \partial u_i/\partial t + \rho u_j \partial u_i/\partial x_j = \rho X_i + \partial P_{ij}/\partial x_j,$$

in which ρ is density, u velocity, and P stress. P is a function of the viscosity resisting shear and the strain rate. X_i represents an external force acting on the fluid. The conditions may represent flow purely under buoyancy convection, or this with the addition of external stress components as by deformation of the "container," for instance by the migrating bulge or wave. Multiplying by the velocity, the viscous dissipation within a volume V (assuming incompressibility and uniform viscosity), is

$$\int_v \rho u_i X_i d\tau + \int_v u_i \partial P_{ij}/\partial x_j \, d\tau.$$

Chandrasekhar notes that in respect to thermal convection, flow commences at the minimum temperature at which a balance can be maintained between the kinetic energy dissipated in the viscosity and the energy released by the buoyancy forces, i.e., at the critical Rayleigh number. The Rayleigh number of the mantle by any plausible estimates is orders of magnitude greater than the critical value, about 1700, and requires the onset of convection in terms of the locus Σ of the marginal state in the space of parameters such as the thermal gradient, prescribing a surface beyond which heat transfer becomes more efficient via bulk flow than via conduction.

On this basis, Chandrasekhar assumes that the diffusion term in the momentum equation (accounting for heat transmission by conduction and viscous dissipation) is negligible compared with the term describing transport by flow, and concludes that viscous dissipation is not significant relative to the overall energy flux.

Tozer (1965) has approximated the ratio of viscous heating in convection to the overall heat flow in terms of

$$A^2 g \beta L / C$$

in which A is nearly unity, β the expansion coefficient, C specific heat, and L a linear dimension. For conditions in the upper mantle the ratio is as Chandrasekhar suggests very small, Tozer estimating 6×10^{-3}.

Re-examining Chandrasekhar's formulation, Gebhart (1962) and Rice (1971, 1972) defined an analogous dissipation number

$$\varepsilon_{visc} = \beta g L / C.$$

Quantifying, these authors showed that it is unsafe altogether to dismiss the role of viscous dissipation, among other cases in fluids having the viscosity of the upper mantle. Perhaps unexpectedly, under thermal convection decrease in the ratio of viscous dissipation versus conduction goes as increase in viscosity.

In the Earth Σ is subjected to disturbance by forces having among other components $T_{\alpha\alpha}$, (table 3.2), reaching values as large as those in thermal convection, and having a secular component. T is imposed on a system already at the point of yield in non-linear flow.

Lithgow-Bertelloni et al. (1993) have separated the toroidal from the poloidal component of "observed" lithosphere motion, plotting it against that obtained in 1000 test models assuming partition under random plate motion (figure 4.15). An implication is that the order-1 toroidal component, representing net lithosphere rotation, is not zero but amounts to some cm/yr.

The net-lithosphere-rotation implied by this and similar investigations (Ricard, Doglioni, and Sabadini 1991; O'Connell, Gable, and Hager 1991; Cadek and Ricard 1992) is difficult to explain in the absence of a force recognised to displace the lithosphere relative to the mantle. Faute de mieux, it has been shown (Ricard, Doglioni, and Sabadini 1991; O'Connell, Gable, and Hager 1991; Bercovici and Wessel 1994) that motion in this style accords with a specified incorporation, under random convection, of viscosity heterogeneities representing lithosphere plates already in existence.

In addition to the fact that this solution leads to potential discrepancies with respect to surface-observed length-of-day and lunar-separation data, the principal direction of lithosphere motion, westward, must then be ascribed to chance. It

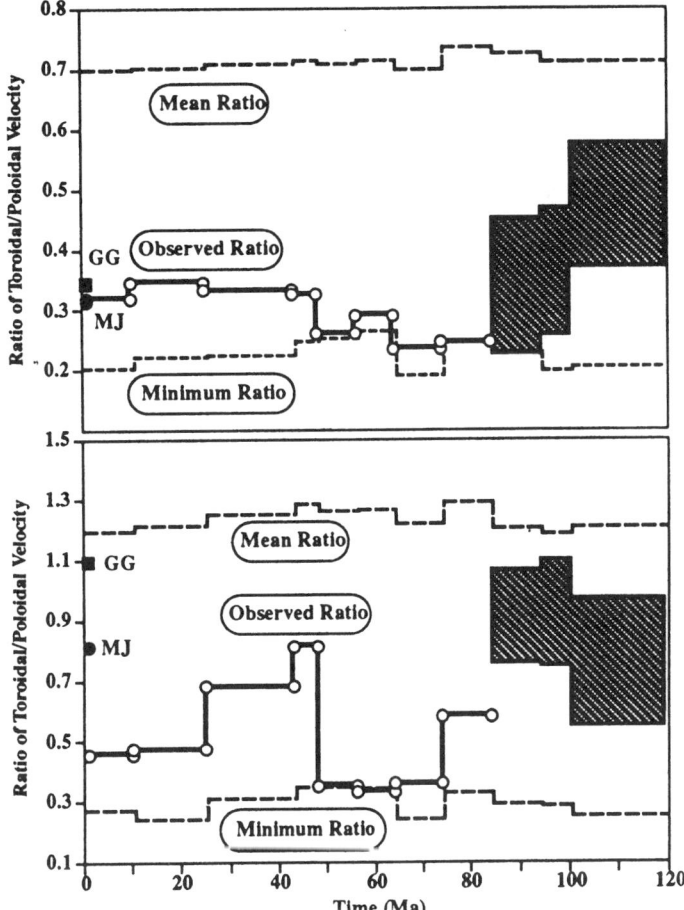

Fig. 4.15. Displacement of partition ratio, toroidal vs. poloidal plate motion, in the case of excluding net-lithosphere-rotation (upper), and permitting net rotation (lower). In each case, the observed ratio is compared with the mean of 1000 random-motion computations. MJ and GG indicate curve placement based upon alternative plate-motion models. (Reproduced from Lithgow-Bertelloni et al. [1993]).

might be preferable to suppose that rather than being fortuitous, the displacement from random of the partition-curve represents the effect of vorticity induction.

Despite Chandrasekhar's conclusion that viscous dissipation is insignificant relative to heat transport in conduction and bulk flow (streaming), to err on the high side let us suppose that in the overall flow as much as 1% of the total heat transference from the interior, approximately twice the estimate of Tozer, takes place via viscous dissipation. The total heatflow is thought to be 44.2×10^{12} W (Pollack, Hurter, and Johnson 1993). The most direct available observations of the "solid" tidal dissipation, independently employing gravimetric data (Melchior 1995b) and VLBI data (Herring and Dong 1994), indicate a dissipation rate of about 5×10^{11} W. Ocean effects and geographic heterogeneities cause there to be present a large potential error in both data sets. Nevertheless, it is apparent that the tidal viscous dissipation, perhaps operating as a bias, could account for a substantial part of the total. Within presently observed values it is possible that mantle flow is strongly biased.

4.4 Dissipation: Material at Critical Point

With reference to term [4], equation (4.4), the tidal stress σ_{ik}^T, besides incorporating rotating stress axes, does not act upon a passive earth having viscosity pertinent to such effects as deglaciation, but on mantle portions of which are maintained by convection at the point of transition, state change, differentiation with evolution of volatiles, and fracture. Several coupling mechanisms exist.

Information as to the existence of low-Q mantle material, related to subduction, was early obtained by Barazangi and Isacks (1971), figure 4.16. Widmer, Masters, and Gilbert (1991), Romanowicz (1994) and others have since mapped the global distribution of Q, using all wave modes and extending coverage by detectors sensitive to low and ultralow frequency, short of dc.

4.4.1 Volatiles and Phase Change

It has become apparent (Peacock 1990; Thompson 1992; Ahrens 1989) that in the upper mantle the role of the volatiles component in convection is important, and may be paramount, in determining physical properties. Pressure-dependent hydration ↔ dehydration transitions are intrinsic. In his address "Oceans in the Upper Mantle," Nolet (1994) has pointed to the large volume of water constantly entrained in this regime. Gasparik (1993), Nolet (1994), Philippot (1993), and others have identified no less than tens of transitions in the mineral assemblage $Mg.SiO_2 - Mg(OH)_3$ and others, as this is subjected to pressure/temperature change in regions of subduction and upwelling.

Figure 4.17, based on an illustration by Gasparik (1993), diagramatically displays regions of downflow, hence instability, under pressure increase (subduction) regions, and those of upflow, hence steady decrease (mid-ocean rift system; mantle plumes). The cycle results in large-scale circulation of water and volatiles such as CO_2 between the surface and the interior.

4.4.2 Subduction (Pressure-Increase) Regions

Computations by Peacock (1990) indicate that of the water entrained in subduction, estimated to be 8.7×10^{11} kg/yr, only 1.4×10^{11} (one-sixth) is expelled adjacent to the foundering slab in arc volcanism. The balance enters the extended, intramantle hydrous cycle. Seismically-mapped attenuation indicates that transitions, with attendant volatiles release and partial melting, are prevalent in regions adjacent to

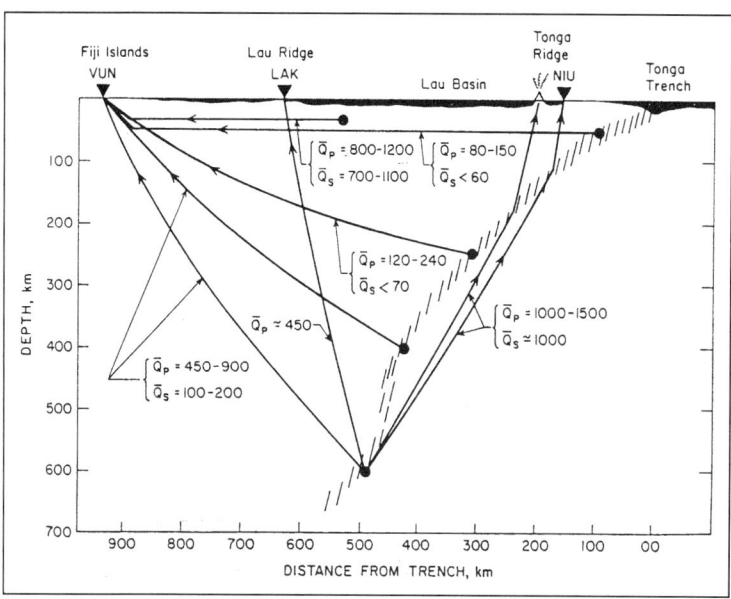

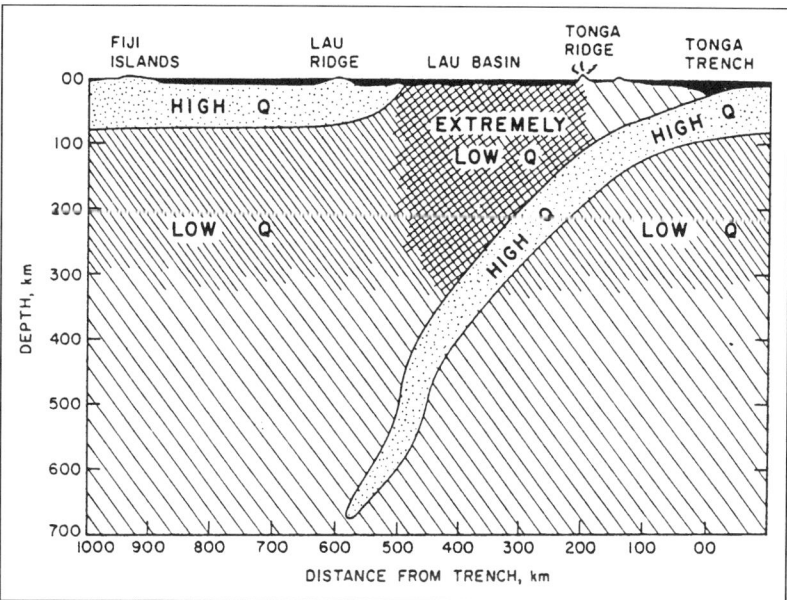

Fig. 4.16. Seismic attenuation in Tonga subduction zone (southwest Pacific). (Top) Raypaths with indicated values of Q. (Bottom) Section, region of low Q and "extremely low Q." The primary data are described by Barazangi and Isacks as "visual observations of large and unsubtle differences in the predominant frequencies and amplitudes of P and particularly S waves." (Barazangi and Isacks, 1971.)

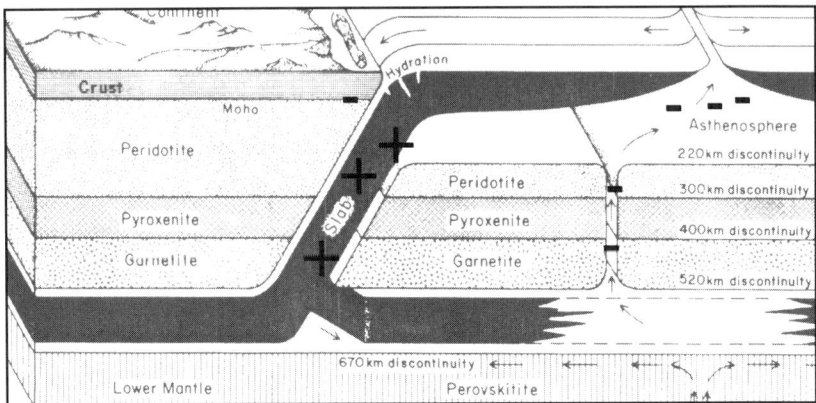

Fig. 4.17. Regions of downflow (pressure increase) and upflow (pressure decrease). The background is from Gasparik (1993, figure 8). Inscribed: representation of pressure-dependent phase transitions: + + + under continual pressure increase; – – – under ongoing pressure decrease. The convection system shown is two-layered, within a stratified upper mantle. Velocity data have been interpreted by many (figure 4.2) to suggest that material enters and leaves the lower mantle, constituting single-layer system, with volatiles circulation extending to all depths (Bell and Rossman 1992).

the foundering slab and in the overlying mantle wedge. Deep-seated dehydration and expulsion processes play a dominant role in transitions, differentiation, and viscosity reduction (lubrication).

The regime varies (Thompson 1992), according to whether the descending slab is mature, cold and thick, or young and relatively warm. In hot, young slab, as represented by the Juan de Fuca plate, water released by dehydration of minerals such as mica and chlorite is recycled directly back into the lithosphere. In old, thick, cold slab, as at the western Pacific margin, dense hydrous magnesium silicates (DHMSs) undergo successive P-T transitions, thus transporting water down to at least the 650-km seismic discontinuity. Transitions are especially pressure-sensitive in the region of steep P-T gradient above 200 km, but persist to great depth. Some transitions take place within a narrow pressure interval. Ito and Takahashi (1989) find that dissociation from magnesian spinel into perovskite and magnesiowustite takes place over a pressure interval less than 0.15 GPa at 1600°C.

4.4.3 Upwelling (Pressure-Decreasing) Regions

Upflow regions are marked at surface by vulcanism in the form of mantle plumes and the seafloor spreading network. Numerical models and field observation indicate that, as compared with downflow regions (foundering slab), those of upflow are less geographically concentrated. Morgan et al. (1995) suggest that upflow in tall narrow plumes, supplying the asthenosphere, accounts for a large proportion of the total. Hill et al. (1992) consider that convection is of two modes, that driving the lithosphere plates, and upflow as deep-seated plumes. S-wave velocity anomalies (Su, Woodward, and Dziewonski 1994), our figure 4.2, accord with geological conclusions as to the existence of deep-seated "super-plumes," or groups of plumes.

Perhaps due to lesser volatiles content and the potentially cascading effect of eruption and depressurization, state change under upflow may play a larger role than in downflow. Pressure decrease under hot conditions is characterized by partial melting, release of volatiles, in particular H_2O and CO_2, and gaseous state changes fostering volcanism. Wide-spread eruption of mantle-derived magma in such regions as southern Africa points to the depth range to which these effects extend. The mid-ocean rift system may be passive, with pressure decrease under lithosphere extension causing depressurization melting (Nicolas 1990) and consequent upwelling. Saunders et al. (1992) and White and McKenzie (1995) point to evidence that lithosphere thinning may lead to voluminous melt generation and the formation of large igneous provinces.

4.4.4 Tidal Modulation: Pressure Coupling

Under the convection, downgoing and upwelling material passes successively through marginal states. A fraction of the material is always at the point of pressure-transition. Thus in addition to the secular (vortical) distortion by tidal action, material in subduction zones is sensitive to the scalar and oscillatory component represented by the tidal pressure pulse.

Referring to term [4] in equation (4.4), the pressure variation in a foundering slab, under the semidiurnal lunar tide (T) plus subduction (C) at a point at 200 km in depth, is illustrated in figure 4.18.

The pressure-increase assuming foundering rate 2 cm/yr, section of average density 3.3 g/cm^3, is:

$$g \times 2 \times 3.3 \times 1/720 \text{ dyn.cm}^{-2}/12 \text{ hr} = 9.17 \text{ dyn.cm}^{-2}/12 \text{ hrs } (C)$$

In comparison, the pressure fluctuation in the course of one tidal cycle (typically ± 0.1 mGl) at the same point is:

$$g \times 10^{-7} \times 2 \times (2 \times 10^7 \times 3.3) \text{dyn cm}^{-2} = 13{,}200 \text{ dyn cm}^{-2}/12 \text{ hrs } (T).$$

The reason for the gross $C{:}T$ disparity is that in (T), although minute the tidal gravity variation acts on the entire 200 km column. The pressure change (C) under subduction in the same interval is represented by the addition of only 2/720 cm^3 of material to the overburden. Under adiabatic conditions, transition under solely the tidal, oscillatory pressure-change would be self-reversing (symmetric). However there is present a continuous increase in the backgound pressure, and dissipation is intrinsic.

The transition process is thus entropic, probabilistic and to a large extent as irreversible as the convection. The probability-gradient within a transition zone under increasing pressure is positive, the probability reaching 100% at the point at which the process is complete. Within a downgoing slab subject to secular pressure-increase, a mineral fraction nearing the upper, unstable pressure limit of phase n experiences transition, jumping into phase ($n + 1$), stable in this region. It is

Fig. 4.18. Pressure variation and phase-transition mimicked by shading, within foundering slab. The pressure at an instant is the sum of the two graphs. During secular pressure increase, the probability of transition to higher-density macrostate increases from 0 until it is 100%. Typically, within slab foundering at 2 cm/yr at a depth of 200 km, within the 1-day interval shown the slope of the subduction pressure-increase, 18 dyn/cm², is scarcely visible. During semidiurnal tidal cycle, the pressure variation (text) is many times greater. In material undergoing transition to denser mineral phase, probability is great that transition will take place to denser, stable phase during tidal pressure increase; within new equilibrium phase, probability is correspondingly small of reversion during decrease part of cycle. Throughout, slab transitions take place predominantly during increase in total pressure. Transition is experienced as decrease in Q_{bulk} and tidal modulation of the transition. The process is phase-inverted, but likewise dissipatory, in upwelling (pressure-decrease) regions.

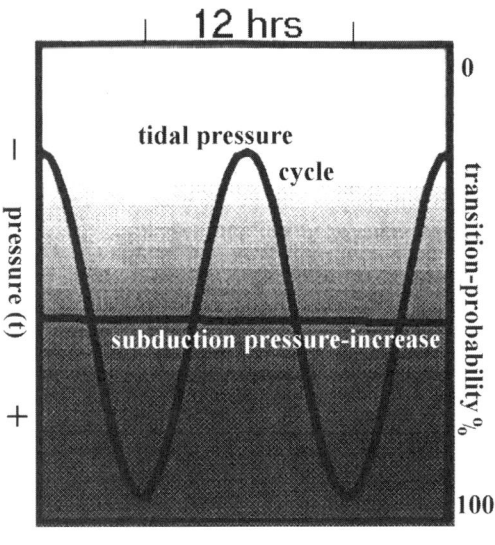

subject to a correspondingly reduced probability of reversion to the n state, less stable at this depth, during the inverse part of the tidal pressure cycle.

The effect is integrated over the thickness of a transition region and the pressure-modulation affects all depths. The total dissipation (irreversible increase ΔS in the Boltzmann entropy) is therefore the same whether a transition is concentrated within a thin zone, or extends over a depth of many tens of kilometers. Within statistical but no less robust confines (see for instance Reif 1965), the probability bias is ineluctable, operative in both pressure-increase (subduction) regions and those of secular pressure-decrease (upwelling). With respect to some transitions, their occurrence is accompanied additionally by state change, irreversible differentiation, and dispersal.

It would be surprising if, within volumes held continuously at transition by the convection, transitions are not more numerous during tidal pressure-change in the direction of stability, than in the inverse part of the tidal cycle. To judge by the amplitude of the tidal modulation, several hundred times the rate of secular pressure increase, it might be suspected that this form of bulk viscosity is responsible for a great portion of the "missing" tidal dissipation, such as that sought by Schwiderski (1985a,b). By the same token, the process constitutes tidal modulation of the advance in the convection. Significant oceanic loading must add to the tidal pressure variation, causing it to be unlikely that a linear, isotropic ocean-loading factor is applicable across regions incorporating phase transition; in re this, see recent papers by Matsumoto et al. (1997, 1998).

Anderson (1980) has pointed out that bulk absorption is required "somewhere in the Earth" to account for the decay of eigenvibrations. It might be supposed that transition under phase instability is relevant in this case also. Transition would be effected more efficiently by the grossly larger, longer-period tidal pressure variation. Stevenson (1983) has extended the treatment by Landau and Lifshitz (1959) of sound-attenuation (dissipation) in multi-phase media to pressure-changes of tidal frequency. In two-phase planetary material having a primary bulk viscosity of about 10^{12} P Stevenson examines the case of a dilute suspension of one phase in another, subjected to an oscillating pressure field, assuming continual thermodynamic equilibrium at phase boundaries. The bulk-attenuation factor is reduced to $Q \approx 10^2 - 10^3$. It is apparent that in the terrestrial case, the tidal pressure pulse acts upon a system perpetually pushed to the transition point by the convection. Within the fraction at transition, bulk viscosity may apparently decline without limit. The simultaneously imposed tidal shear strain adds to the probability of in-phase pressure-transition.

Together with this mechanism, the heterogeneous nature of the asthenosphere supports the possibility under consideration by the USGS for many years (Shaw, 1970; Shaw, Kistler, and Evernden, 1971), that localized, tidal dissipation may play a significant role in tectonic energy supply. Concentrated as it is within less than one ten-thousandth of the Earth's volume present in subducting slab, loss in transition-modulation may entail tidal energy deposition locally exceeding that in radioactivity and in shear friction.

Honda and Uyeda (1983) have examined the well-known paradox, that elevated heat efflux is found in regions of voluminous inflowing cold slab. It has become conspicuous that subduction regions are characterized by secondary heat-flow highs, rather than as is to be expected, being the site of the major global lows. The western Pacific region of subduction is the site of a well-marked high (figure 5.9). Lithosphere cooled from having traversed almost the entire Pacific floor flows into this region, there foundering. Ida (1983) has pointed out that

> "A most serious paradox in geothermal problems of the subduction zones is what mechanism could generate heat to allow partial melting for volcanisms. It is generally accepted that a downgoing slab should cool its environment so exhaustively that neither frictional heating nor other heat sources associated with radiogenic or chemical reactions could raise the temperature of the mantle enough, competing with the cooling effect."

Melchior and colleagues (Melchior and De Becker, 1983; Melchior, Ducarme, and de Becker 1986; Melchior and Ducarme 1991) report anomalies in the bodily tides in tectonically active regions such as the southwest Pacific (subduction region) and northeast Africa (upflow). Melchior (1995a) has observed a geographic correlation between tectonics, heat-flow and anomalies in the bodily tide. Tidal interaction with the subduction is more than sufficient to account for the heatflow anomaly. It may be thought that in subduction regions, enhanced heatflow may result from concentrated tidal dissipation; and vice versa. The process tends to be exponential, in that increase in temperature must be associated with reduced viscosity and localized drop in Q. Tidal modulation of volcanism (Mauk and Johnston 1973; Dzurisin 1980) is discussed in section 7.6. The phasing of earthquake swarms associated with the oceanic rift system (Klein 1976), following, indicates that tidal modulation is also active in upflow regions.

4.4.5 Cotidal Seismicity

Besides the nonlinearities present in phase and state transitions, the lithosphere, including parts of the upper mantle, is frangible (figure 4.8). Beyond a critical strain rate strain nucleation commences, lately examined by Nemirovich-Danchenko (1997). Continuous deformation is replaced by faulting and fracture, if abrupt with the radiation of energy in elastic waves.

4.4.5.1 Initiation. Tidal initiation of seismicity ("triggering") has been investigated for many years (e.g., Stetson 1929), partly as a forecasting tool, generally without success. The problem resides in correlating a combination of numerous tidal stress components, plus regionally-variable ocean loading, with a combination equally intricate and generally unknown, representing tectonic stress.

Kilston and Knopoff (1983) have identified an association of large California earthquakes with tidal reduction of stress across faults. The association found is with the solar (as against the larger lunar) semi-diurnal, the lunar fortnightly, and the 18.6-yr tidal components. The technique correlates seismic events with the instantaneous dipolar (symmetrical) tidal stress field. The results identify "unlocking," namely release by tidal stress of elastic energy already present, as distinct from direct dissipation of tidal energy. Weems and Perry (1989) demonstrate correlation of the gravity tide and the incidence of large earthquakes of the eastern North America littoral. The correlation represents tidal modulation of the dissipation, but the extent to which this represents unlocking cannot be identified.

Klein (1976) has specifically correlated events on the mid-ocean ridge system with the tidal stress component topically adding to the tectonic stress. In this realm the tectonic stress (Sykes 1967) is shear in fracture zones, extensional in spreading segments. Klein separately examines earthquake swarms and discrete large events. The latter characterize the fracture zones. Swarms, distinguished from aftershocks, represent advance in the convection in the form of volcanism and fracturing (Francis 1968, 1974; Kanamori and Stewart 1976). Klein's technique incorporates dynamic rather than static tidal stress and in practice is sensitive to the direction (polarity) of the nonreversing torque, (figure 4.19).

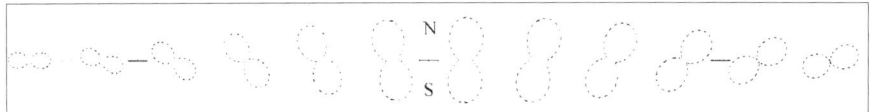

Fig. 4.19. Computed tidal strain at one-hour intervals over one semi-diurnal cycle, at a mid-latitude point in the northern hemisphere, displaying nonreversing dextral rotation. In the Southern Hemisphere, rotation sense is the opposite. (After Klein 1976.)

In this scheme, to act as an initiator a component of the tide stress must act in the direction of the tectonic stress. Thus Klein identifies dissipation of unknown magnitude, in the form of displacement in the tidal direction times tidal stress (figures 4.19, 4.20); cf. [4], Eqtn. (4.4).

4.4.5.2 Far-Field Displacement. Mass displacement in the seismic focal region is relatively small, but regionally a large mass is permanently displaced. Figure 4.21 shows the displacement estimated for an event of moderate size ($M_s = 6.9$). The earthquake took place on the Mid-Atlantic Ridge in the region of cotidal events tabulated by Klein, but was of a later (1974) series.

The distal displaced mass increases at least in proportion to the dissipation, namely in respect to major and great earthquakes by a factor of thousands. Investigations have confirmed that the displacement in great events extends to large portions of the globe (Steketee 1958; Chinnery 1961; Maruyama 1964; Press 1965; Berg and Pulpan 1971; Sterling and Smets 1971), figure 4.22. Press (1965) finds that in repect to major earthquakes, displacements as large as 0.5 cm are to be expected at distances of several thousand kilometers.

Press (1965), Berg and Pulpan (1971), and later investigators (Jiao et al. 1995) have found that for reasons not fully understood, far-field displacement is grossly larger than theory predicts. It might be expected that concentrated within a two-dimensional lithosphere the offset is larger than that forecast for a homogeneous half-space. Displacement propagates within regionally variable crust and mantle. Far-field offsets related to events in the Gulf of California were conspicuous in the trace written by the 1-km Cascades laser extensometer (Bostrom and Vali 1968;

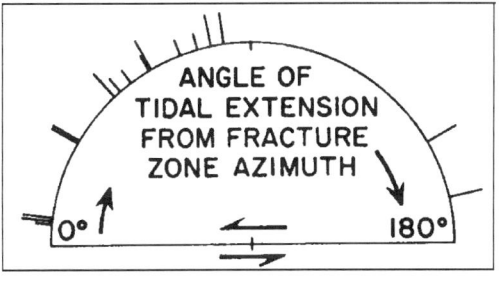

Fig. 4.20. Instrumentally derived stress angles of Atlantic fracture-zone strike-slip earthquakes, 7°N to 74°N, as compared with topical tidal extension. Of the 16 represented mainshocks on FZ's inferred to be strike-slip events, the tidal shear of 14 is in the direction of the displacement. (After Klein 1976.)

98 Tectonic Consequences of the Earth's Rotation

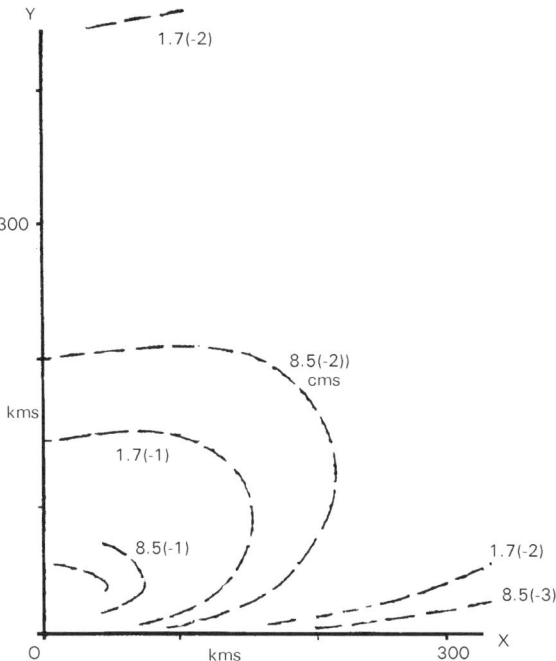

Fig. 4.21. Computed far-field displacement at time of the 1974 Oct. 16 strike-slip event, Gibbs fracture zone, northern Mid-Atlantic Ridge. The formulation here used is due to Steketee (1958) and Press (1965, equation 8), employing the focal displacement, 170 cms, and fault length 70 kms instrumentally determined by Kanamori and Stewart, 1976. I assumed the ratio fault-length to vertical extent to be 10, representative of a crustal fault in this region of thin crust. X axis is in direction of fault plane. 8.5(−2) signifies 8.5×10^{-2} cm.

see also Sterling and Smets 1971). The recent deep 1994 Bolivian earthquake (Jiao et al. 1995) was accompanied by static displacement of some centimeters at stations hundreds of miles from the epicenter.

Deconvolution permits separation in seismograms of source, path and receiver functions (e.g., Priestley, Zandt, and Randall 1988). In respect to major earthquakes, the source structure and history have been found to be far from simple (e.g., Wyss and Brune 1967: Johnston and Langston 1984). Brune (1991) concludes that data have established beyond a doubt that *most* earthquakes have complex sources. Simple ruptures are the exception. Even with respect to a tectonic feature for long closely investigated, "the absolute state of stress along the San Andreas fault remains as enigmatic as ever." Slip seems to be occurring devoid of an identifiable driving force, and in the presence of normal-stress which might be expected to cause continuous locking. In the case of the deep 1994 Bolivia earthquake,

Tidal Action 99

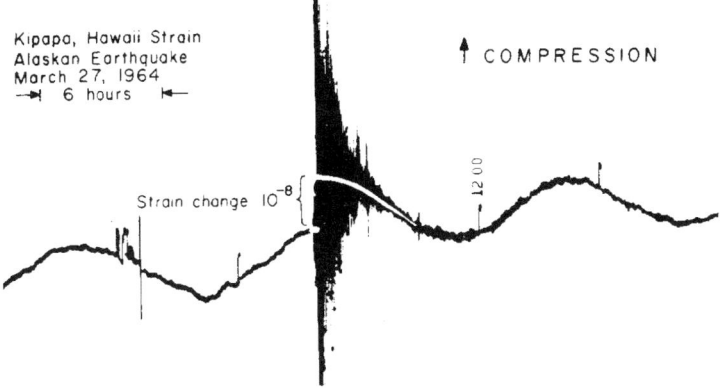

Fig. 4.22. Strain recorded in the central Pacific Ocean (Hawaii) at the time of the 1964 great Alaskan earthquake. As authentication, Press displays the tidal variation recorded for some days before and after the event. This has since been shown (Wyss and Brune 1967) to have had a most complex source-history. (From Press [1965]).

network records have shown (Hara, Kuge, and Kawakatsu 1995) that failure occurred in shear. At such depth, fracture rather than for instance an implosive (isotropic) failure mode is difficult to understand. Digital records of well-instrumented near events (e.g., Anderson et al. 1986) have revealed major source complexity.

4.4.5.3 Energy Release. Energy release at the time of seismicity (Kanamori 1978) consists of: (1) a part expended in the generation of the elastic wave reaching seismometers; (2) a part dissipated in the focal region as heat, not available to wave generation; and (3) a part represented by aseismic strain-energy change.

In respect to the total available, Kanamori (1978) contrasts seismogenic energy with the terrestrial heat efflux, equivalent to 7×10^{27} erg/yr. The average rate of seismic energy release s.s. is 4.5×10^{24} erg/yr, approx. 1/1500 of this figure.

If the stress drop in seismic slip is smaller than the prevailing tectonic stress, the unrecorded aseismic strain energy change (part 3) may be several times greater than the radiated wave energy. It may then represent a more substantial part of the total. The laboratory fracture strength of rocks and the field-apparent magnitude of the tectonic stress, as in the case of the San Andreas, are consistent with the possibility that only a fraction of the strain change may be signaled as waves.

4.4.5.4 Correlation. It will be apparent that almost no information is held as to a correlation of tidal stress components with the displacement in great earthquakes, accounting for most of the seismic dissipation. The correlation problem is bound up with major uncertainty surrounding the tectonic stress and displacement, rather more than with formulation of the tide stress. Besides the source uncertainties identified by Brune (1991), Rudnicki and Kanamori (1981) have distinguished a zone of inelasticity extending beyond the bounds of the seismic slip surface. It might be expected that a zone experiencing slow relaxation is normally present in earthquake belts, but little information exists as to event-dependent deformation or its time-scale. With the advent of accurate electronic survey (chapter 7) this situation may be changing. From the point of view of entropy, it would be remarkable if tidal stress components act globally contrary to the dissipation direction. To judge by the multiyear variation observed in global seismicity (Kanamori 1978, figure 2; Romanowicz 1993) it may eventually be necessary to take into account the very long period (orbital) tidal components. Du (1994, 1996) has concluded that globally, the incidence of all great earthquakes is affected by the 18.6-yr lunar declination cycle.

Figure 4.23 compares seismic and tidal energy dissipation. If even a large proportion of wave-generative seismic dissipation turns out to be cotidal, this path can account for only a minor fraction of the solid-Earth dissipation consistent with gravimetric and VLBI observations. In the extreme, if coseismic strain energy change amounts to several times the expenditure in wave-generation, and the process is strongly cotidal, this dissipation path might account for one-third of the tidal dissipation.

4.4.6 Tidal Coupling: Oceans to Solid Earth

No matter the hydrodynamic mechanism proposed (Schwiderski 1980, 1985a,b; Le Provost et al. 1994; Sjoberg and Stigebrand 1992), a large part of the global tidal dissipation must be marine. A proportionately large part of the torque responsible for day-length increase must be exerted via a mechanical coupling, ocean to solid earth. Schwiderski (1985a,b) and Melchior (1989) have drawn attention to the fact that under the low water velocity there obtaining, bottom friction in the featureless deep ocean must be negligible. Schwiderski finds that oceanic torque acts in direction to *accelerate* the rotation. To produce the deceleration well established astronomically, it has been proposed that coupling is effected via an increment in oceanic pressure on the eastern flank of seafloor topography, such as the continental margins.

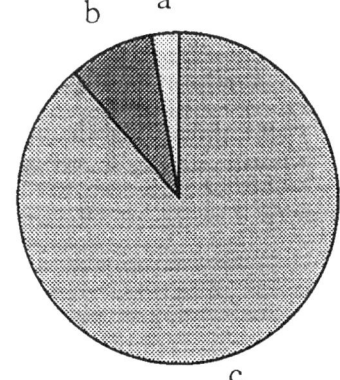

Fig. 4.23. Relative energy expenditure. (a) Elastic wave-generative seismicity, (b) possible coseismic strain energy change, (c) solid-Earth tidal loss. (Sources:) a, Kanamori (1978), his 5-yr running average; b, lower part of range estimated by Kanamori and subsequent investigations (e.g., Brune 1991) of tectonic stress; c, solid-Earth tidal Q based on earth-tide observations by Centre International des Marées Terrestres (Melchior 1995a,b) and VLBI data (Herring and Dong 1994).

The consequent direct tangential torque within the asthenosphere must be minute, averaging less than 10^{-4} dyn/cm^2, equating with the retarding torque, to comply with the astronomically bounded reduction in rotation, and must have negligible tectonic effect in the case of a passive uniform Earth. In the convective Earth, the marine bottom pressure variations are locally comparable in magnitude with those imposed by the gravity tide (figure 4.18), and might be expected to induce modulation of transition via the same mechanism.

Figures 4.24 a,b (reproduced from Le Provost et al. 1994) display the difference between the marine tide computed on a hydrodynamic basis and the satellite-observed total tide M_2 in the seafloor plus overlying ocean.

In the notation of Ray and Sanchez (1989), the figures display the difference in the M_2 tide:

$$\delta\zeta_o[\text{cm}] = \zeta[\text{estimation; hydrodynamic, tide station control}]$$
$$- \zeta[\text{observed; Topex satellite}].$$

Both terms on the right take into account tidal loading, computed on the conventional isotropic linear basis (Ray and Sanchez 1989), and the conventionally estimated contribution of the solid-Earth tide. The latter assumes solely oscillatory spheroidal deformation and seismologic value of Q.

Ray and Sanchez find that displacement under seafloor loading may amount to 10 cm. The authors draw attention to a possible accuracy limitation introduced because of employment of a "homogeneous" (not regionally-variable) load tide. The incorporated solid-tide correction is small, based on the assumption (Zschau

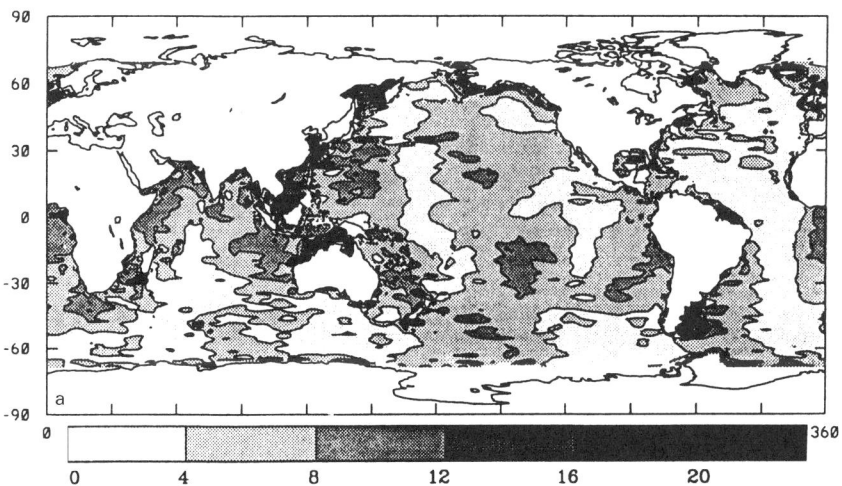

Fig. 4.24. (a) Vector difference (cm), M_2 hydrodynamically computed based on bathymetry and forcing function, vs. M_2 observed via altimetric satellite (Cartwright and Ray 1991). From Le Provost et al., 1994.

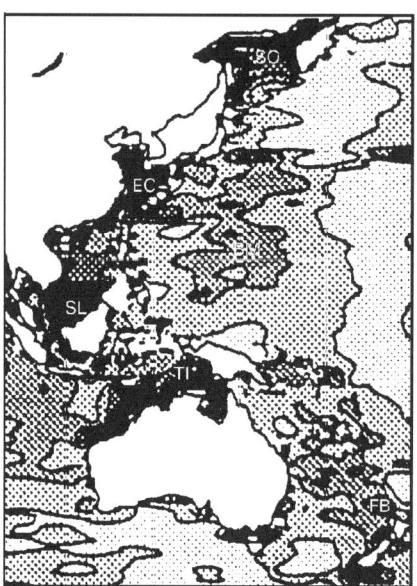

Fig. 4.24. (b) Region of principal discrepancy (western Pacific margin), between computed marine tide and that incorporating Topex data. The discrepancy, up to 25 cm, is systematically associated with subduction zones and back-arc basins overlying them. These are identified as: SO, Sea of Okhotsk; EC, Yellow and East China Seas; BM, Bonin-Mariana subduction; SL, Sundaland,Timor subduction region; FB, Fiji Basin, backing the Tonga-Kermadec subduction. Contours as in figure 4.24(a). Data from Le Provost et al. 1994.

1978) that dissipation takes place in a passive earth, solely in the mode oscillatory shear.

In an ideal that is impossible to reach, the values shown in figure 4.24(a) should be insignificant everywhere, and have no correlation with the bathymetry. Le Provost and co-authors draw attention to misfit of the values east and west of Africa, at the longitude of Australia, and west and east of South America. They note also the difficulty in reconciling estimated and observed values in areas such as the Patagonia shelf.

A mild discrepancy correlates with the topography of the Mid-Atlantic Ridge. Schrama and Ray (1994) have incorporated in a later solution Topex data based on constrained orbital parameters. The latter solution (cf. their Plate 1) does not eliminate the clustering of short-wavelength anomalous values at the west Pacific margin (figure 4.24b). The global discrepancies may be attributed to the following:

1. Dissipation via mixing processes in internal tides generated by bottom topography (Sjoberg and Stigebrandt 1992); the anomaly following the mid-Atlantic Ridge (figure 4.23a) resembles the feature shown in Sjoberg and Stigebrandt's figures 9 and 10;
2. An unrepresentative distribution of marine tide stations, serving as benchmark values in respect to the computed tide;
3. A geographically variable load factor not taken into account; and
4. Use of a fictitious model of the solid tide component.

These and similar factors are taken up in section 7.4.5.

4.5 Dissipation in Pre-Stressed Earth

Summarizing, in figure 4.25 some dissipation paths are tabulated against processes identified in equation (4.4) and succeeding sections. Stylized sketches cannot display the nonlinearities and discontinuities with which the material of the lithosphere and mantle is replete. In addition, in vorticity induction as in thermal convection, self-organization (cell-formation) is indigenous, constituting the most efficient flow mode. It is convenient to visualize both a volume element and Moon as resident in the Equator plane, with coordinates $+\varphi$ east, $+\theta$ north, and $-r$ radial.

Overall, the vorticity path with its various components seems likely to account for the major part of the dissipation. This form of induction, due to its proneness to bias, interact or "cooperate" with the convection, must be predominant in the mantle. Near the surface, having contributions by fracture, phase change, and differentiation (encompassing any form of "viscosity" ultimately definable by a constitutive equation), this mode encompasses what Russian geologists describe as "tectonic flow." Having in mind the magnitude of the dissipation and the pervasiveness of the stress, it seems inconceivable that over an extended period advance in global tectonics can take place in antiphase (speeding the rotation), rather than in phase with the tidal incremental stress. For the same reason a tectonic component, the folding ubiquitous in sediments, apparently must advance predominantly in phase, rather than in antiphase. With respect to the observation problem (chapter 7), the question is "To what extent?"

Some specific components appear incapable of accounting for a major part of the dissipation. Our best evidence is that tidal losses amount to many times the dissipation represented in seismicity; no matter how high the correlation may be, coseismic dissipation can account for only a small part of the phase lag.

Conversely, the quantities measured permit that a portion of global seismicity may represent the release of tidal energy accumulated long-term. If this path is substantial, seismicity at plate margins, as for instance at the western termination of the Pacific plate, may in part represent release of tidal energy accumulating as strain in the elastic Pacific lithosphere. Seismicity resulting from the accumulation would consume only a fraction of the induction. So far as known, only the investigation by Klein (1976), which takes into account the secular rotation factor, is relevant to the problem of correlating tides and seismic incidence. With reference to the seismic incidence reported by Mogi (1974, 1979), Romanowicz (1993), and Du (1994, 1996), it may be worth pursuing further the incidence of seismicity relative to long-term peaks in the induction.

With respect to the mechanics of the dissipation, Truesdell (1954; see also Serrin 1959) defines his geometrical rotationality-measure, describing flow in an extended medium having any physical and mechanical constitution (for instance frangible, elastico-viscous, or containing phase transitions)

$$\mathcal{R} = \{(\Omega{:}\Omega)/(D{:}D)\}^{1/2},$$

where Ω and D are respectively the vorticity and deformation tensors. It has been shown mathematically (Noether 1918), and confirmed physically down to the sub-

104 Tectonic Consequences of the Earth's Rotation

Mechanism, text section	Period	Term equation (4.4)	Deformation mode	
Spheroidal oscillation S. 3.5	1/2 day	[2]	$(e_{\varphi r} + e_{r\varphi})$	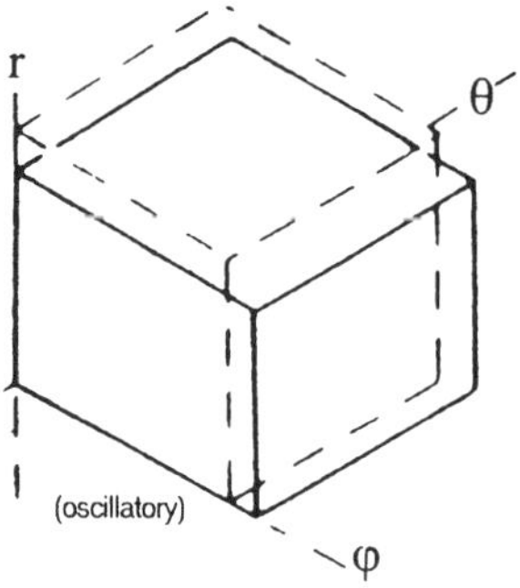 (oscillatory)
Phase-compression, expansion S. 4.4.4	1/2 day	[4]	$(e_{\varphi\varphi} + e_{\theta\theta} + e_{rr})$	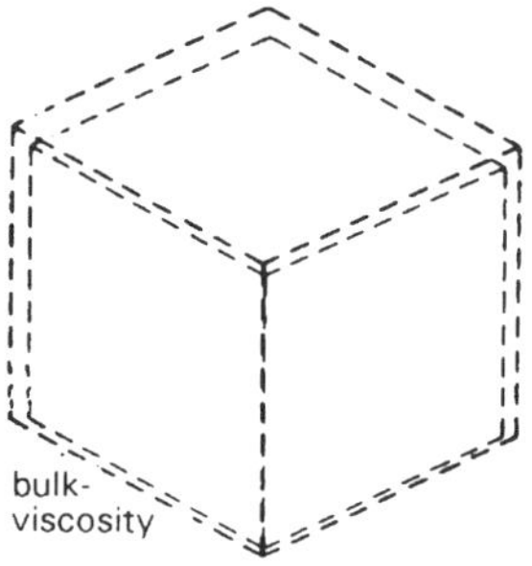 bulk-viscosity
Vorticity induction S. 4.3.2	Secular	[4]	$(e_{\varphi r} - e_{r\varphi})$	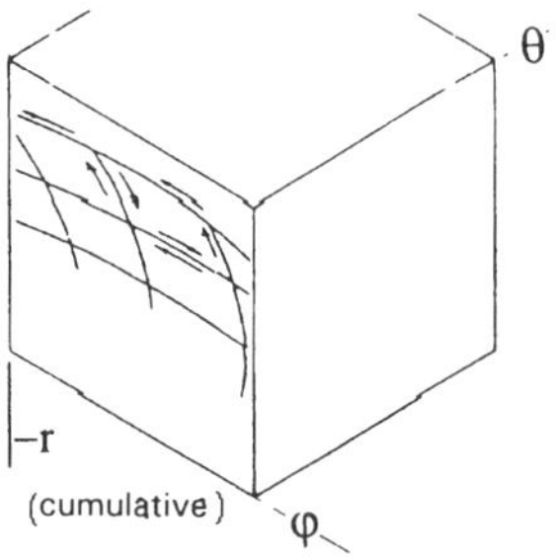 (cumulative)

Fig. 4.25. Dissipation paths: deformation and distortion in tidal mantle. "Term" refers to term in equation (4.4), see text.

atomic level, that linear (translational) and angular momentum are separate entities, independently conserved. One cannot be derived from the other.

In employing orthodox (Darwin/Love) tidal models, by admitting no antisymmetric stress term we fictitiously restrict Ω to the value 0. In actuality, the wave tides being fundamental, the value of $\mathcal{R}$ must be finite and of primary significance. Integrated over time, displacement under imperfect elasticity is vortical and cumulative. In terms of the paradigm used by Kelvin and Lamb to make comprehendible Helmholtz's vortex motion, cycle to cycle the principal tides (M_2, S_2) induce *exclusively* incremental circulation; rather than Truesdell's numerator Ω having zero value, its value must be finite and significant, whereas in the denominator, $D \Rightarrow 0$.

Dimensionally, the domain of Ω is that of the thermal convection representing extant vortical flow. By excluding vorticity from tidal models, a priori we exclude its self-organizing quality, that makes it prone to couple with or form part of the convection.

In chapter 5, an attempt has been made to identify dissipation paths by employing the geological record.

5

Tectonic Record

5.1 Overview

If global tectonics is driven purely by thermal convection, which necessarily is random, it should be devoid of directional bias. Conversely, if tidal action is significant it should appear as latitude-dependence, or east-west asymmetry, in geodynamic processes. To identify phenomena many of which are indetectible by instruments, this review calls upon the tectonic record.

Satellite and oceanic data have now produced a comprehensive view of the Earth. Synoptic images permit extrapolation of a century of geological mapping, deeply analytical, but restricted mainly to "dry land." Within localized maps the action of a broad-scale system, such as the hemispherical tide, is hard to perceive.

In what follows lithosphere motion is first reviewed, commencing with plate-to-plate (relative) motion. An attempt is then made to separate "absolute" motion, namely motion of the lithosphere relative to the inner Earth. In recent years reference to mantle plumes, forming "hotspots," has resulted in models calling for net lithosphere rotation about a high-latitude pole. Gravity-gradient images here summoned point to the existence of a multiplicity of plumes, but also to the limitations in separating lithosphere motion by observation of plume trails. It may be preferable to relate plate motion to tomographic deep mantle objects. As "excess flattening," gravimetry has shown that an unidentified agency has accumulated, as excess in equatorial latitudes, anomalous masses representing several times the potential energy in those of the whole of the rest of the planet.

Paleomagnetics cannot distinguish rotation about the magnetic polar axis, but corroborates an early perception of Runcorn (1962b), that plates tend to rotate clockwise in the Northern Hemisphere, and vice versa. Correspondingly, plate motion during the Tertiary has been disproportionately fast in low latitudes. In the principal regions of seafloor spreading, those of the low-latitude Pacific and Atlantic oceans, motion is predominantly E-W.

An asymmetry has been apparent between east-facing and west-facing Pacific subduction zones. The west limb of the East Pacific Rise is grossly extended, whereas the east limb is diminishing, under replacement by the west limb of the mid-Atlantic spreading center. Subtraction from the eastern Pacific Ocean of cooling lithosphere, its westward trans-Pacific motion, and its accumulation beneath the western equatorial margin, has resulted in the formation of the Earth's most intense geoidal high. An analysis by Wilson (1973), identifying westward extension of the Atlantic realm, increasingly is supported by geophysical data.

An attempt has been made to correlate the tectonic record with the tidal dissipation paths identified in chapter 4. The agency chiefly responsible, both for the dissipation and the tectonic effects, is likely to be vorticity- (circulation-) induction. This agency is secular and produces lateral motion. Its effects therefore are cumulative, but lacking a reference point exceedingly difficult to measure. Without the inclusion of this motion component, it seems not possible that flow models can tally with the tectonic record.

It is noted that the difference between the static lithosphere of Venus and Earth's mobile lithosphere is associated with (but not necessarily caused by), the presence of the Moon, rapid Earth rotation, and correspondingly vigorous vorticity induction.

5.2 Plate Motion

5.2.1 Relative Motion

A recent representation of plate relative motion (figure 5.1, after Gordon [1995]), is based on geological data. To an important extent the motion component in late-Tertiary data accords with that shown by space geodetic observations of current motion. The author points out that whilst the interior of plates essentially is rigid, older views must be modified to the extent that non-rigid boundary regions may occupy as much as 15% of the Earth's surface.

5.2.2 "Absolute" Plate Motion

The motion of plates relative to the Earth's interior has been examined by Minster and Jordan (1978), referring to the trace of mantle plumes, assumed to be quasi-stationary (figure 5.2). The subject is taken up in section 5.4.

5.2.3 Vortical (Toroidal) Motion

When plate motion is dissociated further (Forte and Peltier 1991), into poloidal and toroidal components (figure 5.3), it becomes apparent that counterclockwise rotation is predominant in the southern hemisphere, clockwise in the northern. In respect to alternative models the disparity is greater.

Ricard, Doglioni, and Sabadini (1991), figure 5.4, point out that fundamental, order-1 toroidal motion represents a net westward rotation of the lithosphere as a

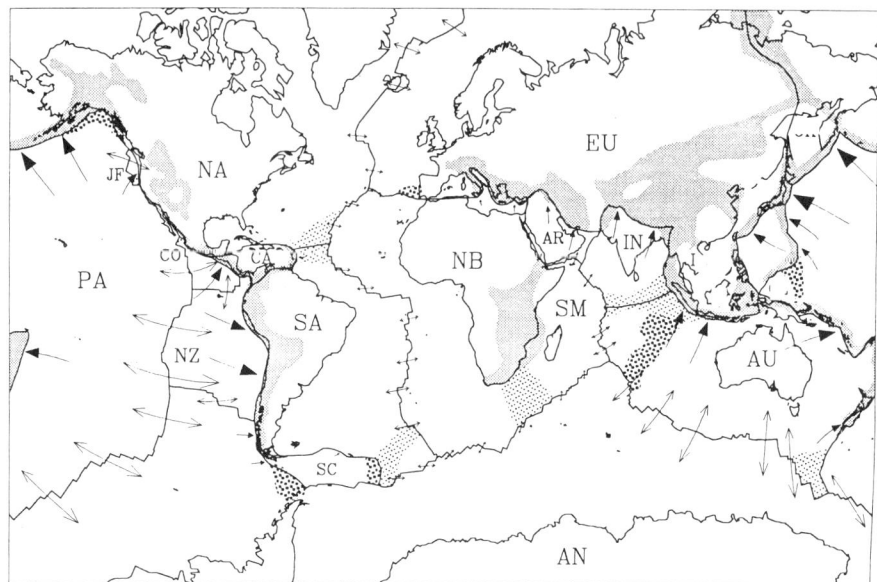

Fig. 5.1. Plate-to-plate relative motion, based principally on geological information. The length of the arrows describes displacement per 25 Ma. Fine, medium, and coarse stippling indicate regions of deformation, identified respectively: on land; at sea on the basis of nonclosure of plate motion circuits; at sea on the basis of seismicity. The data are from Gordon (1995), who identifies additionally a number of smaller, plate-like rigid regions, and cautions that the plate boundaries shown in this representation are idealised.

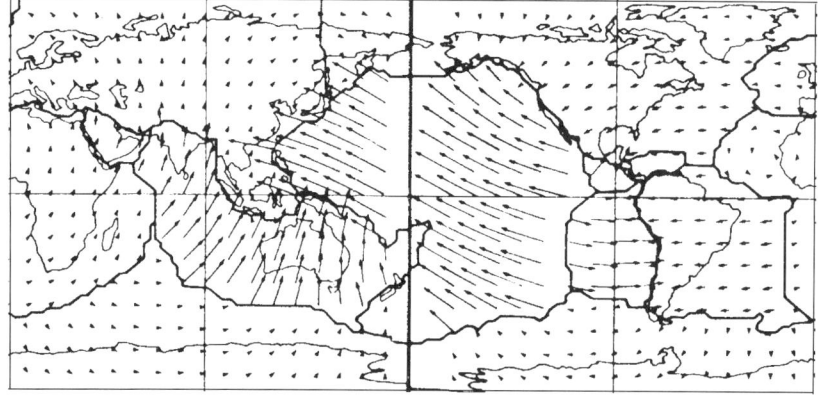

Fig. 5.2. Surface plate "absolute" velocity field, referred to mantle-plume frame. (From Forte and Peltier [1991], based on data from Jordan and Minster, 1978.)

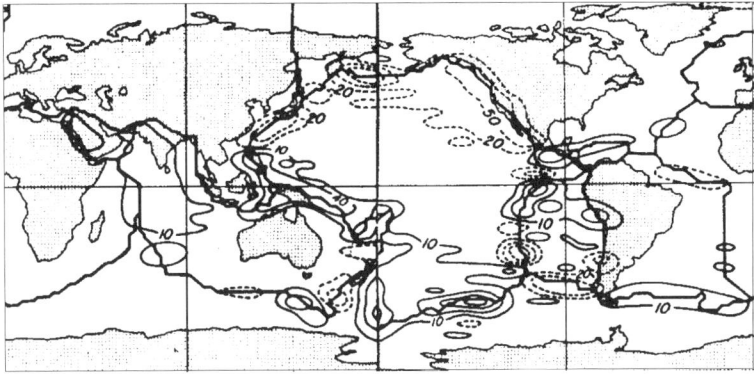

Fig. 5.3. Radial vorticity, detailed to degree and order 32, based on the plate motion model of Minster and Jordan (1978). Solid and dashed contours represent respectively counterclockwise and clockwise circulation. The units are 10^{-9} rad/yr (Forte and Peltier 1991a,b).

whole about the underlying mantle. Dependent upon the absolute-motion model used, the rotation rate lies between 0.11°/Ma (1.2 cm/yr) in respect to older models and 0.32°/Ma (3.6 cm/yr) using NUVEL 1. In the absence of a known torque other than the extremely weak retarding couple, in absolute-motion models the condition has been imposed that the lithosphere experiences no net torque relative to the underlying mantle.

Simple westward motion is not seen in any one plate, at one time, but makes its appearance as an overall bias, figure 5.4a (Gripp and Gordon 1990). A systematic feature of plate motion, identified by Gordon and Jurdy (1986), is that throughout at least the Cenozoic, plates have moved faster near the Equator than when near the poles.

Cadek and Ricard (1992), figure 5.4b, have adjusted no-net-rotation models based on the data of Gordon and Jurdy (1986) in such a fashion as to conform with a priori no-net-torque, and apparent former poles of net rotation. In explanation of the motion, Ricard, Doglioni, and Sabadini (1991), O'Connell, Gable, and Hager (1991), Bercovici and Wessel (1994) and others note that a large fraction of toroidal flow can be engendered in models incorporating lateral viscosity heterogeneities, in flow primarily poloidal (as in buoyancy-convection). Data compiled by Gordon and Jurdy (1986) have been used by Cadek and Ricard (1992) to extend the record of global lithospheric rotation having the same sense back to 64Ma. If the heterogeneity model of plate motion is complete, the principal motion component (westerly) is fortuitous.

Lithgow-Bertelloni et al. (1993) have located net-rotation poles relative to present geography employing independent models of plate motion (AM1–2; the data of Gordon and Jurdy [1986]; and NUVEL 1). It is concluded (figure 5.5) that a significant change in the toroidal/poloidal motion ratio took place at 43 Ma. Since that time there has occurred a decrease in overall plate motion. The toroidal

110 Tectonic Consequences of the Earth's Rotation

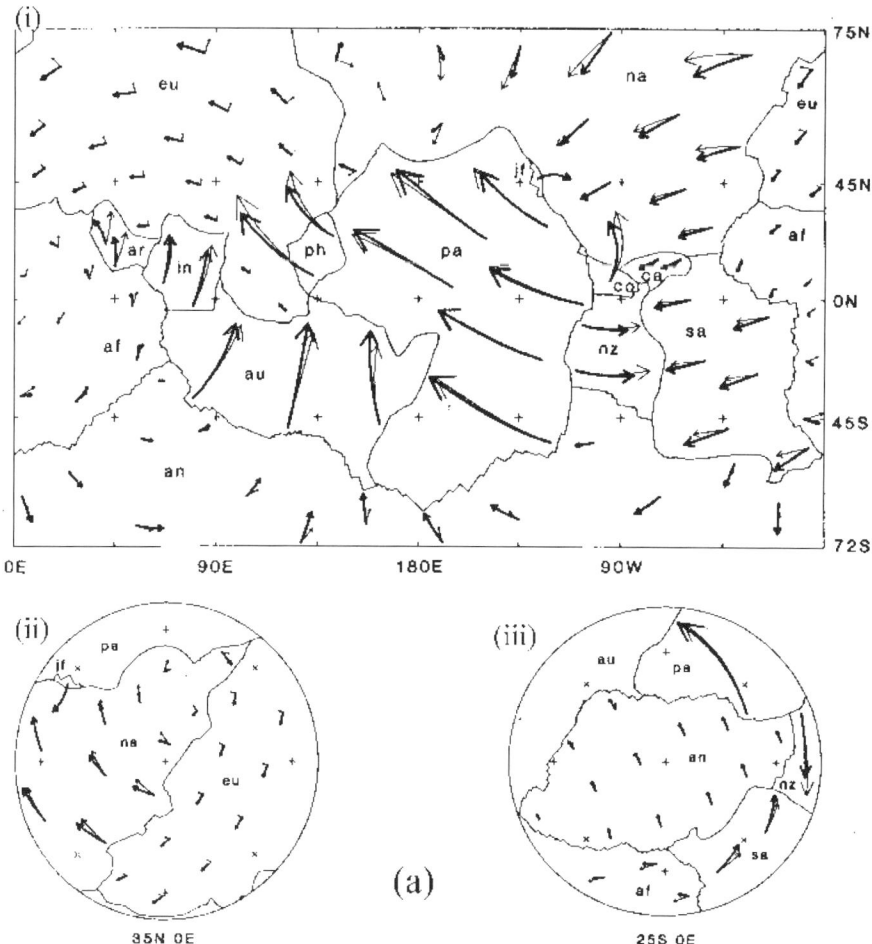

Fig. 5.4. (a) (i) Plate velocity relative to hotspot sets according to model NUVEL 1 (thick arrows) and earlier model AM-2 (Minster and Jordan 1978), thin arrows. The arrow-length signifies distance displaced in 50 Ma. (ii), (iii), North and South polar regions, displaying predominance respectively of clockwise and counterclockwise rotation. From Gripp and Gordon (1990).

component of motion has varied comparatively little. Shifts in the pole of rotation thus based must be distinguished from polar wander based on paleomagnetism.

5.2.4 Magnetics

Plate motion independent of the hotspot record may be sought in paleomagnetism. A major limitation is that order-1 (net-lithosphere) rotation is not recorded in paleomagnetism. Furthermore we make an assumption, that averaged over time the geomagnetic field has existed as a geocentric dipole aligned with the pole of rotation.

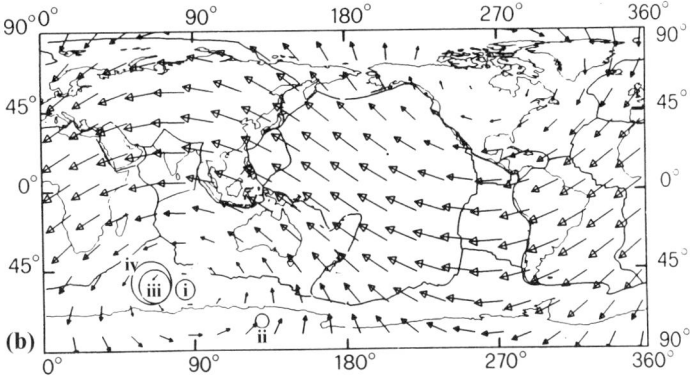

Fig. 5.4. (b) Plate motion, order-1 component (random motion component removed). The net-rotation poles are those of models: i, NUVEL 1; ii, AM1; iii, AM1-2; iv, HS2-NUVEL 1, according with choice of hotspots. From Ricard, Doglioni, and Sabadini (1993). If mantle-reference should be the east-west spreading direction indicated by gravity images of the asthenosphere, the rotation-pole would more nearly coincide with the Earth's present rotation axis.

In his demonstration of what at the time was referred to as continental drift, Runcorn (1962b) established the track of continents reference the paleomagnetic field, figure 5.6; see also Irving, 1964 and Creer, 1964. Since commencement of opening of the South Atlantic at about 80 Ma, the counterclockwise rotation of Africa/South America has almost ceased, and been replaced by fragmentation represented by independent westward drift of South America (compare gravity data, section 5.4.2). Later reconstructions (figures 5.7, 5.8), using the extended record provided by continental paleomagnetism, have confirmed the tendency perceived by Runcorn of plates within the Southern Hemisphere to rotate counterclockwise, as against clockwise rotation in the Northern. In the case of India and Siberia, the sense appears to reverse, from counterclockwise to the opposite, upon drifting into the northern hemisphere. The motion accords with the vortical (toroidal) component of plate motion separated by Forte and Peltier (1991).

5.3 Data-Indicated Flow

5.3.1 Geophysical Parameters

Independent restraints as to the internal flow regime are presented by the heat flow, the form of the geoid, and tomographic mapping of the interior.

5.3.1.1 Heat Efflux. Based on almost 25,000 spot measurements, the total flow of heat from the Earth's interior (Pollack, Hurter, and Johnson 1993) is estimated to be 44.2×10^{12} W. At surface, the efflux is strongly concentrated in the crestal

112   Tectonic Consequences of the Earth's Rotation

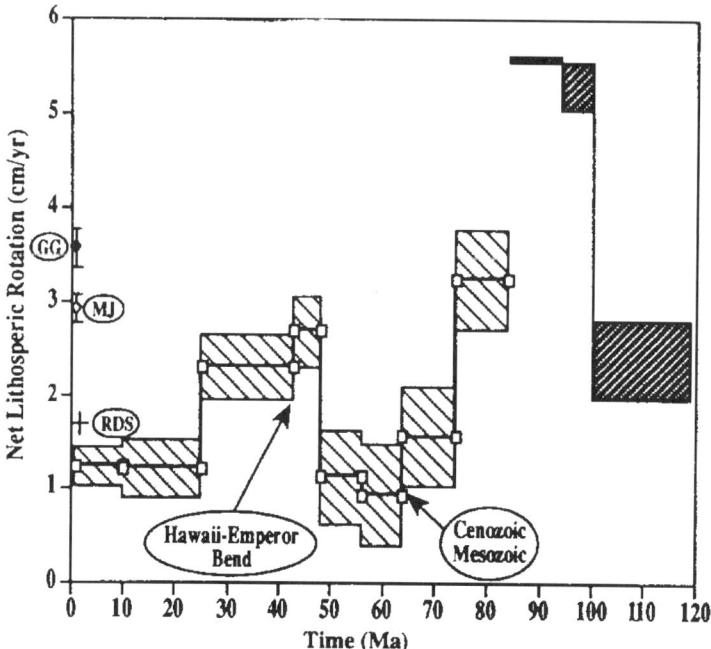

Fig. 5.5. Net lithosphere rotation showing past changes in rate, based on the hotspot identification used in various models: (GG), HS2-Nuvel1; (MJ), AM1-2; (RDS), the model of Ricard, Doglioni, and Sabadini, 1991. Lithgow-Bertelloni et al. identify also the apparent pole of rotation at past times. From Lithgow-Bertelloni et al. (1993).

region of the mid-ocean ridges, figure 5.9. As pointed out by these authors, the pronounced sectoral distribution reflects a conspicuous north-south orientation of much of the ocean ridge system.

5.3.1.2 Velocities. Tanimoto and Zhang (1990) and Anderson, Zhang, and Tanimoto (1992) have employed seismic velocity, as surrogate for density and temperature, to evaluate the distribution of temperature at depth in the asthenosphere and upper mantle, figure 5.10.

The distribution departs from that at surface, losing its strongly sectoral pattern and acquiring a concentration of fast values in high latitudes. At depth, hot areas are widespread, and not confined to the vicinity of surface-mapped hotspots. High and low velocities appear to record an earlier position of trenches and ridges, respectively, implying that these are migratory. Ridges tend to be asymmetrical, the east flank appearing different from the west.

Surface heatflow observations now are representative of both continental regions and the world ocean. A limitation in the use of trans-oceanic heatflow profiles to model flow is that whereas they may be used (Stein and Stein 1994a,b) to identify inadequate models, the assignability of parameters prevents models from

Tectonic Record   113

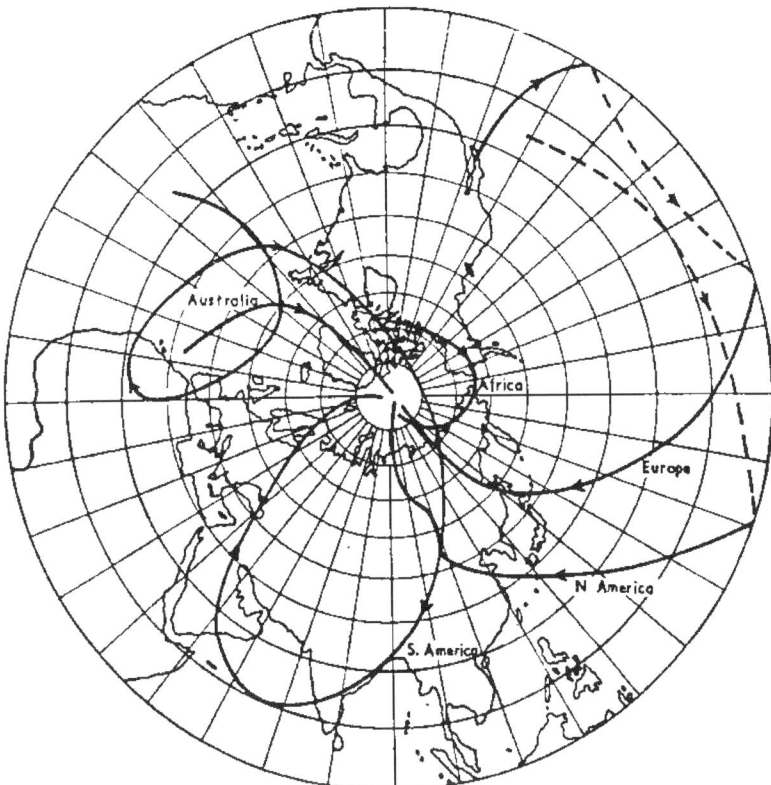

Fig. 5.6. Pole path of individual continents based on paleomagnetism, adduced by Runcorn as evidence of continental drift and posssibly polar wander. Paths indicate clockwise rotation whilst in Northern Hemisphere, counterclockwise whilst in Southern (dashed part of path). Drawing by Runcorn, 1962b.

being unique. The oceanic lithosphere is seldom undisturbed over great distances, and the seismologic results of Anderson, Zhang, and Tanimoto (1992) seem to preclude a cooling formulation applicable to all oceans. Stein and Stein (1994a,b) have used bathymetry as an additional constraint. In model fitting, heat flow and bathymetry represent the the joint effect of several variables, including the thickness and temperature of nascent lithosphere; its horizontal velocity; its continuing influx of heat from the substrate, affected by the existence and form of plumes and secondary convective structures in the asthenosphere (see for example Haxby and Weissel [1986]; Cazenave, Parsons, and Calcagno [1995]); and not least, as demonstrated by the Steins, the conveyance of heat by hydrothermal processes prior to sealing by sedimentation.

Forte, Peltier, and Dziewonski (1991) have used Bayesian inversion to successively refine models of the internal heatflux and viscosity in the light of incoming data. Forward modeling of the flow to be expected, based on seismologic mapping

114 Tectonic Consequences of the Earth's Rotation

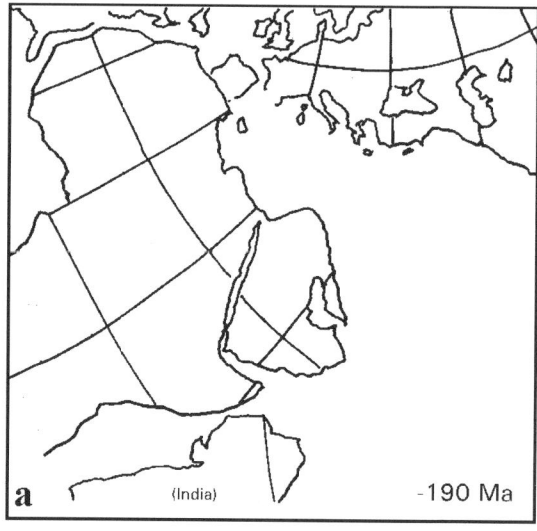

Fig. 5.7. Orientation of Africa relative to the geomagnetic field, as Africa drifted north in the Southern Hemisphere during the breakup of Gondwana. While attached to Africa, South America rotated likewise. The frame is oriented North-South. Re-drawn from Savostin, et al., 1986. (a) At −190 Ma. (b) At −80 Ma. (c) At present.

Tectonic Record 115

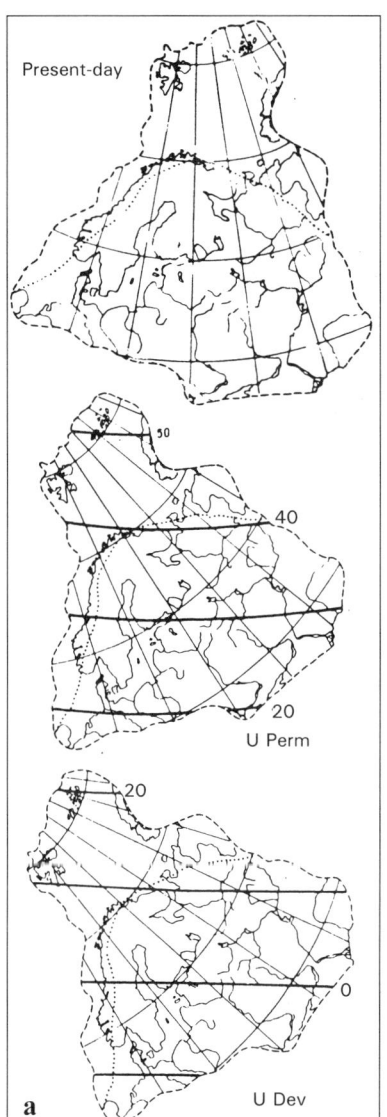
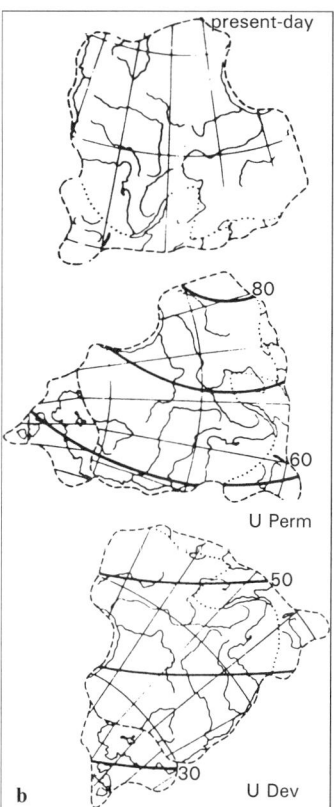

Fig. 5.8. Rotation since Middle Paleozoic of (a) Euro-Russian platform, and (b) Siberian platform Displayed reference to the geomagnetic field by removal of paleodeclination. (Maps from Khramov, Petrova, and Pechersky [1981]).

116   Tectonic Consequences of the Earth's Rotation

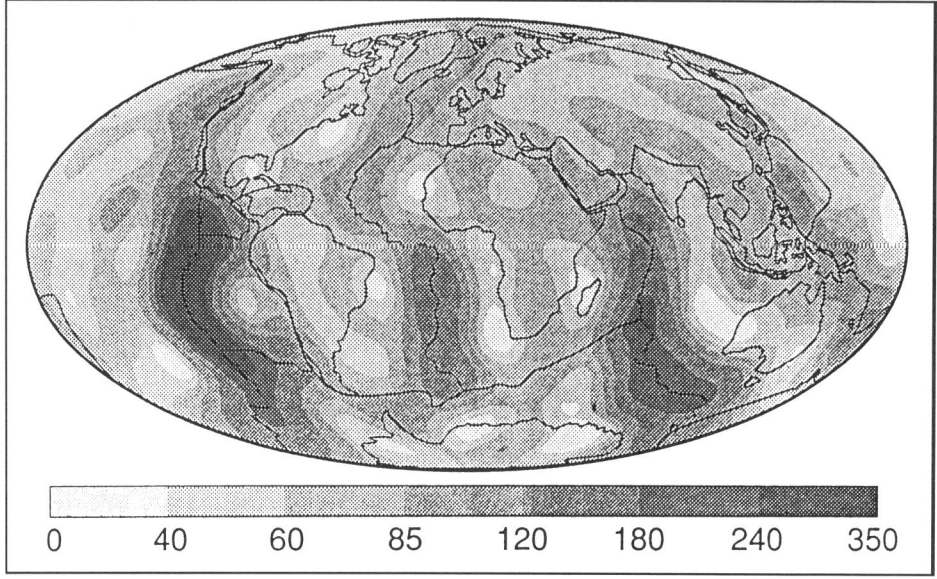

Fig. 5.9. Distribution of global heat flow, to degree/order 12. Units mWm^{-2}. (Reproduced from Pollack, Hurter, and Johnson [1993].)

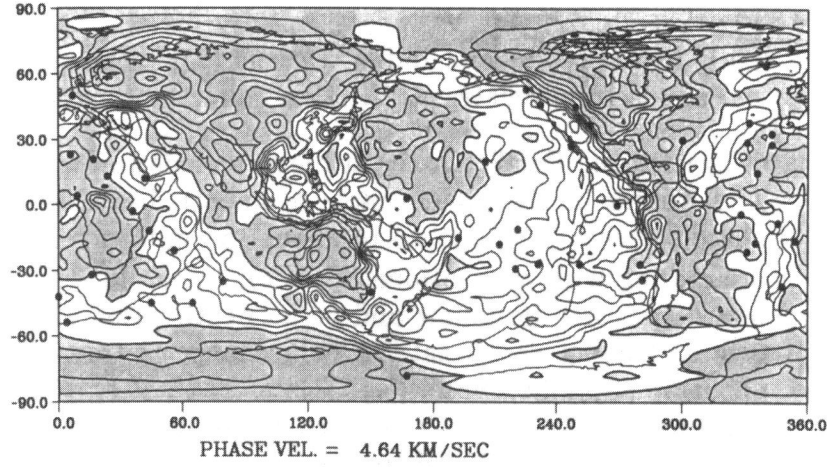

Fig. 5.10. Love wave shear velocity for period 100 sec. The shaded/unshaded areas divide the surface into regions faster (believed to be colder) and slower (warmer) than the average, 4.64 k/s. The black dots are hotspots. (From Tanimoto and Zhang 1990.)

of the quasidensity, has been used by Forte, Dziewonski, and Woodward (1993), Pari and Peltier (1995), and Forte and Mitrovica (1996) to arrive at optimal estimates of the viscosity distribution and heat advection through internal surfaces.

Corrections have been made for the contribution to heatflow of variations in thickness of the radioactive crust and for the effect of compressibility. The geophysical constraints indicate the existence (figure 5.11) of a viscosity low at about 670 km depth, where might be expected the transition spinel to perovskite/magnesiowustite. Their model fits the independently observed hydrostatic (complete) geoid.

5.3.1.3 The Excess Flattening. The observed complete geoid, figure 5.12a, displays an excess flattening of the Earth, in the sense that the Earth's figure is more oblate than is in equilibrium with its rotation (Henriksen 1960; Jeffreys 1963; Nakiboglu 1982; Alessandrini 1989; Denis 1989). The excess in harmonic degree two constitutes by orders of magnitude (table 5.1) the Earth's largest departure from hydrostatically stratified equilibrium.

The "excess" degree-two component is arbitrarily removed in preparing geoids described as "best-fitting." Without its removal, the higher order components associated with tectonic features, such as subduction, are masked.

McKenzie (1968) has pointed out that the bulge in excess of hydrostatic equilibrium represents about five times more accumulated potential energy than the remaining terms put together. Models constructed by Forte, Dziewonski, and

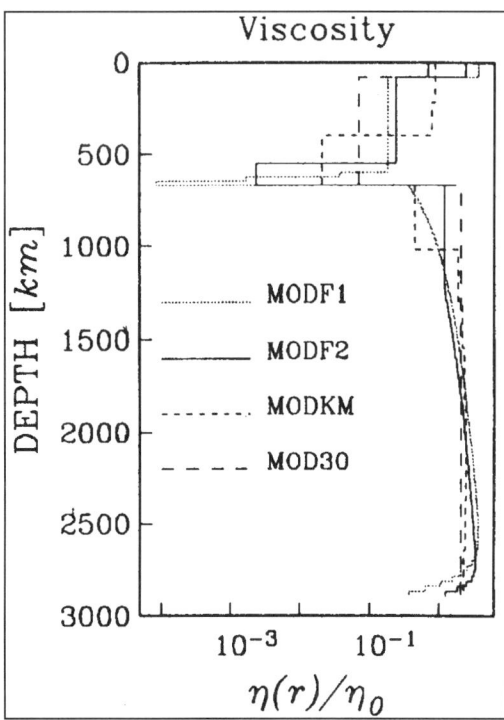

Fig. 5.11. Variation of viscosity with depth, displayed as ratio of various models against reference viscosity η_0. Alternatives range from a layer of greatly lowered viscosity concentrated at the base of the asthenosphere (700 km) to a model in which the drop is less but extends to shallower parts of the asthenosphere. (From Forte, Woodward, and Dziewonski [1994].)

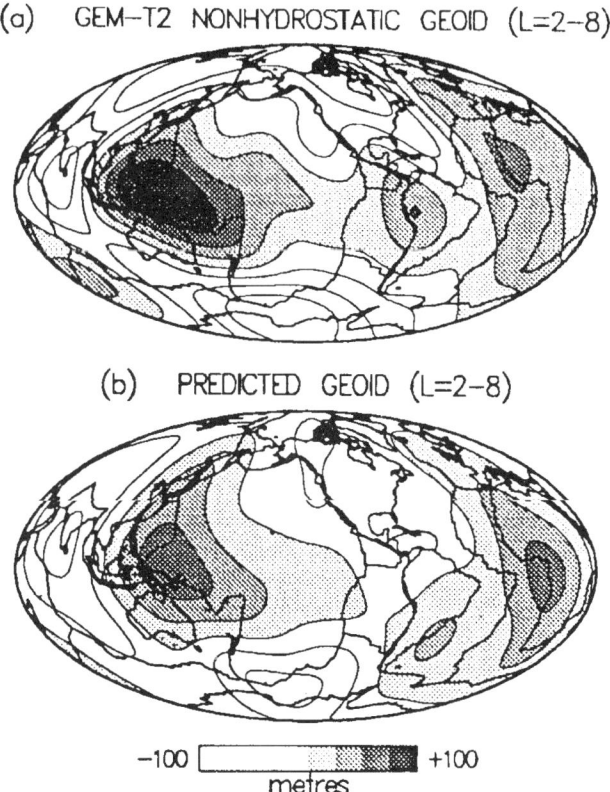

Fig. 5.12. Reconstructions of internal flow structure, as restricted by the geoid, seismological controls, and variations in the viscosity of the mantle: (a) Observed (GEM10B) geoid after removal of the flattening representing rotational equilibrium, the latter as determined by Nakiboglu, 1982. C.I., 20m. The excess flattening becomes evident as the domination of high latitudes by negative values, and the almost unbroken band of highs centered on the Equator. (b) Geoid determined by forward-modeled flow based on seismologic (tomographic) determination of pseudodensity and probable viscosity distribution. (From Forte, Dziewonski, and Woodward ([1993].)

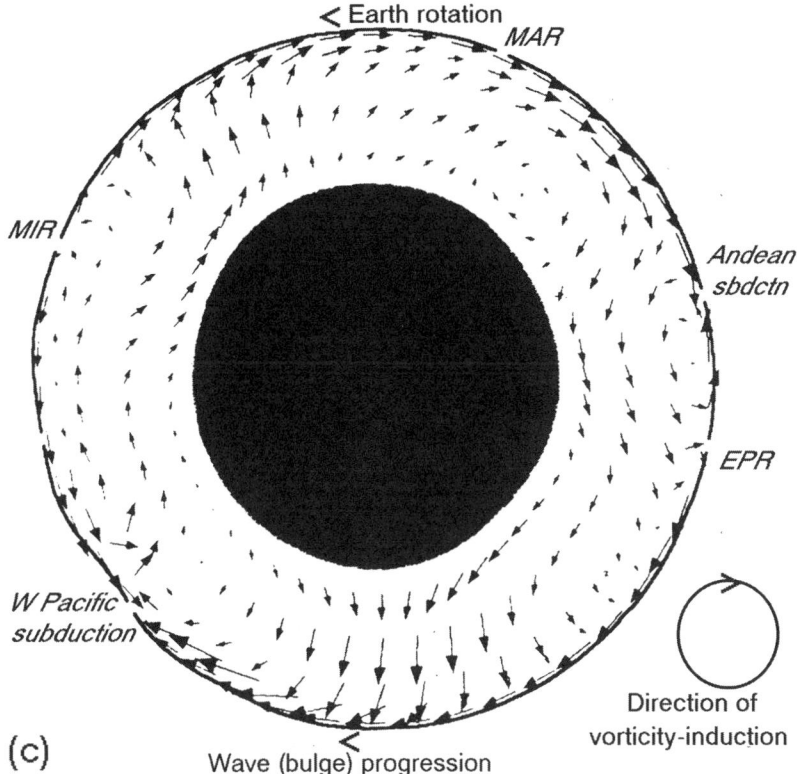

Fig. 5.12 (continued). (c) Mantle flow in the Equatorial plane computed on the basis of independently determined mass anomalies, and viscosity adjusted so as to fit observed plate velocities. The shape of the geoid is indicated by the undulation of the boundary. Direction of circulation induction under M_2 wave is shown for reference; prevalent sense of the circulation is visible to the eye. Note unbroken westward motion beneath Atlantic, as against symmetric spreading from MAR; cf. Atlantic plate motion as derived by Tanimoto and Zhang (1992, fig. 5.20) and Phipps Morgan et al. (1995, fig. 5.23). (Data from Vigny, Ricard, and Froidevaux [1991, fig. 10A]; re-projected from North Pole.) Continued overleaf.

Woodward (1993), Forte and Peltier (1995), and Forte and Mitrovica (1996) demonstrate that forward flow-modeling *starting with the already-developed mass distribution* accords with the observed geoid, including the "excess flattening," figure 5.12. The background question is, what is the geodynamic process producing the Earth's greatest hydrostatic anomaly in the first place?

Vigny, Ricard, and Froidevaux (1991), figure 5.12c, have modeled the flow to be expected based on geologically "observed" plate velocities, the topography, and the internal density distribution as constrained by the low-order geoid and seismic tomography.

A flow model constructed by Bunge and Richards (1996), figure 5.12d, sur-

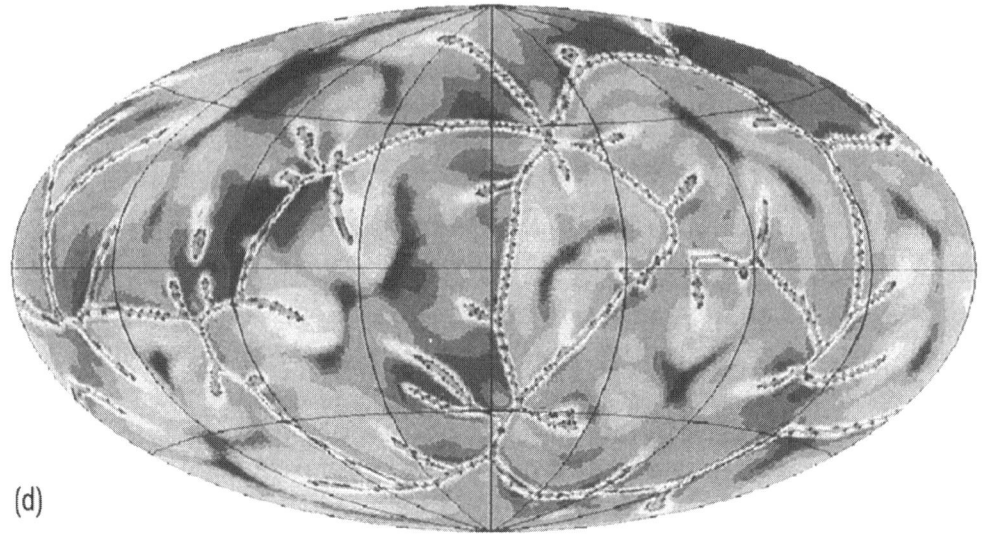

(d)

Fig. 5.12 (continued). (d) Flow model by Bunge and Richards (1996) incorporating high-viscosity lower mantle. At 300 km depth, broad warm upwelling regions develop, separated by narrow bands of cool "subducting" material.

mounts the long-standing difficulty that the Rayleigh number in the mantle is so high ($\geq 10^5$) that one would expect the subdivision of convection into small cells. Employment of a viscosity strongly increasing with depth, accordant with independent restraints such as glacial rebound rates, results in near-surface cells well able to account for plate size.

In a fundamental paper, Goldreich and Toomre (1969) have examined the location of the Earth's principal axes relative to its rotation axis. Within limits of slow tidal transfer of angular momentum, the latter is fixed in inertial space.

Well-established coefficients of moment about the principal axes times 10^{-6} are:

$$C_{20} = -481.2; \ \overline{C}_{22} = 2.4; \ \overline{S}_{22} = -1.4$$

Table 5.1. Components of the Geoid, by Degree and Order

Index	Coefficient $\times 10^6$	Index	C	$S \times 10^6$
2, 0	−484.16602	2, 1	0.00535	.00258
4, 0	0.54063	3, 1	2.03444	.25330
6, 0	−0.14885	4, 1	−.53956	−.47935
8, 0	0.05015	5, 1	−.05430	−.10712
10, 0	0.05330	6, 1	−.07699	.03304

Source: Marsh, Lerch, and Williamson (1985).

The observed coefficient of the flattening, $\overline{C}_{obs}$, consists of the sum of the coefficients of the equilibrium- and the excess-flattening, here dubbed $\overline{C}_{eq}$ and $\overline{C}_{exc}$, so that

$$\overline{C}_{obs} = \overline{C}_{eq} + \overline{C}_{exc} = -481.2.$$

Goldreich and Toomre then "imagine that the earth's rotation is halted," and that "the present rotational bulge, but nothing else, is allowed to subside." The operation is represented by forming

$$\overline{C}'_{20} = \overline{C}_{20} - \overline{C}_{eq.} = \overline{C}_{eq} + \overline{C}_{exc} - \overline{C}_{eq.} = -4.7$$

It emerges that the low-order components are so large that the Earth is more usefully described as a triaxial body than a spheroid. Parenthetically, Pavoni (1985) has pointed to the fact that in addition to the recognized rotational polar axis, geotectonic axes are located in Africa and the tropical Pacific.

It is then pointed out by Goldreich and Toomre that the differences in moment between the residual figure, C_{exc}, describing the "excess flattening," and those about the other axes, are comparable with mass shifts occurring under recognized geological processes. It might accordingly be expected that true polar wander will be commonplace. Their test computations suggest that at times wander may be as rapid or more rapid than plate motion, as a result of random mass shifts under convection, erosion, and sedimentation.

It has since become apparent (Stevenson 1993) that true polar wander (chapter 6), although occurring (Courtillot and Besse 1987; Gordon 1995), is a comparatively slow process. During the Cenozoic, polar motion has been slower than the motion of plates. It has been suggested (Bostrom 1994) that the tectonic elements forming the excess bulge have not in fact formed at random, but are systematically associated with the topical rotation axis. Almost 30 years ago, Dicke (1969) suggested that contrary to a supposition of deep strength in the mantle, "deep convective currents" may support the excess bulge. Outwith the possibility of weak Eddington currents (McKenzie 1968), a mechanism has not been available to explain a latitudinal dependence of convection.

In light of earlier discussion (chapter 4) it might be suspected that tidal circulation induction is responsible both for the maximization of plate velocities in low latitudes observed by Gordon and Jurdy (1986) and for the excess mass accumulation in these latitudes picked up by tomography and gravimetry.

Forming the low-latitude highs, the tomographic density anomalies have been identified with causative tectonic structures (sections 5.3.1.4; 5.4.2) and represent (Forte and Peltier 1995) the physical constitution of the flattening. These features are known geologically to have been in existence since the last reorganization of plate motion, i.e., for at least 44 Ma. If the connection is other than coincidental the induction may act as a stabilizer (since the features controlling the position of the pole are self-renewing, a product of their low latitude).

Eventually, subcontinental heat accumulation and uplift in other latitudes, in accordance with processes such as those outlined by Anderson (1982), must bring about an episode of polar shift, as a result of which the Earth adopts a new orientation of equilibrium. Reconfiguration consequent upon polar wander is taken up in chapter 6.

5.3.1.4 Geoid: Western Pacific Margin. The equatorial western Pacific is the site (at 165E) of a second axis of moment. The geoidal high chiefly responsible for its position (figures 5.13, 5.14) is the locus of a concentration of subduction. The western Pacific margin is the most actively tectonic sector of the planet. A measure is that this region is responsible (Frohlich 1989) for 92% of *all* deep seismicity, marking subduction. Its singularity accords with that of Pacific seafloor spreading, the west limb of the East Pacific Rise (EPR) being grossly developed relative to the east limb.

Tectonism unique to this region has attracted ongoing research. In celebrated papers, Carey (1955, 1958, 1970; see also Coleman and Packham [1976]) identified the 5000 km "Melanesian Borderland" as the site of major left-lateral transpression. This is now seen to be the most concentrated region of sinistral vorticity on the planet (figure 5.13). Its complexity has made difficult a unified tectonic analysis (Packham 1982). Crook et al. (1987) have investigated the mechanics of the incoming Pacific plate. The equatorial westward embayment of the Pacific (Bostrom, Saar, and Terry 1983) may be composed of a separate, small (Caroline) plate (Weissel and Anderson 1978). Powell and Johnson (1980) show that Sundaland has migrated into its present location (figures 5.14a,b) from far to the east. Stevens et al. (1999) show that currently, the Pacific-Eurasian plate convergence is concentrated beneath the narrow, Equatorial Molucca Sea. Packham (1993) has shown that north of the Equator, SE Asia is affected predominantly by *dextral* shearing.

The geological record renders it difficult to suppose that the equatorial embayment of the western Pacific is accidental, or that the concatenation of tectonic features has been subject to significant north or south translation. In this regard, Crook (1978, 1980) has already proposed as the best available explanation that the development of the southwest Pacific is associated with zonal flow as a function of the Earth's rotation.

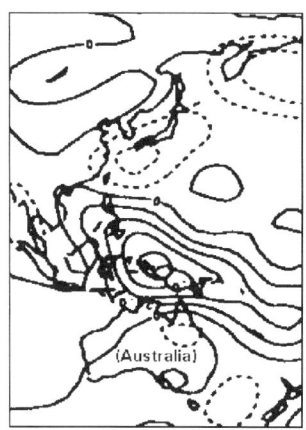

Fig. 5.13. Radial vorticity, equatorial western Pacific. C.I.: 20×10^{-9} rad/yr. Dot contours, clockwise; solid, counterclockwise. (After Forte and Peltier [1991].)

Tectonic Record   123

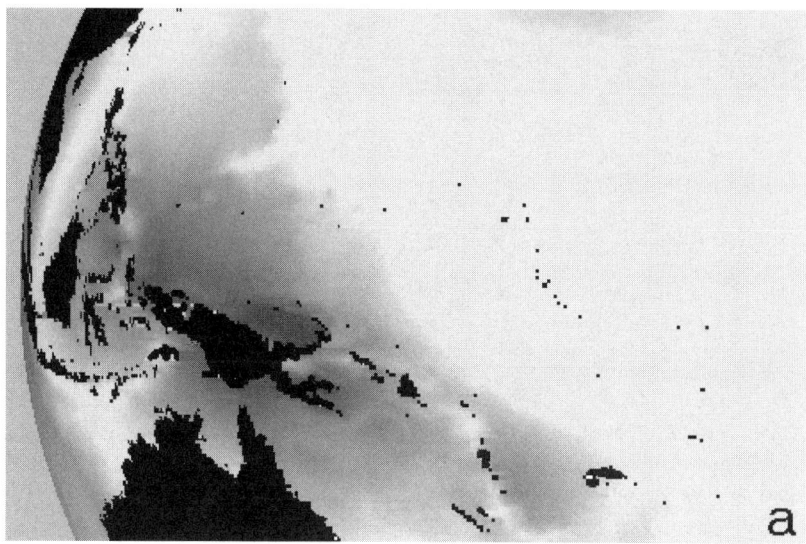

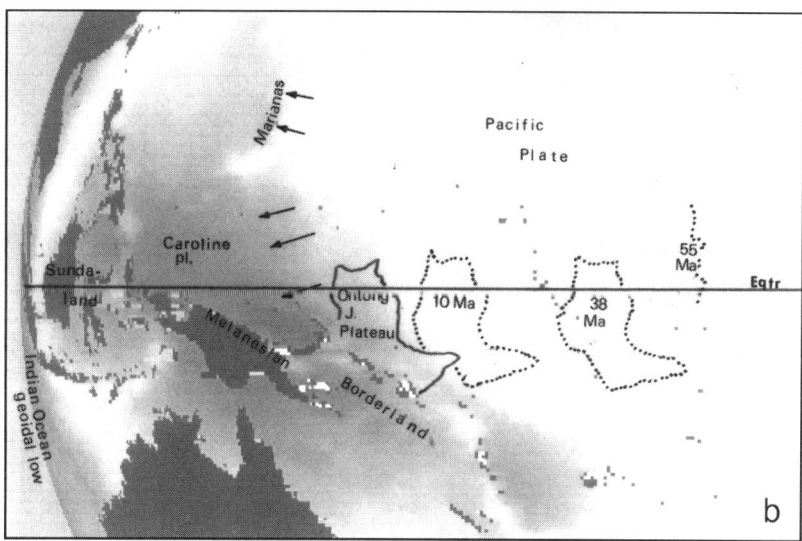

Fig. 5.14. Mass displacement and accumulation, equatorial western Pacific. (a) Excess mass, represented by geoid in which darker shading describes higher values. The geoid here reaches its highest value on the planet (≥100 m). Its steepest gradient occurs on the Equator, where the western Pacific high passes westward into the equally intense Indian Ocean low. (b) Geodynamic elements apparently responsible for mass accumulation. Carey's Melanesian Borderland (Carey, 1955, 1970) is a 5000-km zone of sinistral transpression (see figure 5.13). Powell and Johnson (1980) have traced the westward motion of Sundaland into its present position. Coleman and Packham (1976) have employed the displacement of the Ontong Java Plateau to describe the relative motion of the Pacific plate. Weissel and Anderson (1978) point to the existence of a small separate Caroline plate.

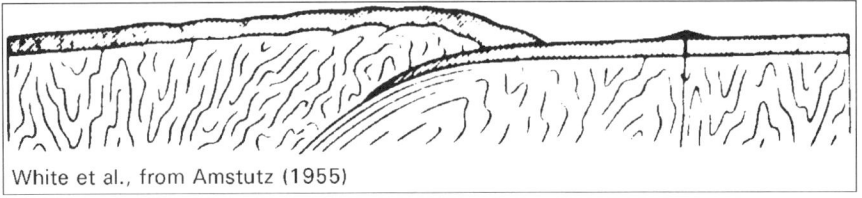

Fig. 5.15. Asymmetrical "Verschluckungs-Zone," in process of developing into alpide belt (from Amstutz [1951], with acknowledgment to White et al. [1970]).

Models constructed over several decades by tectonicists of Japan and Australia, incorporating Carey's "sphenochasms" and "oroclines," identified elements since found intrinsic in plate tectonics. Prior to the identification of subduction and back-arc spreading, and before general admission of the concept "mobilism," it had become apparent to workers in Japan that the local pre-Cambrian basement could only have originated on the Asia mainland, having become separated from kindred Asiatic terrain by the Sea of Japan (cf. Minato 1968). Acceptance of mobilism was preceded by the demonstration by Wadati (1928; c.f. Frohlich [1987]) of mantle activity at depths generally believed (Jeffreys 1928) too weak to accumulate seismogenic elastic stress.

5.3.2 Structural Asymmetry

The asymmetry of individual subduction zones and their precursors in the classical literature, Verschluckungs-Zonen, has been noted since their first identification (Ampferer 1925, 1941; Ampferer, and Hammer 1911; Amstutz 1951; White et al. 1970), figure 5.15. White et al. trace the emergence of the concept as far back as Ampferer (1906) and early alpine geologists.

The asymmetry in early models of convergence has been shown by the constructions of Dewey and Bird (1972) to be the result of lateral buoyancy contrasts, figure 5.16. Modeling by these investigators at once made clear the relation between "plate tectonics" and classical mapping in the alpide belt (Griffin 1970). In

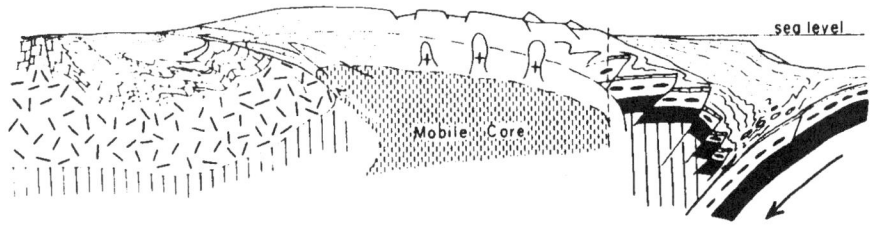

Fig. 5.16. Stage in the evolution of cordilleran (convergence) type mountain belt. Being more dense, the oceanic lithosphere passes below the continental lithosphere, left. (Reproduced from Dewey and Bird [1972], fig. 10D.)

respect still to the individual subduction zone, Wilson (1973) pointed to an additional factor, in terms of the possibility that one plate is relatively stationary.

5.3.2.1 Asymmetry: Global Set. Dickinson (1978) pointed out that subduction zones on the western flank of the north Pacific rim have produced eastward-bowed island arcs, whereas those on the opposite, eastern rim have formed continental arcs (as in the Andean sector). "West Pacific" and "East Pacific" subduction zones characteristically develop structures of the type shown in figures 5.17a,b and 5.18 (after Uyeda and Kanamori 1979). Reviewing convergent plate boundaries, Dewey (1980) identifies rules governing the form of the subduction.

An orthodox explanation of plate-driving forces is that slab-pull under negative buoyancy is fundamental. Difficulty is encountered in supposing that this is operative in regions of small subduction angle. Beneath the northwestern United States the subduction angle is 11° (figure 5.19a). Seismicity yields indication of neither intermediate nor deep foundered slab. Sin 11° having the value 0.19, "slab-pull" as a driving mechanism would seem to operate at a considerable disadvantage, but it is likewise difficult to believe that such a thin overriding wedge can be thrust so far without failure. Induction of the kind examined in connection with the tidal dissipation results in an internally generated crawl or creeping action, as in a pile of papers repeatedly flexed in the same direction, capable of effecting the displacement.

The subduction angle of the lithosphere adjacent to the Chilean Pacific margin

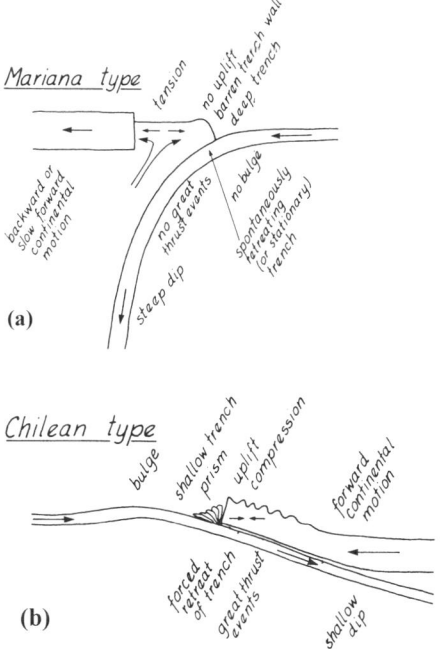

Fig. 5.17. Subduction plate boundaries. (a) Western Pacific (Marianas) type, after Uyeda and Kanamori, 1979. Cool, overdense lithosphere is foundering at a relatively steep angle. This type of subduction bounds the back-arc basins of the western Pacific rim such as the East China Sea and Sea of Okhotsk. (b) Eastern Pacific (Americas) type. Subduction angle adjacent to ocean is shallow. In lieu of quasi-oceanic back-arc basins, the subduction produces convergent alpide mountains such as the Andes. In North America south of Cape Mendocino, this regime prevailed until the Tertiary, during which time the North American plate has reached and is over-riding the crest of the EPR. (Drawings from Uyeda and Kanamori [1979].)

Fig. 5.18. Eastward mantle motion of Uyeda and Kanamori (1979), relative to the crust.

is likewise small. Steep subduction commences hundreds of kilometers inland, in the case of slab-pull presenting the problem of long-distance tensile coupling, via fractured material having zero tensile strength.

Being a gravity phenomenon, tidal action is minutely pervasive, affecting structures at all scales. In one extreme, observations by Zurn, Emter, and Otto (1991) introducing their ultra-short strainmeter confirm finding that "Tides are in the smallest cracks." Locally, the Earth's mechanical inhomogeneity strongly concentrates the effect. By the same token it may be surmised that secondary magma accumulations, hypothesized by geologists to be associated with the crest of the ocean ridges marking mantle upwelling (figures 5.19 b,c), cannot escape the effect of vorticity induction. Its action is to bias flow processes. The mid-Atlantic spreading-axis makes its discontinuous 1500 km jog or jump at the Equator, south of which the South Atlantic developed at the time of separation of Africa and South America. Systematically, for some 13,000 kms in latitude centered on the Equator, continued spreading generates quasi-parallel east-west fracture-zones separated by short north-south spreading-segments. At high latitudes fracture-zone orientation loses its east-west bias.

Earlier Pacific images (figure 5.19d), constructed without benefit of altimetric satellites, did not reveal the symmetry about the Equator of the present axis of seafloor spreading. The Equator evidently forms a boundary in the orientation of the surface axis of the East Pacific Rise (figure 5.19c). North of the Equator, from meridional the axis azimuth veers gradually counterclockwise.

The veering pattern is symmetrical to that south of the Equator, but laterally inverted (mirror-image). With increasing south latitude veering increases, until represented by the large high-latitude fracture zones of the South Pacific. Secondary flow structures are regimented accordingly.

To the north, the EPR acts similarly, but is disturbed through having been

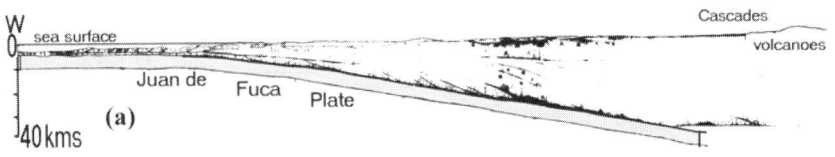

Fig. 5.19. (a) Overriding of Juan de Fuca lithosphere slab by North America plate. (From Cowan and Potter [1991].)

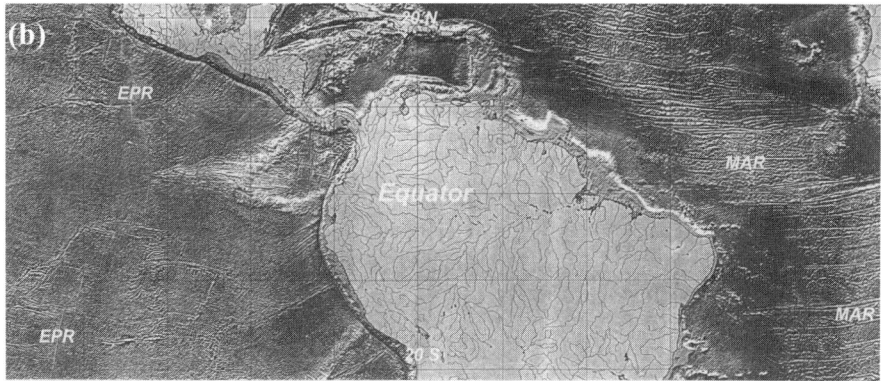

Fig. 5.19. (b) Latitude dependence and orientation of contemporary low-latitude flow structures. EPR, East Pacific Rise axis; MAR, spreading axis, of Mid-Atlantic Ridge. The equatorial Atlantic came into existence at the time of separation of South America from Africa. (From image of seafloor topography by Smith and Sandwell [1997].)

overridden by the North American plate. It has been shown (Mammerickx, Naar, and Tyce [1988]) that relative to the moving North American plate the EPR axis formerly occupied a position further west, on its oceanward margin, before "popping up" within the Gulf of California. That the nonrandom features in the Pacific are not the expression of sedimentation, dominated by the oceanic circulation, is attested by their coincidence with deep- and shallow-seated gravity anomalies, see below.

Time-wise, Pockalny et al. (1997) have demonstrated that the segment of the EPR adjoining the Clipperton and Siqueiros fracture zones (latitude $8°N-11°N$), hence the spreading-direction, has veered counterclockwise $4°-8°$ in the last 2–3 Ma, with corresponding disturbance in the generation pattern of the fracture zones. This represents the most recent stage in ongoing reorganisation, reorientation, and eastward "jumping" or stepping of seafloor spreading in the eastern Pacific (Mammerickx, Herron, and Dorman [1980]; Bird et al. 1998), figure 5.19c.

5.3.3 Motion of the Americas

The forces acting in plate convergence, explaining the locus of the greatest earthquakes, would be greatly increased by independent, Pacific-ward motion of the Americas. Such motion is indicated by data obtained by Montagner and Tanimoto (1991), figures 5.20a,b and Tanimoto and Zhang (1992), based on velocity and anisotropy in the asthenosphere. Using identification of the flow form, Tanimoto and Zhang (1992) postulate that the opening of the Atlantic was asymmetrical (figure 5.20b).

Similar flow geometry is apparent in western North America (figure 5.21), as a result of the North America plate having overridden the crest of the East Pacific Rise. Tanimoto (1992), Tanimoto and Zhang (1992), Zhang, Tanimoto, and Stolper

Fig. 5.19. (c) Zoomed image, equatorial part of the seafloor spreading in the eastern Pacific Ocean. The current extension axis is indicated by small white x's.

The manner in which the low-latitude EPR maintains its eastern location, abandoning those earlier, may be illustrated in the vicinity of the Easter (E; 25°S) and Juan Fernandez (JF; 33°S) "microplates" (Bird et al. 1998 fig. 6, a–i). Eastward stepping at these latitudes, relative to the EPR at higher latitudes, is mirrored at 2°N, where abandoned spreading-axes such as the Mathematicians axis (M) seem to occupy analogous positions.

Enigmatic "cone-in-cone" structures centered on the Equator (figure 5.31a) extend westward from this locality for several thousand kilometers, to the western Pacific. (Data from Smith and Sandwell [1997].)

Tectonic Record 129

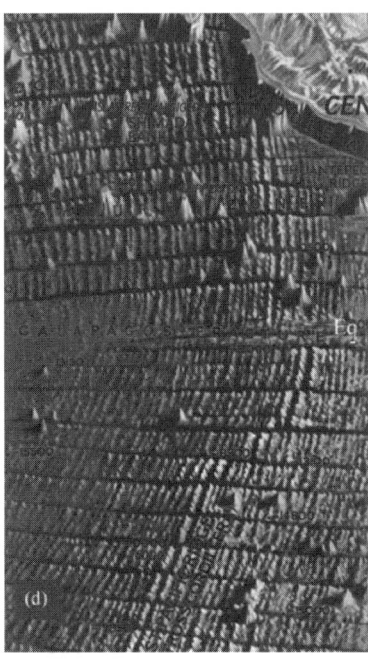

Fig. 5.19. (d) (Left) Pre-satellite version of area shown in adjoining figure, 5.19c. (From Heezen and Tharp [1969].)

(1994), and Phipps Morgan et al. (1995) have obtained velocity data relating to the structure of the Atlantic and Pacific basins (figure 5.22a).

The velocity anomaly across the central Pacific is compared with gravity in figure 5.22b. The high characterizing the east flank of the gravity trough associated with the East Pacific Rise is the expression of faster, denser material abruptly entering velocity profiles at this location. That the EPR is the site of a well-marked gravity low (in contrast to the topographic high) may indicate that upwelling rather than being active is the lagging, passive response to lithosphere extension.

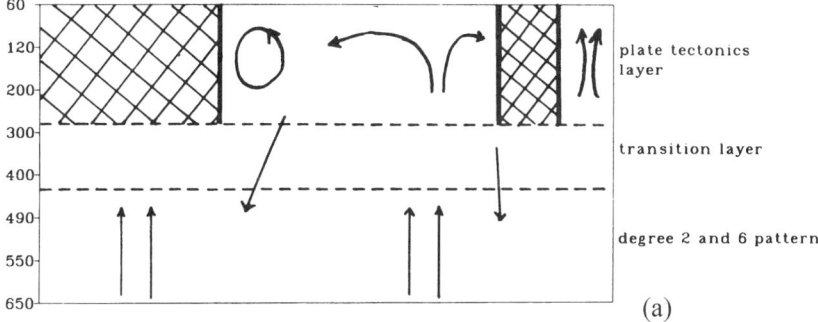

Fig. 5.20. (a) Form of the flow in the upper mantle, as indicated by seismic velocity anomalies and anisotropy. (From Montagner and Tanimoto [1991].)

130 Tectonic Consequences of the Earth's Rotation

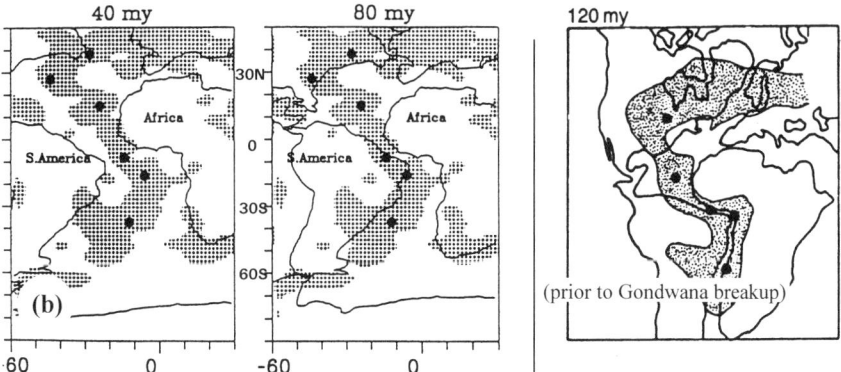

Fig. 5.20. (b) Development of Atlantic since Mesozoic times. In this interpretation the Mid-Atlantic Ridge marks the site of plumes, indicated by black circles. (From Tanimoto and Zhang [1992].)

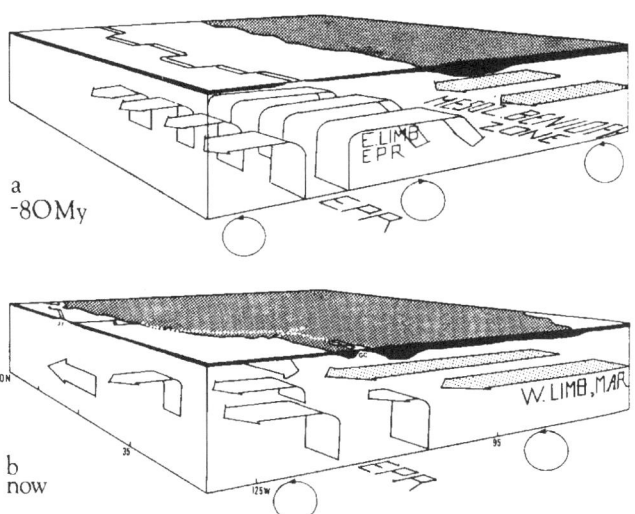

Fig. 5.21. Overriding of East Pacific Rise by North America plate (Bostrom 1981c), resting upon west limb of Mid-Atlantic spreading center (MAR). JF: Juan de Fuca Ridge; GC: Gulf of California (Sea of Cortez). Wilson (1973) postulated that the Basin and Range tectonic province is underlain by EPR. The tectonics of western North America in the Cretaceous resembled that of western South America at present. (Based on Atwater 1970 and Wilson 1973.)

Tectonic Record 131

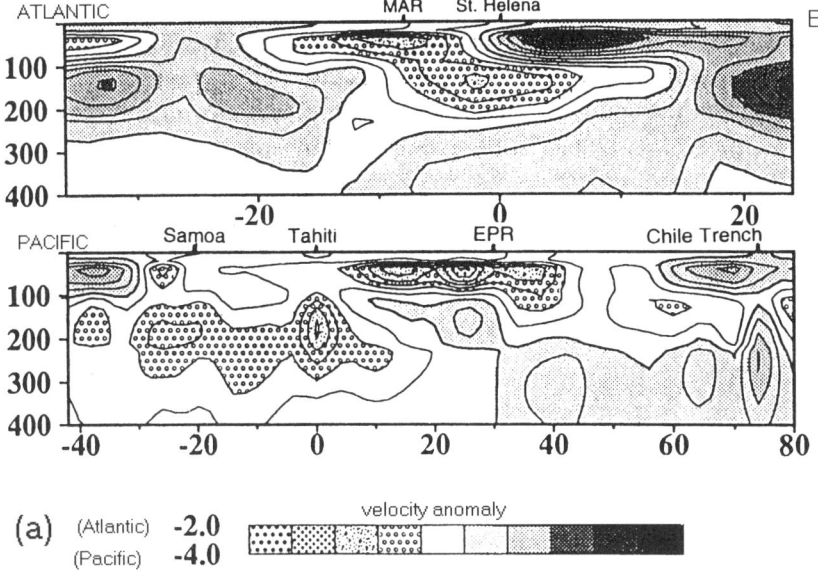

Fig. 5.22. (a) Atlantic (top) and trans-Pacific (bottom) profiles relative to axes of spreading (MAR, EPR), so as to display east-west asymmetry. The positive/negative velocity anomaly is believed to represent dense and less-dense, hence cool and less-cool material in the upper mantle. Note vertical exaggeration. The ticks on x-axis represent intervals of 20° angular distance along great circles. Profile locations: see figure 5.22b for Pacific, figure 5.28 for Atlantic. Profiles from Phipps Morgan et al. (1995); see also Tanimoto and Zhang (1992).

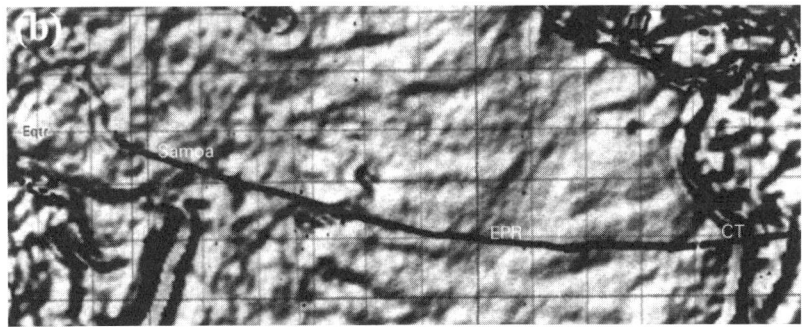

Fig. 5.22. (b) Location of Pacific velocity profile (cf. figure 5.22a; EPR location marked). The gravity-gradient data (background) are from Rapp (1988), see section 5.4.2.

132 Tectonic Consequences of the Earth's Rotation

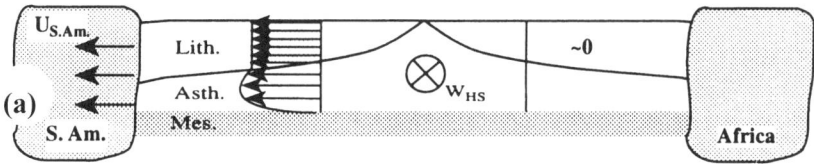

Fig. 5.23. (a) Asymmetrical motion of the Americas, away from the Africa plate, based on asymmetry of the Atlantic bathymetry and geoid in this region. (From Phipps Morgan et al. [1995].)

Phipps Morgan et al. (1995) find that flow in the sub-Atlantic asthenosphere takes the asymmetrical form shown in figure 5.23a. The geoid is negative behind westward-migrating, passive continental margins. The east-west asymmetry of the velocity-anomaly in this region is apparent in figure 5.22. Similar asymmetry is seen in Pacific velocity profiles. The apparently less active east limb of the EPR, on which the South American plate is encroaching, is sublain by faster material.

5.3.4 Asthenosphere Flow System

To accommodate difficulties evident in recent data, as a working model Phipps Morgan et al. (1995) have proposed that instead of upwelling occurring in bulk, as the limb of a large deep-mantle convection cell, upflow takes place principally in the form of numerous deep-seated plumes, figure 5.23b (see also Davies 1984). These authors point to a mobility of seafloor spreading axes, which they identify as passive and induced by extension, rather than driven from below. Additionally

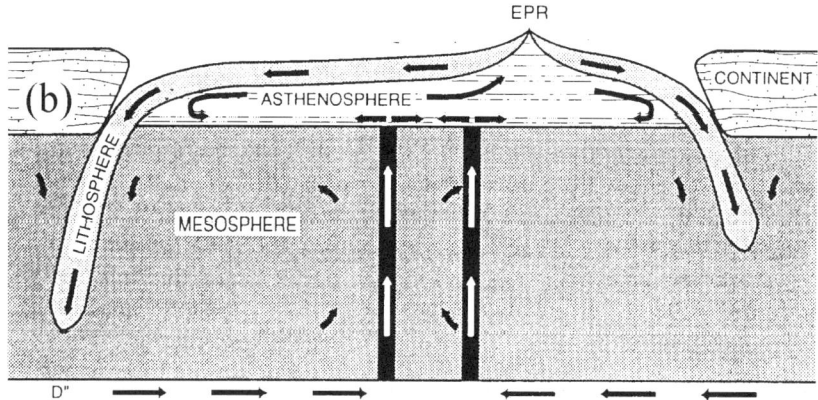

Fig. 5.23. (b) Stylized drawing by Phipps Morgan et al. (1995), describing mantle flow if an upper asthenosphere layer acts to decouple surface plate motions from the interior. The asthenosphere layer is supplied by numerous discrete plumes having a deep origin (as distinct from a whole-mantle convection uplimb).

they point to the remarkable weakness of coupling of the lithophere to the underlying asthenosphere. To judge from their small deformation in traversing large distances, thin oceanic lithosphere segments move as freely as floating ice.

5.4 "Absolute" Motion

The development of the lithosphere in the Pacific, including the East Pacific Rise, has been worked out using the seafloor magnetic pattern together with satellite altimetry (Menard 1966, 1984; Menard and Atwater 1968; Herron 1972; Herron and Tucholke 1976; Macdonald et al. 1988; Mammerickx, Herron, and Dorman 1980; Engebretson, Cox, and Gordon 1984; Mammerickx, Naar, and Tyce 1988; Mayes, Lawver, and Sandwell 1990; Mammerickx 1992; Watts et al. 1988; Lonsdale 1988).

Mammerickx, Herron, and Dorman (1980) conclude that a major reorganisation and reorientation of seafloor spreading in this region took place between 20 and 18.5 Ma. At that juncture, the present direction developed (c.f. figures 5.19c, 5.25a). A conspicuous feature of the direction change is that it took place coherently, at widespread parts of the EPR. The latitude of fastest total (areal) spreading, represented by the sum of the 140 mm/yr E-W on the EPR and 70 mm/yr N-S on the Galapagos axis, (Macdonald et al. 1988, fig. 1), is the equator.

5.4.1 Hotspot Record

To relate lithosphere motion to the underlying mantle, estimates have relied upon the concept of geologically expressed surface "hotspots," thought possibly to overlie deep-seated plumes (Wilson 1963; Morgan 1972a,b; Okal and Batiza 1987; Gordon 1995).

Motion estimates are dependent on the degree of stationarity of hotspots relative to the mantle as a whole or to each other. Identifying a relatively large number of features believed to be hotspots, many in the Eastern Hemisphere, Burke and Wilson (1972), Burke, Kidd, and Wilson (1973a,b) and Molnar and Stock (1987) have produced evidence that within groups hotspots may be stationary relative to each other, whereas groups may be subject to migration. Molnar and Stock point to evidence that Pacific hotspots may migrate relative to each other as fast as the Atlantic is opening (although not in the corresponding direction). Conversely, Duncan (1981) and Morgan (1983) point to evidence indicating quasi-stationarity of mantle plumes.

Fundamental uncertainty exists in identifying "hotspots." As stated by Phipps Morgan et al. (1995), "Apart from 12 or so major hotspots, any hotspot list is fairly subjective." Little doubt exists as to the reality of the Hawaiian hotspot, but other features have a more equivocal identity. Whereas it would be of value to estimate plate motion rates in terms of the volcanics trail left by hotspots, in regions of slow motion a trail is seldom clearly evident. Age dating of apparent trails (Schlanger et al. 1984; Epp 1984; Keating 1987) has been beset by apparently nonsequential dates and recurrent vulcanism. Published vulcanism series have not always included outliers, and dates that are apparently not sequential.

In consequence of the uncertainties, the "hotspot reference frame" employed in estimates of absolute plate motion has commonly been based upon a relatively small number of hotspots believed to have been reliably identified, at times referred to as the "standard set" (Table 5.2) (Minster and Jordan 1978; Gripp and Gordon 1990).

5.4.2 Gravity Data

Gravity coverage has become available globally to degree/order 360 based on orbital elements combined with surface observations (Rapp and Pavlis 1990) and exceeding 1000 over the world ocean, based on radar altimetry (Sandwell and Smith 1996).

Rapp's projection (Rapp 1988) of the gravity gradient at an elevation of 160 km uniformly covers continents and oceans. At this elevation, the function is sensitive to density heterogeneities in the asthenosphere, while attenuating the expression of short wavelength shallow features due to local topography. Inactive regions have become smoothed by the action of isostasy.

The Pacific region (figure 5.24) displays numerous "circular features." These tend to occur in lines. Some are coincident with known plumes, hotspots or "hotlines" (Bonatti et al. 1977; Lonsdale 1988), suggesting that there exists additionally a number that are unmapped. Within the older, western Pacific, plume-like gravimetric features are numerous. None are equivalent to the Hawaii feature in relief and magnitude. Many lineaments are subparallel with the Hawaii plume trail, but several other directions exist. The importance of establishing a fiduciary referred to the inner Earth has justified much effort in examining the mid-Pacific hotspots. Images of this area, sensitive respectively to asthenosphere inhomogeneities and near-surface structure, are shown in figures 5.25 a,b.

The NNE alignment of the Tonga-Kermadec subduction appears in deep-source gravity some hundreds of kilometers east of the trench line. The distant subduction is apparently responsible for surface fractures shown on the altimetric image. In this region four or five structural directions are superimposed. Some,

Table 5.2. Standard Hotspot Set

Hotspot	Coordinates	Motion direction	Uncertainty (°)
Hawaii	20N155W	N64W	±10
Marquesas	11S138W	N445W	±15
Tahiti-Mehetia	18S148W	N65 W	±15
Macdonald	29S140W	N55W	±15
Pitcairn	25S130W	N65W	±15
Juan de Fuca	46N130W	N54W	±15
Galapagos	1S92W	N45E	±10
Galapagos	1S92W	SS85E	±10
Yellowstone	45N110W	S60W	±20

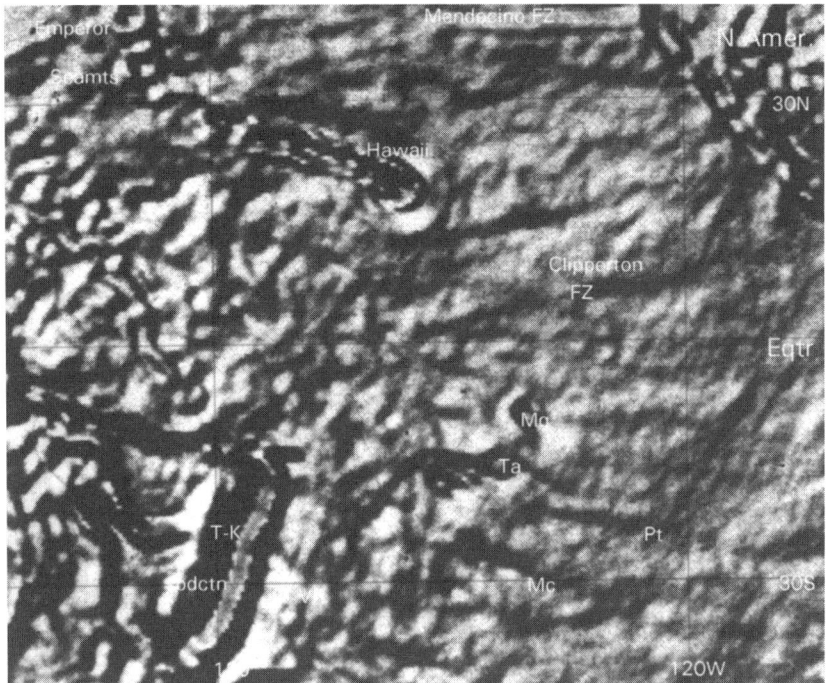

Fig. 5.24. Gravity gradient at 160 km elevation; illuminated from NW. "Known" plume trails such as the Hawaiian and earlier Emperor trail appear as lines of circular features, as do those in the Atlantic region, discussed later. Mq, Ta, Mc, Pt: site of "standard" hotspots; see figures 5.29, 5.30. T-K subduction: Tonga-Kermadec zone of subduction. (DeMets et al. 1990; Phipps Morgan et al. 1995). The least ambiguous trail-generating plumes, such as the Hawaiian, listed in table 5.2, are in the Pacific. The gravity data are from Rapp [1988].)

such as the relict or fossil Galapagos fracture zone, are thought to have little connection with hotspots. The Line Islands lineament is subparallel to the Emperor Seamounts chain, but is associated with a volcanism sequence hard to resolve (Henderson and Gordon 1982; Sager and Keating 1984; Schlanger et al. 1984; Epp 1984; Keating 1987). To judge from its weak deep expression and its left-lateral offsetting of the Galapagos FZ, this linear feature represents shearing deformation bounding the region which has become dominated by the East Pacific Rise.

Desonie, Duncan, and Natland (1993) find that the orientation of the Marquesas hotspot trail varies by 30° from correspondence with the "standard" hotspot direction. With respect to the standard set, deep gravity suggests the existence of lineaments extending N75W to N85W through all "hot spots." However, the latter do not demarcate terminations of lineaments. The MacDonald lineament is part of a line extending for several thousand kilometers, from the north end of the Tonga-Kermadec subduction zone, ESE to the gravity trough associated with the axis of the EPR. To acquire representative date-samples of such an extended structure is

136 Tectonic Consequences of the Earth's Rotation

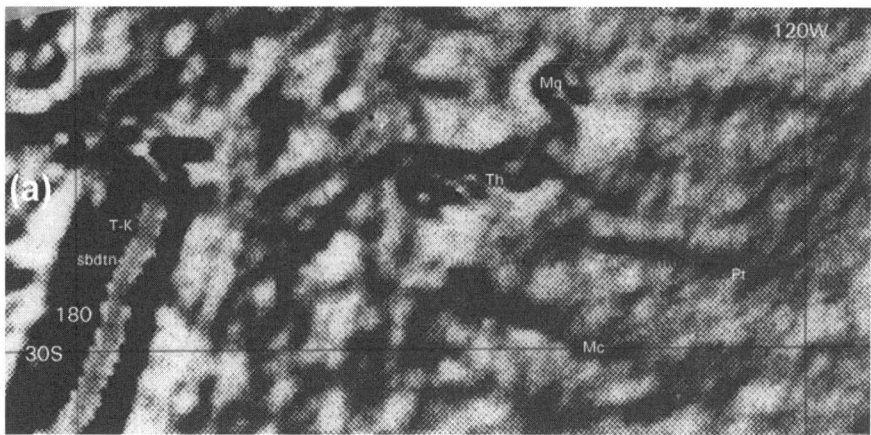

Fig. 5.25. (a) Gravity-gradient at 160 kms elevation; image illuminated from NW. "Known" plume trails such as the Hawaiian and earlier Emperor trail appear as lines of circular features, as do those in the Atlantic region, discussed later. Mq, Ta, Mc, Pt: sites of "standard" hotspots (table 5.2). T-K subduction: Tonga-Kermadec zone of subduction. (Data from Rapp [1988].)

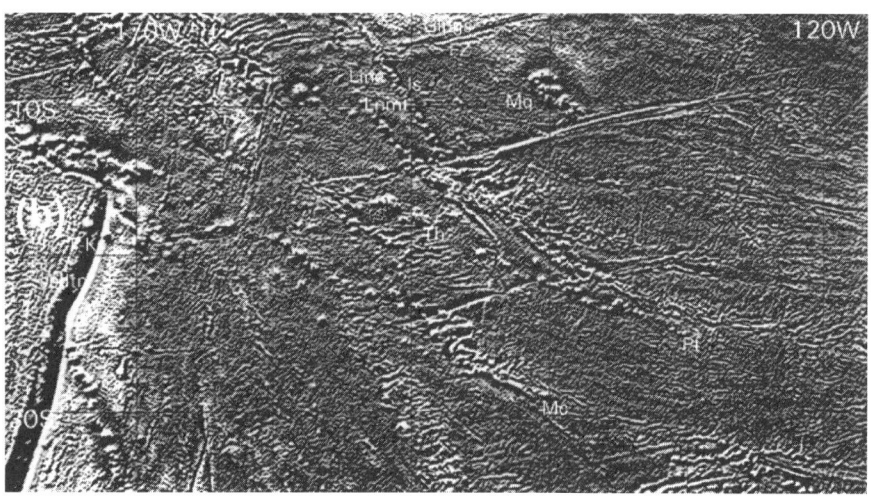

Fig. 5.25. (b) Radar-altimetric gravity image; area as in figure 5.25a, but sensitive to short-wavelength shallow structure. (Data from Sandwell and Smith [1996].)

a formidable task. In form the lineament resembles the structures forming the Louisville "hotline" (Watts et al. 1988).

The eastern Pacific (figure 5.26) is dominated by the presence of the currently-active EPR. Despite its topography, the EPR is the site of a gravity trough, indicating, perhaps, its passive origin, as by lithosphere extension. Gravity features outline its location at earlier times. On the basis of magnetics, Mammerickx, Herron, and Dorman (1980) and Mammerickx, Naar, and Tyce (1988) have identified the location of abandoned axes, e.g., the Mathematician Ridge. The surface location of the present-day spreading axis seems to be located systematically east of the sharply-defined gravimetric axis.

At the latitude of fastest spreading, approximately 20°S, plume lines may cross the EPR axis. It may be significant that at this latitude, Cormier and Macdonald (1994; see also Levi [1998]) have found the EPR axis is stepping or "jumping" west, (effectively, reassigning material from plate to plate). "Plume lines" representing the present spreading direction extend westward beneath lithosphere formed during the earlier spreading direction, as at 15°S.

Plume trails seem to cut fossil fracture zones such as the Clipperton. Whereas fracture zones such as the Clarion and Mendocino in the Northern Hemisphere are coincident with a line of circular features, perhaps a product of transform faulting, the asthenosphere beneath fossil fracture zones south of the Equator, such as the Marquesas and Austral, has been subject to reorganization, erasing their deep expression. The latter fracture zones seem to survive only at surface.

The change in surface spreading-direction from W by N to W by S in crossing the Equator northward (Macdonald et al. 1988, fig. 1) is reflected in asthenosphere trends. However these extend for some thousands of kilometers on the west flank of the EPR, beneath lithosphere having the older trends. As a result, asthenosphere lineaments symmetric about the Equator converge toward it as far as 160°W.

The directions of the principal geoidal lineations in the Pacific have been statistically separated by Baudry and Kroenke (1991), Maia and Diament (1991), Cazenave et al. (1992), Wessel, Becovici, and Kroenke (1994), and Cazenave, Parsons, and Calcagno (1995). Wessel et al. find that medium-wavelength lineations are aligned in the direction of absolute plate motion, assuming standard hotspots. Baudry and Kroenke (1991) and Cazenave et al. (1995) find elongated geoidal anomalies colinear with the standard Polynesian hotspot swells, but extending east of the EPR. Within statistical constraints, correlation maxima are present at wavelengths of 750 and 1100 km, suggesting a location within the deep asthenosphere. Avoiding prior assumption as to plate motion based on hotspots, and taking precaution against introducing filter artefacts, Cazenave et al. (1992) found that medium wavelength geoidal anomalies are preferentially oriented in the E-W direction. As in the case of the high-level gravity image, within the region formed during post-19 Ma spreading (Mammerickx, Herron, and Dorman 1980), lineations are oriented E-W (Cazenave et al. 1992 pl. 3). Based on earlier discussion, it is apparent that lineations of longer wavelength, stemming from the asthenosphere, are little affected by conspicuous shallow features such as the Marquesas and Galapagos relict fracture zones.

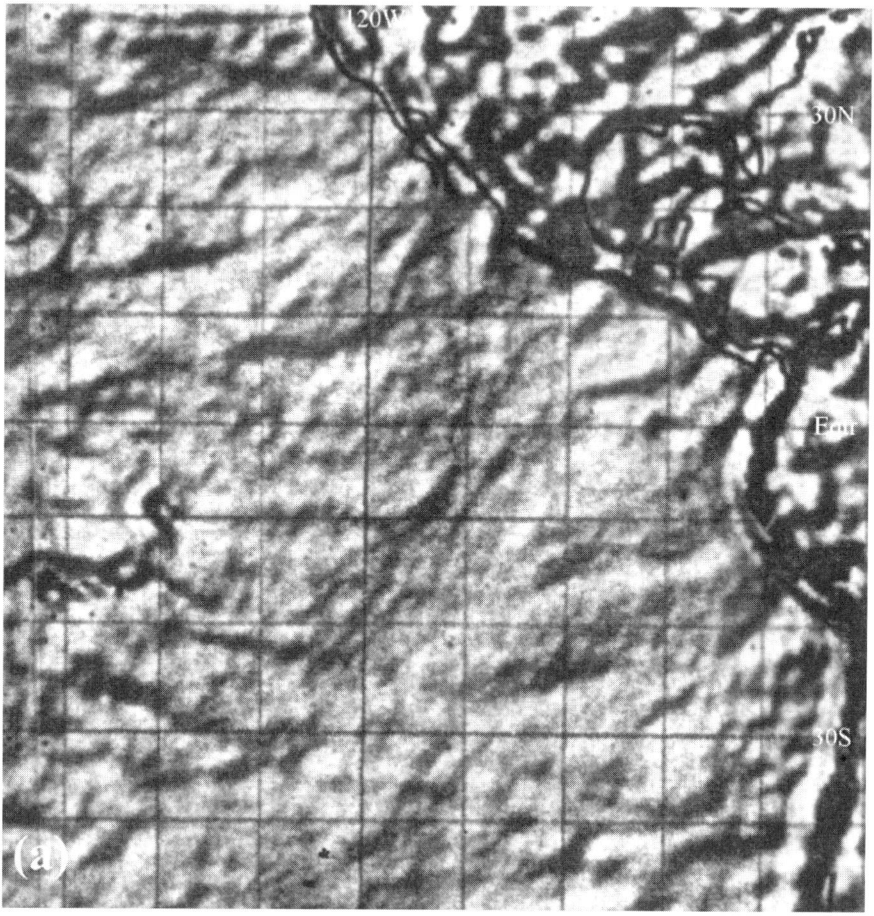

138 Tectonic Consequences of the Earth's Rotation

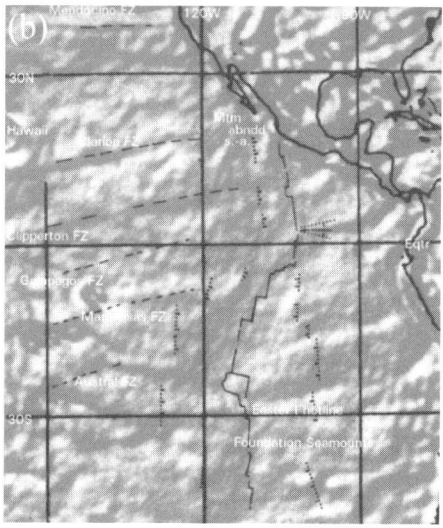

Fig. 5.26. (a) High-level gravity gradient, eastern Pacific; lit from NW. The projection of the field at elevation 160 km is due to Rapp (1988). (b) Tectonic elements. Black-line: current surface spreading axis. Dot ornament outlines lithosphere area within which spreading segments may have been re-aligned NNE since 26 Ma BP (Mammerickx, Herron, and Dorman 1980). Prior to this, segments may have been aligned NW, with spreading direction NE (refer also to Joseph 1993). *Mtm abndd s.a.:* site of Mathematicians spreading axis, abandoned since axis "jump" (Mammerickx, Naar, and Tyce [1988]). Compare with altimetric data, figure 5.19c.

A plausible explanation of the existence of the large number of "plumes," due to Richter and Parsons (1975) and Bonatti and Harrison (1976), may be that they form at the upwelling join of parallel convection rolls (figure 5.27). This mechanism would explain the difficulties encountered in searching for age sequence in volcanics. Age is then prone to be random along lineaments, and unfortunately not diagnostic of sequential motion. Gravity-gradient images suggest that at least 2 categories of objects identified as hotspots or plumes may exist, namely lines formed via the Richter/Parsons mechanism, and structures native to Hawaii and the Atlantic region (section 5.4.2.1), representing upflow from the deep mantle. Additionally, it may be the case that the vertical-axis, vortical motion present in transforms is recorded as the circular gravity features that conspicuously follow the trace of fracture zones, as at Mendocino (fig. 5.26a).

5.4.2.1 The Atlantic Ocean. Deep-seated latitudinal (east-west) features are also conspicuous in the Atlantic Ocean (figure 5.28). These become dominant in equatorial latitudes. Plume-like circular features are common.

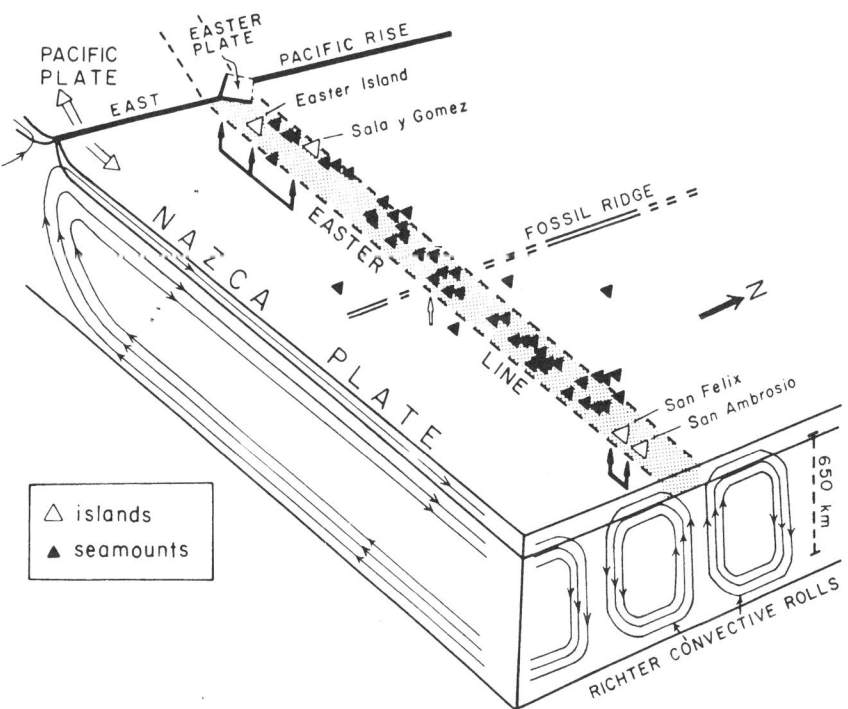

Fig. 5.27. "Rolls" forming downflow from convective spreading axis (arrows) and volcanic structures (triangles) outlining Easter Island hotline in equatorial Pacific. Bonatti and Harrison propose that "spouts" or plumes form at the upflow join of conjugate convection rolls. (From Bonatti and Harrison [1976]; see also Bonatti et al. [1977].)

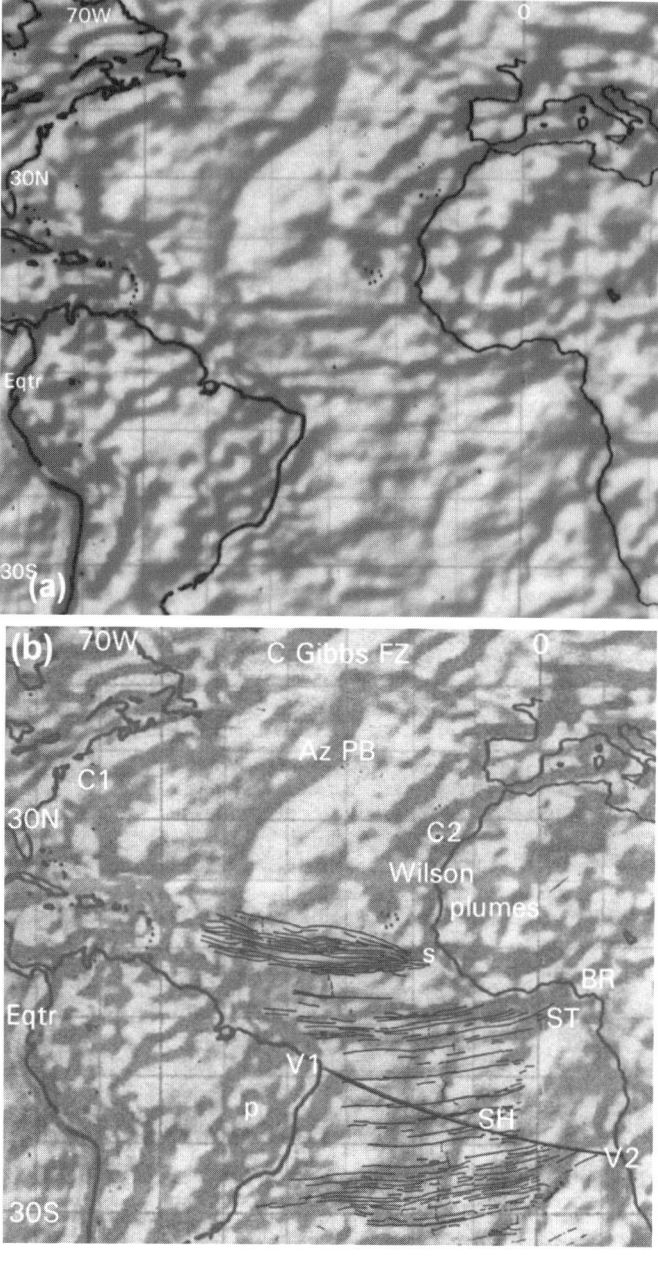

Fig. 5.28. (a) Gravity-gradient at high level, Atlantic Ocean. (b) Tectonic features. *South Atlantic:* Thin black-line: fracture zones and mid-ocean rift. BR, Benue rift system; P, a Parana plume; S, "source" of C. Verde linears; SH, St. Helena; ST, Sao Tome. *North Atlantic:* Az PB, Azores plate boundary; C1, C2 original and present location, Canary Island "plume." The thick line V1–V2 is location of velocity profile (figure 5.22a). The gravity data are from Rapp (1988).

Many of the circular features are coincident with African and other hotspots identified geologically 25 years ago by Burke and Wilson (1972) and Wilson (1976). Rather than supposing that the multiple African features represent hotspot trails, these writers suggest that since −25 Ma, Africa has been stationary relative to the mantle.

A typical deep E-W structure (figure 5.28) underlies the peculiar braided fracture zones of the equatorial Atlantic. A "bundle" of fracture zones emanates from a point located near the small circular feature at 5°N 20°E, diverging WNW from this point across the Atlantic. It would be illogical to suppose that the fracture zones in this group start at the west Atlantic margin (oceanward of the Antilles), converging eastward so as to vanish at a point. It must be supposed that the trans-Atlantic surface flow at this point is westward, commencing near the Africa margin. The "source" marked S is proximate to analogous circular features, perhaps the expression of hotspots, on the Africa mainland. The deep-seated feature extending east-west across the Atlantic ending in the active volcanism of the Benue region, is also possibly the trace of a plume.

In the South Atlantic, the eastern termination of fracture zones (figure 5.28) is a ridge marked by a gravity swell extending SW fom Sao Tome through St. Helena, and a similar feature paralleling the Walvis Ridge. The age of the South Atlantic fracture zones is well known, as a result of sampling and in terms of the numbered seafloor magnetic anomalies.

Rather then paralleling the ridges at which they originate, the fracture zones now extend WSW across the Atlantic. The opening sequence resembles that outlined by Fairhead (1988, figure 6), in which at first opening took place SW-NE relative to Africa, then veered to its present E-W direction. The "plume line" aligned along 14°S seems to represent the trail of a plume that sourced the Parana basalts (cf. VanDecar, James, and Assumpcao 1995), of which the original surface expression is now located at 15°S 45°W.

For these relations to be satisfied (and for there to exist no east-lying zone of subduction), rather than originating at what is now the location of the mid-Atlantic, spreading originating at the Africa margin must have taken place by means of westward translation of the spreading axis, leaving behind it Africa and material forming the east limb of the spreading. The Atlantic islands such as St. Helena lie several hundred kilometers east of the apparently migratory MAR spreading axis, displayed as a velocity anomaly (Zhang, Tanimoto, and Stolper 1994) and in gravity. An analogous sequence has been modeled independently by Pavoni (1993), based on geological evidence.

The system indicated accords with the sub-Atlantic westward flow modeled by Phipps Morgan et al. (1995) figure 5.23a, and with the opening of the Atlantic as perceived by Tanimoto and Zhang (1992), figure 5.20b. Circular features as at 3°N 30W, coincident with the Mid-Atlantic Ridge, may be the expression of plumes contributing to the flow.

The North Atlantic opening style appears similar to that of the South Atlantic. A deep-seated zone of disturbance bears W by N from the actively volcanic Canary Islands plume, C2, extending to a similar but defunct feature C1, centered at 36°N

70°W, southeast of New England. Some suggestion exists that the plume trace persists west of New England. The trace demarcates the New England Seamount line and its trans-Atlantic extension identified by Duncan (1984, fig. 4; see also Sleep [1990] and Duncan [1982]), in examining the opening of the Atlantic. It would be logical to suppose that, as in the South Atlantic, the North Atlantic is opening via westward extension. The mid-Atlantic spreading center with its negative gravity sign is then westward-mobile and passive. The mode accords with the conclusion of Wilson (1973) that the surface motion of this region is predominantly west.

5.4.2.2 The Indian Ocean. As in the Pacific and Atlantic Oceans, structures already identified as plumes are the site of distinctive gravimetric features (figure 5.29). Plumes have been viewed as the source of features such as the 90°E Ridge, a trail beginning at the site of the Rajmahal Trap effusion. The northern sector of the 90°E Ridge has been deflected by the westward displacement of Sundaland (Powell and Johnson 1980), figure 5.14. Discussion has centered upon whether the present plume site is Crozet Bank (Curray and Munasinghe 1991) or Kerguelen Island (Kent et al. 1992); see also Royer and Sandwell (1989). The gravity image suggests that a source may be a weak circular feature now at 43°S 87°E, east of Kerguelen, not far from the present spreading-axis of the SE Indian Ridge. Its axis is the site of plumes that are barely discernible at present resolution.

Discussing jumps in spreading axes, Mammerickx and Sandwell (1986) concluded that Broken Ridge and the Kerguelen Plateau are conjugate features flanking the Southeast Indian Ridge, the current spreading axis. The southern part of the 90°E plume trail underlies lithosphere produced by the SE Indian Ridge. Several other lineaments invite comparison with the conspicuous Deccan Traps and 90°E plume trails. The direction of fracture zones seldom reflects that of plume lines. Apparently, once formed lithosphere is freely displaced.

Within the ocean basin, the curvature of tectonic elements increases westward. Westward of the N-S Deccan Traps plume line, the present seafloor spreading-axis is aligned N-S at the Equator, concave toward the African continent. Fracture zones being produced here are aligned NE-SW, the direction to be expected in an environment in which the Indian (eastern) flank still is experiencing south-to-north displacement, toward the Himal, whereas the African flank is stationary. The sharply-curved Mauritius-Seychelles plume line parallels the counterclockwise rotation of Africa recorded in the paleomagnetic declination (figure 5.7). The plume array marking the separation of Madagascar from Africa formed in the Mesozoic and rotated with Africa.

A plume group north-bounds the Indian Ocean triple plate junction. The enigmatic post-7 Ma tectonic "ripples" affecting the lithosphere south of Ceylon (Sandwell and Smith 1996) appear as faint E-W aligned striae on the high-level gravity map. The Africa continent is marked by the current east-west disruption of the East Africa (Gregory) Rift system, the site of a line of plumes ending in the Afar feature, identified as a plume by Burke and Wilson (1976).

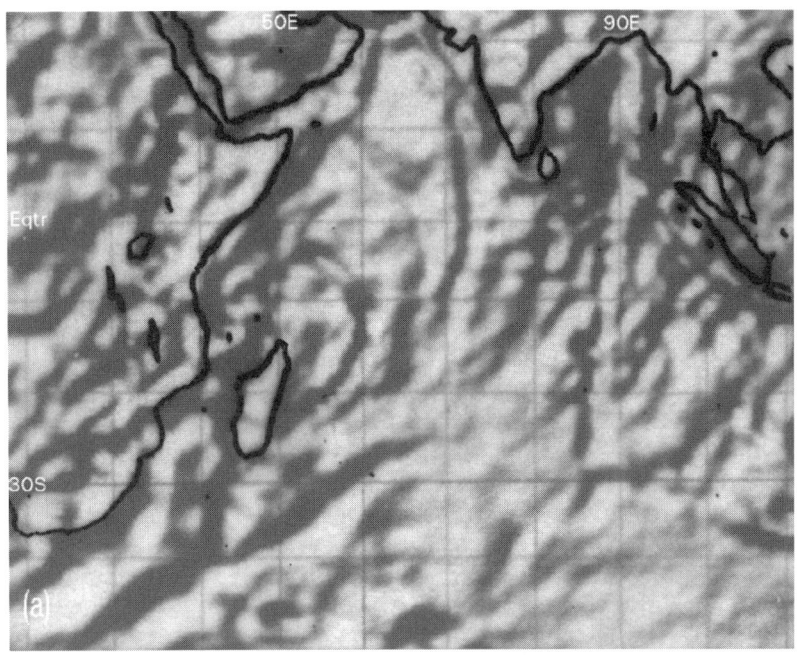

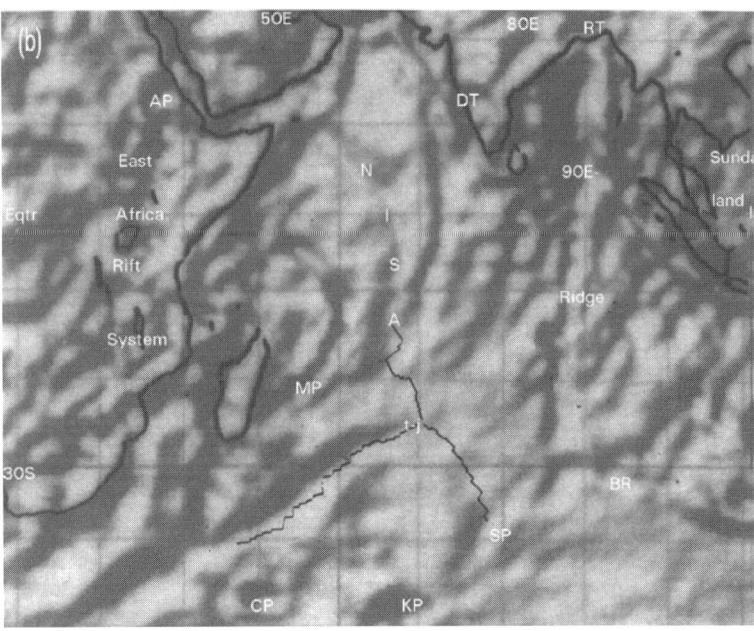

Fig. 5.29. (a) High-level gravimetric image (Rapp 1988), Indian Ocean and periphery. (b) Localities: AP, Afar plume; BR, Broken Ridge; CP, Crozet plume; DT, Deccan Traps; KP, Kerguelen plume; MP, Mauritius plume; NISA, North Indian Ocean spreading axis; RP, Rajmahal Traps; SP, Amsterdam plume; t-j, triple plate junction. Black-line: surface spreading axis, central Indian Ocean.

144 Tectonic Consequences of the Earth's Rotation

5.4.3 Surface Flow Relative to the Deep Mantle

Deep-mantle objects have been tomographically mapped (Su, Woodward, and Dziewonski 1994), figure 5.30a. "Fast" objects, thought to be cooler than average, ring the Pacific, the product of dated subduction zones.

The age of the Pacific plate (figure 5.30b) displays a pattern symmetrical about the Equator, with westward increase. Defunct spreading axes now in the western Pacific record seafloor extension about centers *which, while operational, were located in the eastern Pacific.* The EPR is the site of the current spreading. The parameters of the *system* are:

(1) It has been stationary relative to the objects in the deep mantle. (2) The locus of the spreading axis has, apparently systematically, been the eastern Pacific.

Whereas the surface-westward limb of the EPR, carrying the Pacific plate, has continually extended across the Pacific, its counterpart, the surface-eastward limb carrying the Nazca, Cocos, and Farallon plates has been correspondingly small. It seems to be being replaced by the extending west limb of the Atlantic spreading (figure 5.34, below). The Atlantic-Pacific realm accounts for two-thirds of the global low-latitude belt. No such flow pattern can be seen in the remaining one-third.

5.5 Lithosphere Mobility

5.5.1 Random vs. Nonrandom Motion

Theoretically, mantle flow driven by unrestricted thermal convection must result in purely random motion. This condition applies whether or not one assumes the incorporation of high-viscosity "plates" in the convection (Ricard, Doglioni, and Sabadini [1991]; Bercovici and Wessel [1994]; Puster, Hager, and Jordan [1995]). The Taylor number of the mantle (Chandrasekhar 1961) is too minute to affect the convection pattern, as are Coriolis forces in material of such high viscosity. Eddington flow (Dicke 1966, 1969; McKenzie 1968, app. B) is likely also to be minor or insignificant.

5.5.2 Latitude Dependence and Asymmetric Motion

As evident in review, the tectonic record displays a preferred east-west alignment of structures in many low-latitude spreading regions. The effect is not present in the Indian Ocean. Subduction, overwhelmingly most voluminous at the western Pacific margin, (figures 5.14) takes place preferentially E-W.

To compensate for prevalent E-W flow in the Pacific-Americas-Atlantic two-thirds of the Earth, it might be proposed that exactly canceling meridional flow is dominant in the remaining one-third. Before the availability of high-resolution gravimetry, Wilson (1963) postulated the existence of convection as a single cell encompassing the Earth. On present evidence (figure 5.31), the flow direction in the Eastern Hemisphere offers little promise of cancellation, and some indication of adding to the imbalance with respect to the Western. In low latitudes, only the India/Australia plate is affected by meridional motion, whereas there is evidence

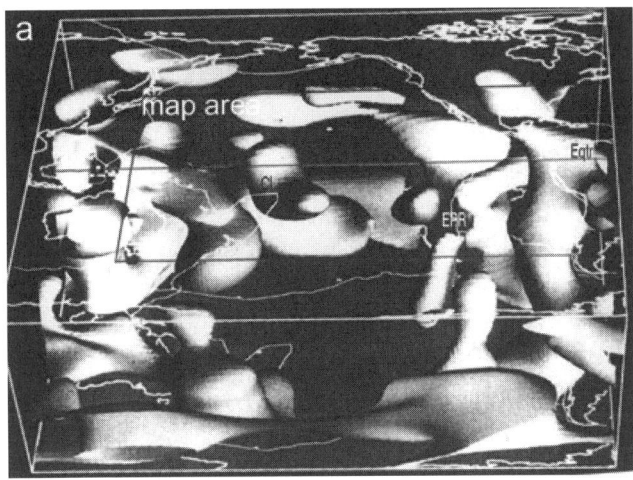

Fig. 5.30. (a) Tomographic "cool" objects located in the lower mantle. "map area" outlines adjoining map, fig. 5.30 (b), of seafloor age. For cross-reference, in each figure "CI" (Canton Island) is one of several sites of defunct seafloor spreading on mobile Pacific plate (Mayes et al. [1990]). The 3-dimensional projection is from Su, Woodward, and Dziewonski (1994).

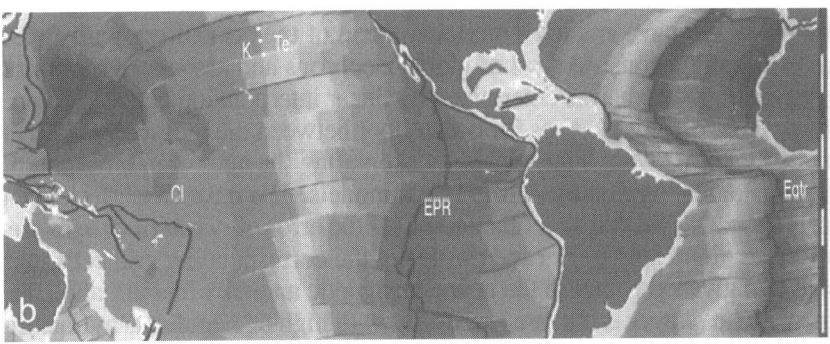

Fig. 5.30. (b) Age of lithosphere (shaded bands), from seafloor magnetic anomalies. In Pacific, age increases westward, from zero (EPR) to Mesozoic in far west. Canton Island (CI) (Joseph 1993) and other sites mark present location of trace on plate of defunct spreading axes. Atwater et al. (1993) trace U. Cretaceous eastward axis "jumps," using fracture zone geometry and the placement of the Quiet Zone boundary (K : Te) in the north Pacific. The age-boundaries pattern describes a persistent veering and eastward jumping of the spreading axis, suggesting that it has acted so as to remain in the eastern Pacific, as in case of present day EPR (figure 5.19c). Atlantic spreading likewise is asymmetrical, no subduction being located to its east.

To judge by location of immobile deep-mantle mass accumulations (previous figure), marking the site of prolonged subduction, and present location of defunct spreading axis traces, the Pacific convection system or cell has been located more or less as at present since the last "plate reorganization," at −43 Ma. (Base: colored global map by Muller et al. [1997]).

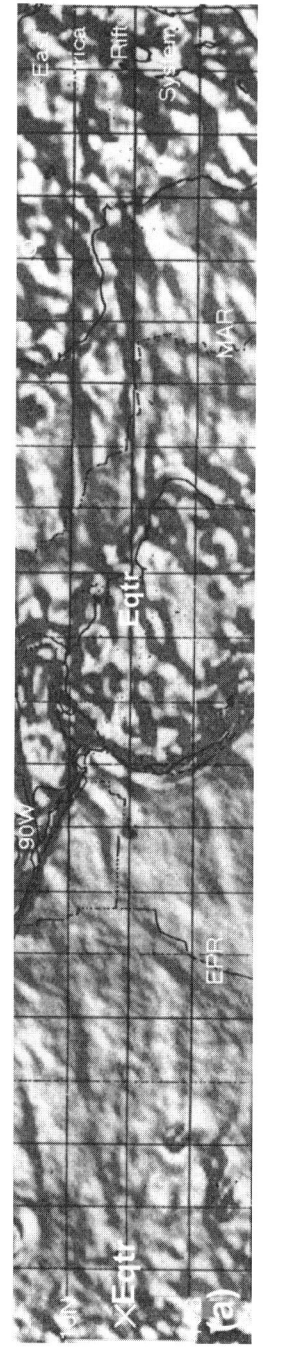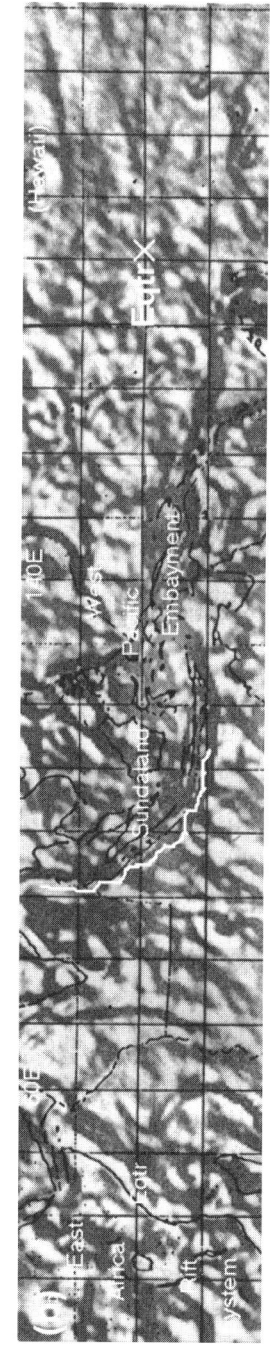

Fig. 5.31. Orientation of gravimetric tectonic features in 35° band centered on the Equator. The gravity data are from Rapp (1988). "x" (Equator, western Pacific) is common reference point. (a) Western Hemisphere. (b) Eastern Hemisphere.

that Sundaland has been displaced some thousands of kilometers west. In the figure, a white "×" marks the westward termination of the peculiar "cone-in-cone" pattern that characterizes the gravity expression of the equatorial Pacific. Referred to hotspot sets, albeit hard to define, overall plate motion has suggested to some a net-lithosphere-rotation about the inner Earth. Net rotation of the lithosphere has been evaluated by vectorially adding relative plate motion as a system to best estimates of absolute motion, indicated by selected plume trails (e.g., Ricard, Doglioni, and Sabadini [1991]).

McKenzie and Parker (1974) have drawn up a convenient representation of plate motion as a polyhedron, the "geohedron" (figure 5.32), referred to angular-velocity axes. To comprehensively define global plate motion, the geohedron should form a closed solid. Component data quality has been reviewed by Gordon (1995). The asthenosphere flow system may be an entity independent of deep mantle convection except for plume penetration, a possibility examined by Phipps Morgan et al. (1995), Davies (1988), and others; this is lent weight by the abundant asthenosphere structures evident in gravity gradient images.

As there is available no explanation of the "cone-in-cone" pattern of gravity in the equatorial Pacific (figure 5.31), if real the pattern indicates neither a surface-

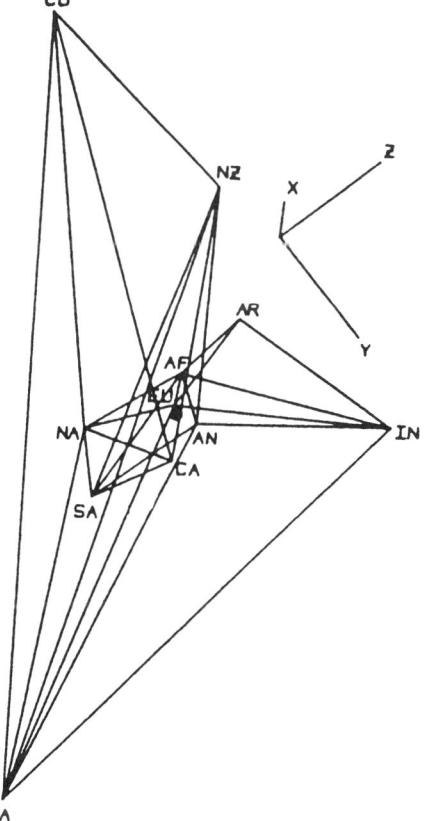

Fig. 5.32. Lithosphere motion represented as the "geohedron" of McKenzie and Parker (1974). The structure is a polyhedron in angular-rotation space. The z-axis is referred to the Earth's rotation, x-axis to the Greenwich prime meridian. The vertices then represent the rotation of individual plates (standard abbreviations). Joining them, the edges represent plate boundaries. Displacement of the origin represents rotation of the lithosphere as a whole ("net-lithosphere-rotation" about the inner Earth). Based on a hotspot set incorporated in model AM1-2 selected by Minster and Jordan (1978), the displacement plots as the black dot, representing a major westward component of rotation and a minor northward component. If net rotation should instead be referred to the spreading direction indicated by gravimetric images of the asthenosphere and deep-mantle objects, the origin may be displaced by several times the amount shown, in the z-direction. (Figure from Minster and Jordan [1978].)

eastward nor a surface-westward preferred component of motion. Perhaps not coincidentally, this feature terminates to the west in the concentration of displacement, subduction and mass accumulation associated with the great westward embayment of the Pacific. Magnetics, surface (radar-altimetric), and deep-seated gravity are insistent that flow is preferentially latitudinal in the low-latitude EPR and MAR. Such features as the asymmetry noted by Phipps Morgan et al. (1995), figure 5.23a, suggest not only that the flow is zonal, but that the sense is surface-westward. It is difficult to suppose that most material is added to the lithosphere in the E-W direction relative to the mantle, without supposing also that E-W rotation is added to the geohedron.

With respect to the rotation of individual plates, McKenzie and Parker (1974) point out that algebraically, if we neglect the sense of observed motion a mirror-image geohedron exists in a quadrant of the angular velocity frame. Excluding reference to hotspot trails, weak other evidence suggests that a mirror-image interpretation is untenable. Plates in the Northern Hemisphere seem to rotate clockwise, the converse of the situation in the Southern Hemisphere. It is found (Forte and Peltier 1991) that counterclockwise vorticity is slightly prevalent in the Southern Hemisphere (figure 5.03), based upon both the sense of transforms and forward flow modeling using the tomographic density distribution. If we assume stationarity of the geomagnetic dipole, paleomagnetic data (figures 5.07 and 5.08) tend to corroborate the sense of the rotation.

In respect to absolute motion, it may be difficult to determine to what extent plumes, if once identified, are moving with asthenospheric flow, or are "up-rooted" or tilted by it. Such motion is not necessarily detectable as inter-hotspot motion (c.f. Burke and Wilson 1972; Burke, Kidd, and Wilson 1973a,b; Schlanger et al. 1984, fig. 2; Okal and Batiza 1987; Molnar and Stock 1987). No way is apparent of directly determining whether the Hawaii plume trail, if sourced in the deep mantle, is affected by an independent asthenospheric flow system, biasing it. For lack of a better procedure, hotlines having conflicting directions generally have been selected, to tally with the Hawaii or Emperor orientation. Recently, O'Connell and Steinberger (1998) have employed models of the flow in the deep mantle to estimate the contribution of this to hotspot motion, and its implications as to true polar wander. It may be the case that lithosphere motion should be referred to the flow seen in robust gravimetric and magnetic images, rather than to motion indicated by plume trails barely identifiable.

Flow referred to deep-mantle objects would best account for the mass distribution required by the Pacific geoid. In absolute terms, "an agency" is subtracting material from the eastern Pacific, creating a mass deficiency marking the EPR; moving the material in the form we identify as the Pacific plate across the Pacific; and depositing it at the western equatorial margin, there to form the greatest geoidal high (figures 5.14).

5.5.3 Tidal Terms?

The E-W asymmetry of the Pacific spreading, and latitudinal symmetry about the equator of the upwelling marked by the East Pacific Rise, became evident in 1978 during the International Geodynamics Conference, Western Pacific, held in Tokyo. At this, Dickinson (1978) pointed to the fact that asymmetries conspicuous in the Pacific Hemisphere could be explained were there to exist a relative eastward displacement of the upper mantle relative to the lithosphere. A flow bias attributable to the feeble retarding torque is entirely inadequate to explain this form of asymmetry.

The spreading-rate of the EPR, recorded by the geographic distance ridge-crest to Paleocene-Eocene boundary (figure 5.33), well displays the low-latitude maximum in surface-west motion, whether coincidentally solely under thermal convection, or under convection biased by an independent agency. On-land and marine investigations had already shown that the surface-eastward limb of the EPR is being replaced by the surface-westward limb of the mid-Atlantic spreading axis. The concentration of subduction at the west Pacific margin was seen as due to convergence between the west limb of the EPR and less mobile continental material of Australasia and eastern Asia.

Heezen and Fornari (1976) based the empirical relations shown in figure 5.33 mainly on magnetic and bathymetric data. Objective and comprehensive coverage, by altimetric satellite and ship survey (Smith and Sandwell 1997), has on the whole confirmed the identity of the features perceived twenty years ago, and revealed a

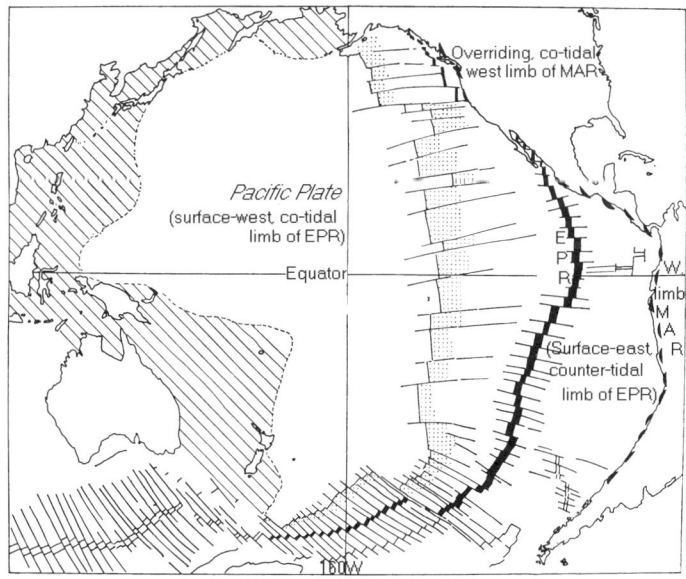

Fig. 5.33. Pacific Hemisphere as seen in 1977, at the time of a Tokyo conference considering the large-scale tectonics of the Pacific. Dot pattern marks the location of the Paleocene-Eocene boundary, hence indicates rate of post-Paleocene spreading; in this projection horizontal scale is the same at all latitudes. (Data from Heezen and Fornari [1976]; see also Bostrom [1984, fig. 5]).

wealth of detail. Parenthetically, if the geometry of spreading (figure 5.19b,c) relative to the contemporary equator is other than coincidence, the antecedent history of spreading, worked out by Mayes, Lawver, and Sandwell (1990), might record the relation of the Pacific plate to the terrestrial rotation-axis during much of the Tertiary (Gordon 1983).

If the global gravity data of Rapp (1988), figure 5.26, are to be heeded, the gravity trough (as distinct from a ridge) associated with seafloor spreading raises a difficulty, with respect to supposing that "ridge-push" (Forsyth and Uyeda 1975) plays a role in driving plates. Why should new lithosphere drift away from a region of shallow mass deficiency? In respect to plate motion, the gravity relation substantiates the possibility that upflow and spreading may represent a response to lithosphere extension or removal, rather than a driving agency. Data now obtained by Project MELT (Levi 1998) corroborate the asymmetry of the EPR, and the likelihood that the upwelling is passively initiated, by the separation of plates. The Pacific plate on the west flank of the EPR is moving west faster than the Nazca plate, on the opposite flank, is moving east.

Forsyth and Uyeda showed that of the forces slab-pull vs. ridge-push likely to be causing plate motion, the former is larger by an order of magnitude. Tidal induction of vorticity, independently activating the asthenosphere, would eliminate a celebrated difficulty. The latter is inherent in transmitting slab pull or tensile stress via lithosphere, clear across the Pacific Ocean, from subduction zones located at its western margin—while overcoming the basal drag everywhere imposed by an inert asthenosphere.

In contrast to imposing drag, activated asthenosphere (namely, that not so perfectly elastic as to be immune to passage of the 40 cm M_2, S_2 waves), would promote the motion of what have become the largest plates on the planet. The latter compose the surface-westward limbs of the MAR and EPR. If the mantle flow system is of this type, convection is not driven solely by vorticity introduced locally and randomly, by foundering slab; but this biased by tidal vorticity induction (TVI), everywhere in phase. The effect is maximum in the low-Q asthenosphere.

To identify the tidal dissipation paths which may be significant, those inventoried in chapter 4 may be compared with the tectonic record. The matrix likely to contain significant terms is (equation (4.4), term [4])

$$\sigma_{ik}^T v_{i,k}^C$$

identifying cross-products of the tidal stress σ and the velocity gradient v under the convection. The dissipation must be increased by the extreme nonlinearity of mantle processes; material is perpetually at yield or "failure" point. As evident earlier, the lowest estimates of tidal energy dissipation are in a position to account for a major flow component.

Reviewing section 5.4.3, if surface flow is referred to objects in the the deep mantle (figure 5.30), it describes an asymmetry in global subduction; the proclivity of the Atlantic to extend itself westward; and the tendency of the west limb of a mobile EPR to become dominant, at the expense of the diminishing east limb. Accumulating geophysical information (Tanimoto and Zhang [1992]; Phipps Morgan et al. [1995]) seems to enhance the likelihood that Neogene flow has been of the form postulated by Wilson (1973), figure 5.34, inexplicable in terms of symmetrical, random convection.

This being the case, it is tempting to relate the observed tectonic biases to the continuous rotation of the wave-tide strain axes relative to particle lines. A component of the mechanism by which modulation may be effected, besides direct vorticity induction, is the isotropic pressure tide in subduction regions (section 4.4; figure 4.18).

It seems inevitable that, overall, the advance in global tectonics must occur in phase with tidal stress increments, entailing the dissipation.

5.5.4 Comparative Planetology

The planet Venus may offer an informative comparison with the Earth (figure 5.35a). Remarkably similar materially (Venus/Earth mass, volume: 0.815, 0.88), astronomically the "sister planets" strongly differ, in that Venus lacks a moon and rotates only slowly (day-length = 243 Earth-days). Nimmo and McKenzie (1997) have pointed to the enigmatic disparity between heat production and efflux in the Earth and Venus.

The tectonic regime on Venus (figure 5.35b) is not one of lithosphere mobility (see for example Phillips, Grimm, and Malin 1991; Turcotte 1993, 1996; Lenardic and Kaula 1994; McGill 1994; Strom, Schaber, and Dawson 1994; Fowler and O'Brien 1996; Solomatov and Moresi 1996, 1997). Instead, beneath a stagnant

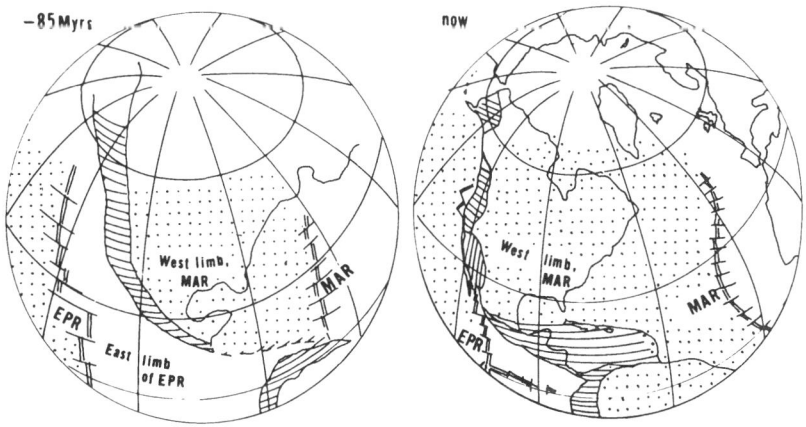

Fig. 5.34. Increase through time of the area underlain by surface-westward circulation. The east limb of the EPR is being replaced by the west limb of the Atlantic spreading. The greatly developed, cotidal west limb of the EPR extends to Asia. Burke, Kidd, and Wilson (1973a,b) and Wilson (1973) suggest that the east limb of the Atlantic spreading center has become stationary during Tertiary times. (Bostrom [1981c], based on Wilson [1973] and Atwater [1970].)

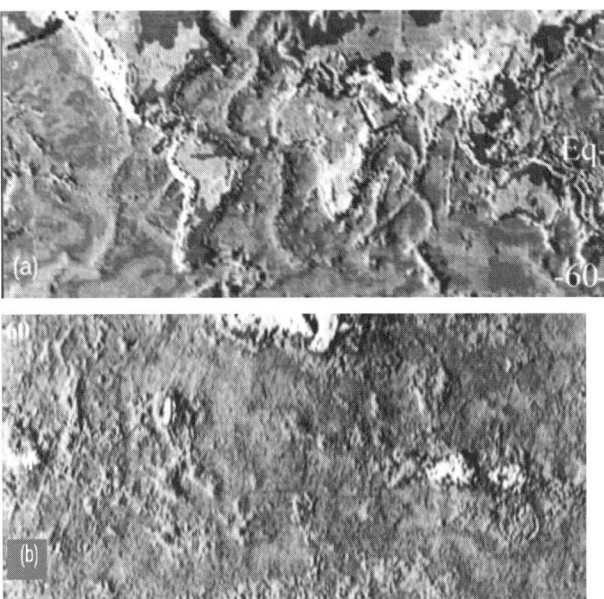

Fig. 5.35. Images of (a) Earth and (b) Venus, here reproduced at equivalent resolution from NASA data. The vertical relief of the Earth is four times that of Venus. Venus is devoid of the extended linear features, such as the mid-ocean ridge system and mountain chains, characteristic of Earth. Nimmo and McKenzie (1997) point out that the Earth's internal heat production seems less than its efflux, which takes place disproportionately at the site of the ridge system. In contrast, the heat efflux of Venus is less than its expected production (implying internal warming), and probably is effected via conduction, transport in static plumes, and episodic mantle overturn.

lithosphere, heat accumulation in the Venusian mantle leads to a static plume regime, with widely separated episodes of mantle overturn causing "replating."

Being moonless and only slowly rotating, Venus lacks the Earth's unidirectional 0.4 m wave tides. It may conceivably be the case that the lunar accompaniment of the Earth is intrinsic to the Earth's regime of steady convection, with constant heat removal and, inevitably, subduction. In an early paper, Campbell and Taylor (1983) pointed to the fundamental role played by water, granitization, and subduction in the case of the Earth, as against their absence on Venus. By ensuring a regime entailing subduction, the action of wave tides may be essential for the maintenance of the hydrous cycle of Nolet (1994), his "Oceans in the upper mantle."

Concomitantly, subduction ensures that the other major volatile, CO_2, is cycled within the mantle, in sediments, and in the world ocean, rather than becoming resident as CO_2 in a thick greenhouse atmosphere. I have speculated elsewhere

(Bostrom 1998b), that absent its Moon and rapid asynchronous rotation, the Earth would be endowed with a Venusian regime, void of life.

In his classical survey of the solar system, Kuiper (1955) categorizes the Earth-Moon System as a double planet. The orbital mass center lies in the shallow part of Earth's core or in the deep mantle. Observing the physiographic contrast (figure 5.36), it is perhaps legitimate to ask whether, without its primordial accompaniment by the Moon, the features of the Earth would be in any way familiar. A strongly differentiated planet, having a lithosphere overlying an asthenosphere orders of magnitude less viscous, may be peculiarly susceptible to the coherent, self-organizing character of vorticity-induction. Whereas structures (cells or plumes) formed by self-organization under thermal buoyancy are the result of local radial forces, the lateral forces conjured by vorticity act in phase throughout shells circumscribing the Earth. In this vein, a tendency to form a "pacific plate" extending over a major portion of the contemporary equatorial belt may be inherent in terrestrial history; the tendency for coherent induction to form "plates," across which stress is additive due to stress-diffusion, is absent on Venus.

Although possessed of a differentiated and probably molten ferrous core (Zhang and Zhang 1995), Venus lacks jointly a central magnetic field (Russell 1993) and the vorticity induction affecting the Earth at all depths, including the core-mantle boundary layer, D″. The mode and sense of wave-tide induction is dimensionally that of the asymmetrical Coriolis force (Bostrom 1998c), commonly believed intrinsic in development of a central magnetic field. Zhang and Zhang point out that as the Taylor number is extremely large, Venus's rotation-slowness

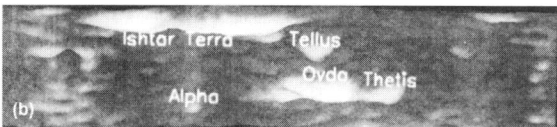

Fig. 5.36. Venus physiography. (a) Radar image draped over topography (black artifacts: data lacunae). (b) Lithosphere density anomalies. Down-warped regions are believed to represent thickened crust. (Perspective images 80°N to 50°S by Herrick [1994], reproduced by permission.)

The Venusian lithosphere displays no evidence of the horizontal mobility characteristic of the Earth (Solomatov and Moresi 1997), but suggests the effect of mantle plumes acting upon a static lithosphere (Turcotte 1993). Strom, Schaber, and Dawson (1994) conclude that "Venus experienced a global resurfacing event about 300 m.y. ago, followed by a dramatic reduction of volcanism and tectonism." Present lava production is 0.01 to 0.15 km³/yr, as against 0.33 to 0.5 km³/yr in terrestrial intraplate vulcanism (mainly, in the form of seafloor spreading). Herrick suggests that during an earlier tectonic regime the Venus lithosphere may have been as mobile as the Earth's at present.

cannot alone be responsible for the absence of a central magnetic field. Accordingly, it would be of interest to compare changes in the inclination of the lunar orbit referred to the Equator plane (Mignard 1981; Laskar 1994; Touma and Wisdom 1994), with changes including reversals in the geomagnetic field, so excellently recorded in seafloor sediment cores; cf. also Hide and Malin (1970).

5.6 Tectonic Record

To the extent that hotspot traces are reliable, the field record suggests that net-lithosphere-rotation is prevalent, in the amount of a few centimeters per year. Unless by large-scale coincidence, viscosity variations cannot explain its dominant westward component. If lithosphere motion should be referred to robust gravity data and deep-mantle tomographic objects, rather than to plume trails hard to resolve, net rotation is greater. In either situation, the associated asymmetric flow in the mantle might account both for the preferential growth of the west limb of seafloor spreading axes, and the various forms of structural asymmetry.

The tectonic record suggests that numerical models excluding the tidal component exclude features of some importance. These include cellular asymmetry, excess of plate motion and subduction in low latitudes (apparently resulting in the gravimetric excess flattening), and secondary features such as the "regimentation" of fracture zones. If correctly posed ab initio (that is, from a random distribution of mass anomalies rather than from the extant anomalous distribution), tideless dynamic models may stabilize at a Venusian static-lithosphere point, denying the existence both of Earth's mobile plates and organized subduction.

Observation of lateral absolute motion is beset with difficulty. Reconciliations of altimetric satellite data with computed ocean tides, summarized by Andersen, Woodworth, and Flather (1995), have assumed a sea-floor component which excludes the primary tidal action. Developments in the field of space-based geodesy, newly available instrumentation, and realistic models, reviewed in chapter 7, may offer a possibility of better measurement.

6

True Polar Wander

Figure Adjustment

6.1 Overview

The Earth is now treated as rotating but isolated, devoid of the tide-causing gravity of the Sun and Moon. Whereas tidal distortion and loss are susceptible to instrumental observation, true polar wander (TPW) is observable mainly in the geological record. The arguments here followed are no less speculative than the reconstruction of lithosphere plates based on tectonic data. The progression followed is from the demonstration by Goldreich and Toomre (1969) that the Earth may be subject to TPW, to its possible explanation of events in Gondwana.

Erosion and tectonic motion redistribute masses forming parts of the Earth. Displacement geographically reorients the axis of maximum moment, about which the planet is compelled to rotate. Simultaneously, conservation of momentum requires that the rotation-axis remains fixed relative to the cosmos. To comply with both requirements, the Earth's rotational bulge migrates geographically, imposing major deformation. At a point on the surface the motion is perceived as true polar wander. Its extent is unknown.

Those parts of the Earth's surface moving toward the equator experience a wave-like displacement. A question arises as to how the Earth's reconfiguration is accomplished. In effecting such large transference of material, is the mechanism deep-seated, as by flow within the deep mantle, or confined to the more mobile asthenosphere? What tectonic effects may be expected?

A hypothetical model earth is viewed as lithosphere and asthenosphere metastable in respect to pressure and temperature, overlying mantle affected by convection. TPW is rate-limited by the heat generation producing convection, rather than by the viscosity. The convection then conforms with the figure jointly determined by rotation and the contemporary location of the moment axes, resulting in such effects as the formation of new ocean basin.

In this account, external coordinates and an earth-fixed system based on the axes of figure are set up, permitting referal of geographic change to an external system. Geological and geophysical data are next reviewed, to assess as best possible whether TPW has taken place at a significant rate. A significant rate is one entailing tectonic phenomena.

Such data as exist suggest that rather than being a steady-rate process, TPW is uneven. "Rapid" intervals, requiring correspondingly fast transfer of deep-seated material, might correlate with the mantle circulation changes that have caused reorganizations of plate motion. It must be noted that several analysts have found TPW, if this is correctly identified, to be slow in comparison with that to be expected under action of the agency identified by Goldreich and Toomre. The possibility exists that a stabilizing or impeding mechanism exists, rendering the action strongly episodic.

Field evidence for TPW is confined mainly to the Eastern Hemisphere, and in particular Africa and the Indian Ocean. As well as the disruption of Gondwana, south to north motion of its parts has become accepted by most geologists, based on paleomagnetic and paleoclimatological data. If this is a manifestation of TPW (rather than individual motion of plates), within post-Paleozoic times figure-adjustment has required the displacement somewhere within the subsurface of material at a rate equivalent to that in mantle convection, as measured by seafloor extension and subduction.

With reference to Africa, material maintaining the bulge was in some fashion added to the continental subsurface, and that of a nascent Indian Ocean. Concomitantly, during northward motion the continent experienced a latitudinal decline in gravity, hence in subsurface pressure, together with areal extension (distension). The distension, although perhaps only coincidentally, was approximately equivalent to the area occupied by the intrusions to which the region became subjected. During the Paleozoic the region had been resident in high latitudes, experiencing only spasmodic and widely separated eruptions.

The geological record, well established as a result of mining activity in the area and extensive laboratory experiment, indicates that the African craton was underlain by upper mantle saturated at ambient pressure with dissolved volatiles, hence in marginal pressure/temperature equilibrium. Commencing with these conditions, perhaps also by coincidence the peak in volcanism took place during the maximum rate of lithostatic pressure decrease and distension. This occurred during the latter Mesozoic. At that time southern Africa, having departed from the Antarctic, was approaching mid-latitudes.

If northward motion of Gondwana fragments took place as individual plates (for arcane reasons sliding over fixed asthenosphere), as distinct from being a manifestation of TPW, absence of the emplacement of new, deep material would much limit the tectonic effects.

During Tertiary times, latitude change in Africa, and its counterclockwise rotation, have become minor. Continued break-up took place in the form of opening of the South Atlantic and westward motion of the South America fragment of Gondwana.

Latitudinal drift since the Mesozoic has been confined principally to India and Australia. For whatever reason, during this period spreading throughout most of the Indian Ocean became uniformly northeast-southwest.

6.2 Displacement in Latitude

Physically, lithosphere segments may be viewed as moving for unspecified reasons across a stationary asthenosphere, alternatively as being carried by the latter during polar wander. The effects are fundamentally distinct.

6.2.1 Displacement As Individual Plates

During variation in latitude, the pressure under a constant lithosphere column varies slightly with the variation in gravity (axifugal component of gravity), figure 6.1a. In respect to a lithosphere plate moving north, as is believed to have been typical during the early stages of Gondwana's break-up, the pressure decrease at constant drift rate reaches a maximum in mid-latitudes (figure 6.1b). This form of motion entails no addition to the underlying mantle in the form of figure adjustment.

6.2.2 Displacement under Polar Wander

TPW requires an adjustment of the Earth's figure, amounting to local change in radius with latitude. As the interior in no ways resembles a body of homogeneous fluid, consideration must be given as to displacement in a layered and unstable subsurface.

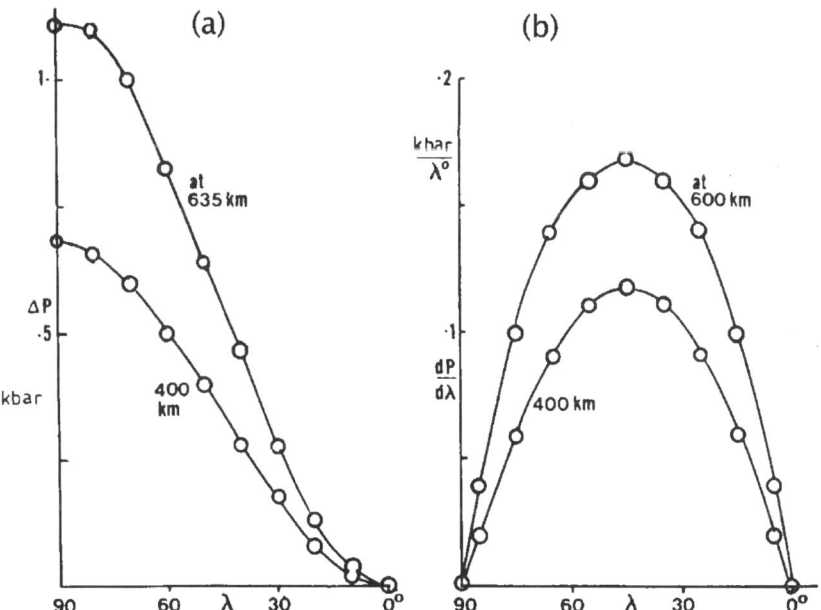

Fig. 6.1. (a) Variation of pressure with latitude at constant depth assuming column average density 3.3 g/cm³. Effects of isostasy do not intervene. (b) Rate of pressure change with change in latitude at specimen depths. The maximum is at 45°.

158 Tectonic Consequences of the Earth's Rotation

In figure 6.2, X_i are axes that are stationary with respect to an external (inertial) frame, X_1 coinciding with the long-term direction of the Earth's axis of rotation. "Long-term" denotes orientation after removal of periodic variations such as precession. x_i are the topical axes of figure, rotating with the Earth, x_1, x_2 and x_3 being those of greatest, intermediate and least moment of inertia respectively.

The moment of inertia of a composite body such as the Earth about a polar axis x_l is represented by

$$I_l = \sum_i (x_2^2 + x_3^2)M_i \qquad (6.1)$$

in which M is a mass element.

Satellite observation of the geoid (Marsh et al. 1990) confirms that the Earth currently is in fact rotating about its principal axis, x_1, x_2, and x_3 project into the eastern and western Pacific. Conservation of angular momentum

$$H = I_1 \Omega$$

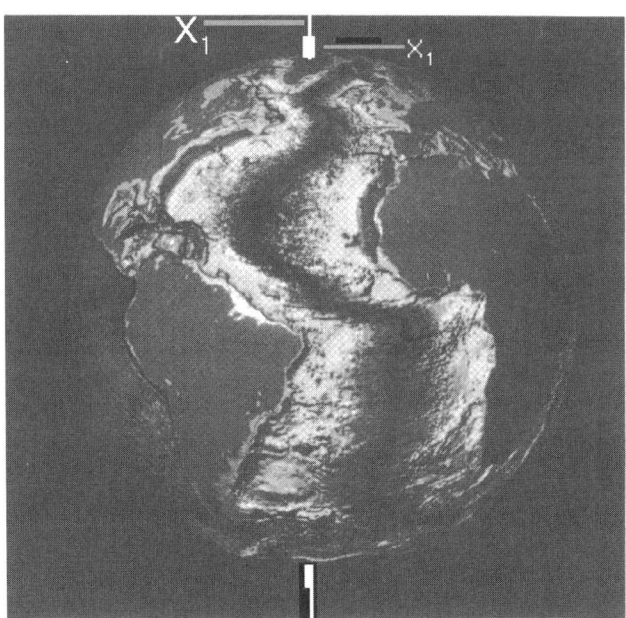

Fig. 6.2. Location of axes X_1 reference to external (inertial) frame, and x_1 describing the location of polar axes of moment. X_1 is space-stationary, complying with conservation of angular momentum. x_i are to be regarded as mobile with *respect to the Earth*, moving in accordance with geological mass shifts. As X_1 and x_1 must nevertheless remain coincident within a small invariant, the body of the Earth migrates with respect to X_1, resulting in true polar wander (TPW).

(Ω, current rotation rate) requires that spin furthermore continues about X_1, almost coincident with x_1, in the absence of torque externally applied.

Newton pointed out that were the Earth spherical, "a heap of new matter" in mid-latitudes would cause the poles to wander about its surface (Cajori, as cited by Munk and MacDonald [1960a]). Almost contemporaneously, Maclaurin (1746) showed that the figure of a body such as the Earth is in fact an ellipsoid, posessed of an Equatorial bulge due to its rotation.

Subsequently, the strength and "moveability" of the 21 km bulge of rotation under mass-displacement have dominated debate as to polar wander.

Based on the conclusion of Kelvin (1863b) that the Earth has the rigidity of steel, but permitting a Maxwellian viscosity, G. H. Darwin (1879a) concluded that polar wander may have occurred, but of insignificant amount. Allowing for a finite viscosity would permit large long-term displacement. Apart from the question of strength of the bulge, Darwin's analysis has been shown (Lambert 1931; see Jeffreys [1952]) to contain an algebraic error annulling his conclusions.

A major impediment to polar wander, and incidentally to lower-mantle convection, seemed evident in the early 1960s, at the time of satellite-based determination of the Earth's rotational flattening. The flattening is observed to be 298.220 (Nakiboglu 1982), after correcting for deglaciation and the permanent tide. In contrast, the hydrostatic (equilibrium) flattening is estimated by Nakiboglu to be 299.638; see also Denis (1989). The unexpected "excess flattening" was hypothesized by McKenzie (1966; see also Kaula 1963, 1967; Caputo 1965; Munk and MacDonald 1960a) to represent lagging adjustment of figure to tidal decrease of the rotation. The present-day flattening is equivalent to the hydrostatic figure of approximately 10 Ma BP, and represents a large quantity of stored energy (section 5.3.1.3). Agreement has not been reached as to the reason for its existence.

6.2.3 Internal Displacement

Sudden displacement of mass within a body such as the Earth, as by a seismic event, displaces the axis x_1 of greatest moment. Because the products of inertia are then non-zero, ongoing rotation contains a wobble, observable astronomically. The wobble decays with adjustment of the Earth's figure to one of equilibrium about the new geographic location of x_1. Adjustment implies reconfiguration of the Earth with respect to the earlier axis, consequently with respect to the previous position of the bulge of rotation. A continuation of the process may be expected to result in cumulative displacement of the pole.

Lambeck (1988) has derived expressions describing the excitation of wobble and its subsequent decay. Wobble is caused by the development of a small angle between the rotation axis and that of inertial moment (Smith 1977), caused by mass shift or change in core-mantle coupling. Observation of wobble provides insight as to internal structure. Hypotheses as to its excitation include mass shift associated with seismic events (Mansinha, Smylie, and Chapman 1979), less abrupt events such as phenomena of the atmosphere and ocean (Wahr 1983), and variable electromagnetic coupling of mantle and core (Runcorn 1982). Lambeck makes clear that although the Chandler wobble is well observed, the nature of its excitation and decay have not been established. Rather than being impulsive, excitation might consist of atmospheric events more prolonged than the decay time. Dissipation may be a result either of bodily elasticity deficiency, of marine tidal dissipation (pole tide), or both. If the excitation is atmospheric rather than related to abrupt mass shift, there is less reason to anticipate cumulative polar wander. Wahr and Bergen (1986) discuss anelastic interaction of wobble and the core.

As might be expected in view of the disparity in frequency, as a measure of viscosity Q in respect to wobble-dissipation (period ≈ 14 months) does not tally with Q in respect to seismic shear (Dehant 1991; Groten, Leonhardt, and Molodensky 1991), having periods of 1hr down to «1s. Q_{wobble} has commonly been estimated to be between 50 and 175 (Ooe 1978; Okubo 1982; Lambeck 1988), in contrast to 300–1000 associated with shear-wave dissipation. It has been argued earlier (chapter 4) that both Q_{wobble} and Q_{seis} are inapplicable to tidal dissipation containing a major secular term. Estimates of the displacement of x_1 relative to X_1 due to an observed large seismic event range from 0.0111 arcsec (O'Connell and Dziewonski 1976) to 0.028 arcsec (Smith 1977).

Slow geologic mass displacement may be expected to produce slow polar migration unaccompanied by wobble. "Slow" is defined as transfer that is slower than the rate at which damping and figure adjustment take place. Attributing the damping of the Chandler motion to low viscosity in the outer Earth, Gold (1955), supported by Inglis (1957), pointed out that, in a spherical earth, the wander of even such small masses as his "beetles" would result eventually in polar wander.

The major limitation of approaches to pole shift based on damping of the Chandler wobble, or viscosity including the effects of the asthenosphere and ocean, is that the values to be employed are grossly uncertain (e.g., Peltier 1989; Hager and Clayton 1989; Jeffreys 1972). Avoiding this, Goldreich and Toomre (1969) have evaluated polar wander rates to be expected in terms of mass displacement and extant convection (see also discussion section 5.3). Mass anomalies are represented by well-delimited geoidal anomalies, such as those associated with subduction zones.

As a preliminary, these authors formulate the overall angular momentum of the Earth, H(t), and components h(t), contributed by internal motion as a Hamiltonian H(q, $p;t$). In configuration space, the generalized coordinate q and momentum p are functions of latitude and longitude. A conclusion is that the value of the action definite integral $\oint p \, dq$ is invariant with respect to an arbitrary time range. This is interpreted to mean that, to within a small adiabatic invariant, the principal axis of figure, under slow displacement by geological processes, remains almost coincident with the space-fixed axis of rotation.

Well-measured quantities to which Goldreich and Toomre then draw attention include:

(1) the small size of the difference between the nonhydrostatic principal moments:

$$C' - B' = 6.9 \times 10^{-6} \, Ma^2$$
$$\text{and } B' - A' = 7.2 \times 10^{-6} \, Ma^2$$

relative to the mass shifts suggested by the geologic record (A', B', C', the principal moments after subtraction of that due to the hydrostatic rotational bulge; M, the Earth's mass; a, its mean radius); and

(2) the large magnitude of the nonhydrostatic moments about x_2 and x_3 relative to the moments about any other equatorial axes; the Earth should be viewed as a triaxial ellipsoid rather than a spheroid.

In a numerical experiment, these authors examine space-referred re-orientation of the Earth (TPW), tracking the axes of figure, based on random, geologically expectable mass shifts. As a result of the smallness of the difference between principal moments, everything being equal it is to be expected that TPW must be prevalent but highly erratic, its velocity at times exceeding the velocity of individual plate motion.

On the basis that plate motion requires mantle convection, the analysis of Goldreich and Toomre brings into relief that the Earth's bulge of rotation is a dynamic not static feature. It has no permanent "strength"; it is an expression of convection patterned by the orientation relative to the Earth of gravity including the rotation-potential, thus imposing no barrier to TPW. Although the fact has not been universally recognised, the analysis by these authors establishes that in a convective Earth, geologic mass shifts must be measured relative to the masses describing the departure from hydrostasy, rather than to the enormously greater mass of the hydrostatic bulge as heretofore.

A numerical experiment based on a formulation akin to that of Goldreich and Toomre was conducted by Keondzhyan and Monin (1977). A value of the viscosity s.l. was not assumed, but their results permit selection of a Maxwell visco-elasticity, 5.9×10^{23} P, best fitting accepted plate motion. Keondzhyan and Monin take advantage of the fact that $(x_2^2 + x_3^2)$ (Eqtn. 6.1) beneath the continents, rising above ocean level, is greater than that beneath ocean basins, despite isostatic compensation.

The disparity suggests that the location of plates that are largely continental might be expected to localize the axes of figure, as earlier pointed out by Munk and MacDonald (1960a,b), in calculating axes defined by "continentality." The ocean function tested by these authors leads to a principal axis presently situated in the north-central Pacific near Hawaii, in contrast to the axis now gravimetrically established, coincident with the rotational pole. The satellite-derived geoid, corroborated by seismic data (Tanimoto 1989), indicates that internal mass concentrations not related to the location of the continents determine the location of the principal axes.

Anderson (1982) has computed that a 50-m geoidal anomaly forms in 100 Ma in a stationary continental area, due to heat accumulation. Having in mind the constraints outlined by location of the axes of moment, it might then be expected that continental areas tend to be stationed in equatorial latitudes. Breakup under continuing heat accumulation would accord with the breakup of Gondwana and observed aspects of continental geology, such as the episodicity of transgressions and magmatism.

Ricard, Sabadini, and Spada (1992) have modeled TPW resulting from tectonic processes based on a mantle having Maxwell rheology, chemical stratification, and mineralogical phase changes. The latter result in large pole shift, countered by hydrostatic compensation at the base of the upper mantle in the case of chemical stratification. Ricard, Spada and Sabadini (1993) conclude that internal mass motion (as distinct from the motion of surface plates) can account for TPW at "observed" rates. However, to restrict rates to those observed, a lower-mantle viscosity of at least 10^{22} P is required.

Besse and Courtillot (1991) and Loper and McCartney (1986) have discussed the possibility that TPW affects the fluid flow in the core, changing the frequency of magnetic reversals, and causing plumed uprise of hot material from the D″ level.

6.3 Geophysical Evidence as to TPW

The foregoing describes the possibility of TPW having occurred. Early suggestions as to polar wander were based on the paleoclimatic record. A reconstruction scarcely different from many today was published in 1924 by Koppen and Wegener (1924; reproduced in Holmes [1978]). Whereas earlier, geologically based, analysis encountered physical objections seemingly insurmountable, acceptance of mantle convection at its "observed" rate renders at once more feasible reconstructions by Koppen and others complying with the geological record.

6.3.1 Paleomagnetics

Establishing a record of TPW faces the challenge that evidence has to be sought within the x_i frame, itself attached to the Earth. Investigation is made feasible using an assumption, that the geomagnetic dipole has tended to be aligned with the axis of the Earth's rotation. The field recorded in rock paleomagnetism is then referable to X_1. However, by the same token, the rock-magnetism record contains directional information only as to rotation of crustal segments and their motion in latitude, and is devoid of longitude information. Livermore, Vine, and Smith (1984) concluded that with respect to paleomagnetic observations, an element of spurious TPW may be contributed by time-dependent departure of the magnetic field from a purely symmetric geocentric dipole. This conclusion is supported by the work of Merrill and McElhinny (1977).

Gordon and Jurdy (1986) have constructed a record of plate motions during the Cenozoic in a mean-lithosphere frame, referred to the location of hotspots (mantle plumes). The best agreement with the paleomagnetic data is provided by an assumption that the Atlantic hotspots are almost stationary relative to those in the Pacific. The hotspot and mean lithosphere reference frame appear to have been in continual slow relative motion, but at a discontinuous rate, throughout the Cenozoic. The rotation of the lithosphere relative to the hotspots is right-handed, approximately 7° about a pole at 46°S 87°E. Plate motion has systematically been fastest in low latitudes. The paleomagnetic axis has shifted 5°–10° relative to the present axis, and less relative to the mean lithosphere. Sager and Pringle (1988) conclude that an element of TPW would resolve discrepancies in a model of Pacific plate motion during Mid-Cretaceous to early Tertiary times.

Courtillot and Besse (1987) estimate that TPW has been at a rate 5 cm/yr, i.e.,

0.45°/Ma, over the last 200 Ma, to be compared with an estimate by Runcorn, 0.33°/Ma (Runcorn 1968). Using data from three major plates, Besse and Courtillot (1991) transfer all magnetically derived data to a common base and compare this with the hotspot record. Referred to the present pole, displacement is somewhat greater than 20° in 200 Ma. Being erratic, the trace or angular path of the pole was much longer, requiring faster TPW. TPW appears to have been episodic, encompassing a quasi-stationary interval between 180 and 110 Ma BP.

Van der Voo (1994) has pointed to the fact that the apparent polar wander paths for all the the continents except northeast Asia (unresolved) in the interval Late Ordovician–Late Devonian are of similar shape. TPW may have attained a larger velocity than individual plate motion during that time. Speculatively, TPW may have been as large as 75° in toto over a 75 Ma interval. The average rate during the mid-Paleozoic was an order of magnitude greater than that during later time. Separating poloidal from toroidal plate motion, Lithgow-Bertelloni et al. (1993) have identified the pole positions of whole-lithophere rotation (as distinct from TPW) since the later Mesozoic.

6.3.2 Gravity Data

Gondwana is believed to have been intact at −300 Ma (figure 6.3, right), and to have lain in such a position that Antarctica was located as now at the South Pole. Africa, India, and Australia were contiguous, in high latitudes. A record of the events in this sequence may be represented in the high-level projection of the gravity gradient by Rapp (1988), figure 6.4 (see also figure 5.29).

The primary tectonic events in this region are widely recognized:

Regarding Africa, this continent moved north from Antarctic latitudes during the Mesozoic. Since then northward motion has slowed, perhaps coming to a halt in the later Tertiary (Burke and Wilson [1972]). The displacement is supported by

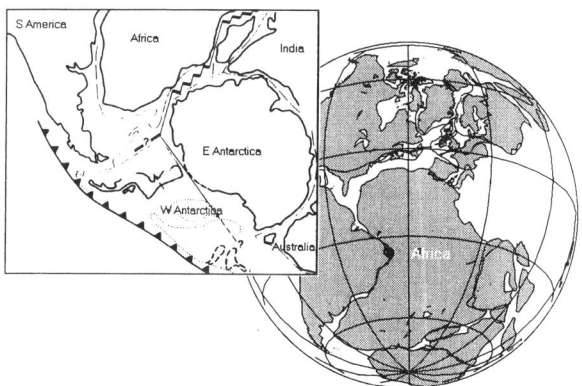

Fig. 6.3. (Right) Location of Gondwana at 300 Ma (McKerrow and Scotese 1990); (Left) Region of convergence adjoining W Antarctica (details from Elliott [1987]).

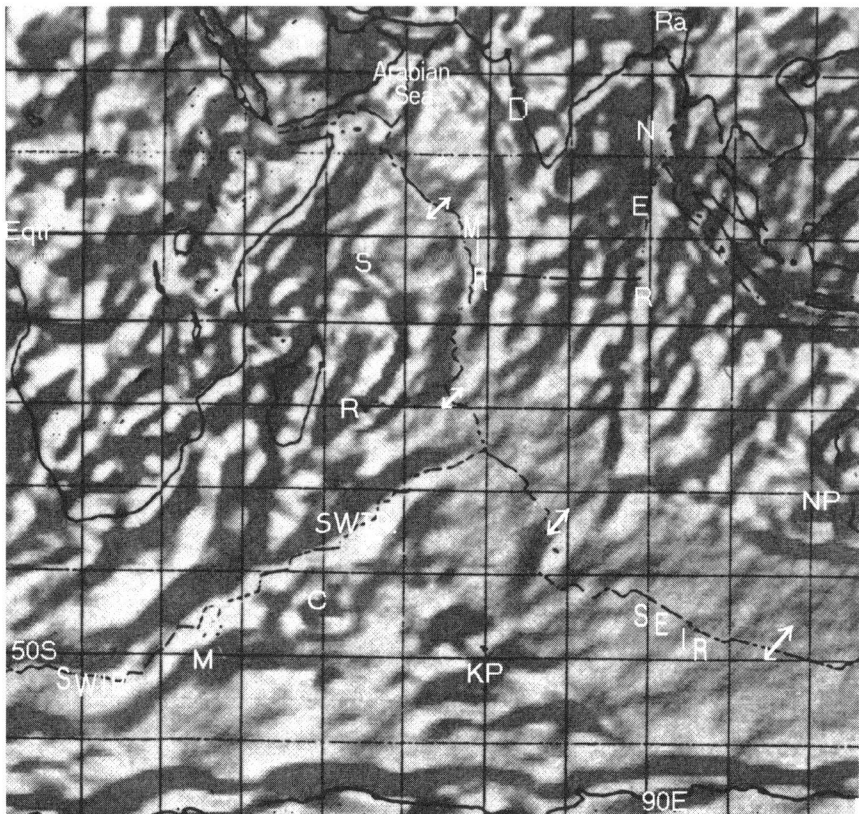

Fig. 6.4. Gravity gradient at 160 km elevation, Indian Ocean and surrounds. Thin black line: surface axis of sea-floor spreading. The circular objects appear related to mantle plumes (cf. Storey, 1995, his fig. 4). C, Crozet Is.; D, Deccan Traps; KP, Kerguelen Plateau; M, Marion Is.; MIR, Mid-Indian Ocean Ridge; NP, Naturaliste Pl.; NER, Ninetyeast Ridge; R, Reunion; Ra, Rajmahal Traps; S, Seychelles. SEIR, Southeast Indian Ocean Ridge; SWIR, Southwest Indian Ocean Ridge. Arrows: spreading direction during Tertiary (NE-SW), in central and eastern Indian Ocean. (The gravity data are from the projection by Rapp [1988].)

mutually corroborative evidence, in the form of the paleomagnetic record on land (figure 5.7) and the dated magnetic record of spreading of the sea floor south of Africa (see for example Mullet et al. 1997).

Regarding the eastern Indian Ocean, in contrast, firmly documented seafloor spreading (Royer and Sandwell 1989; Mullet et al. 1997) demonstrates that rapid northward motion of Australia began in the Eocene, when that of Africa was slowing. Spreading centered on the Southeast Indian Ocean Ridge (SEIR) is well demarcated in the gravimetric map, figure 6.4. For whatever reason, the orientation of spreading (northeast-southwest) during the Tertiary became uniform from the

Arabian Sea in the northwest across the central Indian Ocean, throughout the length of the Mid-Indian and Southeast Indian Ocean spreading axes.

At face value, if the events in the eastern hemisphere are to be interpreted as an expression of TPW, during the Mesozoic tilt of the Earth took place about an axis in the space-fixed equatorial plane (cf. figure 6.6b) which displaced Africa northward. Subsequently, however, during the Tertiary, TPW has been about an axis oriented so as to produce northeast-southwest spreading of the eastern Indian Ocean. The spectacular outburst of spreading responsible for the SEIR, resulting in separation of Australia from Antarctica, ensued from this process.

The original locus of Gondwana fragments and their motion-direction has been determined by Lawver et al. (1987). India is believed to have been separated from Antarctica/Australia in a process carrying it with extraordinary rapidity to its location north of the Equator. During the early stages of break-up, convergence, in contrast to extension, characterised the opposite (western) flank of Antarctica (figure 6.3).

Studies employing petrological, fracture-zone and seafloor spreading data (e.g. Martin and Hartnady 1986; Harris et al. 1987) have established the relation of the Antarctic borderland to adjacent fragments of Gondwana, and the time of the disruption. Many gravity features, such as the line marking the Ninetyeast Ridge, terminating in the age-dated Rajmahal Traps, are coincident with what is viewed as the trace of plumes (Kent et al. 1992; Storey 1995). Opinion has varied as to whether the present site of a plume responsible for the Rajmahal Traps is Crozet Bank or Kerguelen Island (Curray and Munasinghe 1991; Kent et al. 1992).

Subduction is not taking place along the border of eastern Antarctica. In consequence, if the gravimetric "round objects" in fact represent the site of plumes, it would seem that the Southeast Indian Ocean Ridge, and associated plumes, are moving away from Antarctica. If the Karoo Basalts are related to a plume now located at Marion Island (Richards, Duncan, and Courtillot 1989), the plume may have moved north in latitude.

The geometry of the situation seems to require that:

(1) a group of mantle plumes, normally envisaged as quasi-stationary, is moving relative to the mantle, or
(2) spreading axes are moving independently northward, away from Antarctica, or
(3) plumes are as elsewhere almost stationary relative to the mantle, in which case their motion registers tilt of the Earth about an Equatorial axis (TPW).

The last-named motion resembles the explanation tentatively put forward by early investigators of paleomagnetic elements in Gondwana (Day and Runcorn 1955; Creer, Irving, and Runcorn 1957; Irving and Robertson 1969; Creer 1970), namely, that these are an expression of TPW. In the case of TPW, the rapid displacement of India would be comprehendible as motion away from the spreading ridge added to a contribution representing TPW.

166 Tectonic Consequences of the Earth's Rotation

As in all reconstructions, a major unknown is the tectonics at this time of the complementary, Pacific hemisphere.

6.4 Figure Adjustment: Requirements of Bulge Migration

Referred to a point on the the Earth's surface, TPW is manifest as a wave-like passage of the rotational bulge about an axis in the plane of the Equator.

Turcotte and Oxburgh (1973), Turcotte (1974), and Oxburgh and Turcotte (1974) have examined the adjustment in convexity ("membrane tectonics") and resulting fracture as a mechanism in rift formation. Burke and Dewey (1974) concluded that the elastic stress generated cannot endure for significant periods. Sengor and Burke (1978) have concluded that the supporting evidence employed by Oxburgh and Turcotte incorporates K/Ar ratio data which are open to question, and misinterpreted paleomagnetic results.

6.4.1 Areal Adjustment (Distension)

At surface the skin or lithosphere is subjected to areal extension, being proportional to the square of the change in radius with decrease in latitude. Using ellipticity $\varepsilon = (.003353)^{-1}$, as $r = a(1 - \varepsilon \sin^2\phi)$ it will be found as a guide that in experiencing half the pole-to-Equator radius increase, the lineal extension required in all directions is 1.57ms/km (r,a are local and mean radii; ϕ, latitude). Equivalently, each square kilometer of crust must accomodate an extra 3,142 square meters at surface; each areal unit of 100 km square, an extra $3.142 \times 10,000 \approx 3 \times 10^4$ square meters of extra terrain. The specimen area increment is that which would be occupied by a pipe-form intrusion having diameter 200m. Although perhaps by coincidence, the pipes and vents, in excess of two thousand, which appeared in southern Africa during Mesozoic times occupy an areal fraction of the same order.

6.4.2 Volumetric Adjustment

At depth, under TPW three-dimensional adjustment is demanded, requiring the emplacement in some fashion, beneath regions migrating towards the bulge, of material many kilometers in thickness. Rotation of the bulge about an equatorial axis requires the displacement (Bostrom 1990) of

$$4\pi/3 \cdot 1/360 \cdot [(r_{eq}^2 \cdot r_p) - (r_p^2 \cdot r_{eq})] = 10.4 \times 10^6 \text{ km}^3$$

of material per degree of TPW (r_{eq}, the Equatorial radius, 6377 km; r_p the polar radius, 6356 km).

Assuming the TPW rate of 5 cm/yr estimated by Besse and Courtillot (1991) figure 6.6a, based on paleomagnetism, the displacement requirement is 4.7 km³/yr. As an indication of the displacement in mantle convection, on the basis of the spreading rate of the world ocean floor, it has been estimated (Kanamori and Press 1970; Garfunkel 1975) that convection is responsible for the rise of material from within the mantle of 16.7 km³/yr.

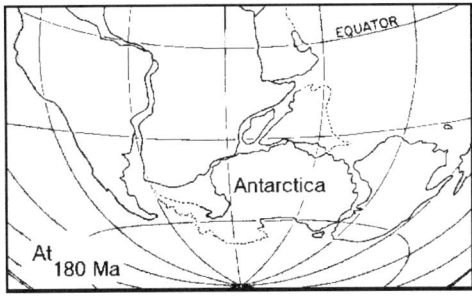

Fig. 6.5. Reconstruction of Pangea at 180 Ma based on paleomagnetics. (After Morel and Irving [1981].)

It may be recalled that rotations are not commutative in addition. The *order* in which TPW episodes take place determines the final outcome. The direction of episodes under geological mass shifts is likely to have been highly erratic (figure 6.6a), with possible "hairpin curves." Under TPW in a single direction, as during one segment of the pole path, it is convenient to view the figure adjustment necessary to effect bulge migration as dividing the Earth into quadrants (figure 6.6b); compare with Spada (1993, figs. 3 and 4).

6.5 Figure Adjustment: Geological Record

Many geologists believe that the continents which comprised Gondwana have been subject to displacement to lower latitudes. The theoretical requirements of figure

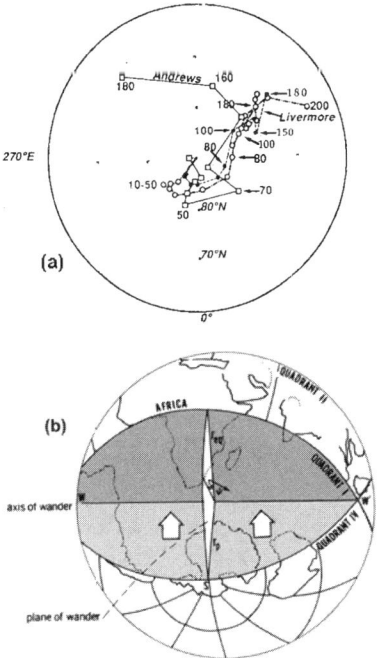

Fig. 6.6. (a) Reconstruction by Besse and Courtillot (1991) of paths of polar wander in the last 200 Ma, incorporating research by Andrews (1985) and Livermore et al. (1984). To smooth erratic data, these investigators employed a sliding 20 Ma window. (b) Division of earth-surface into quadrants during episode of pole shift. W-W': axis of TPW, in (space-fixed) equatorial plane. In respect to migration of bulge in the direction indicated, if this was real the odd-numbered quadrants (Gondwana, no. I) required addition of material from depth, as by Indian Ocean development. The even-numbered quadrants required assimilation of material in the interior, as, perhaps coincidentally, by the outbreak of subduction in the Mediterranean (Tethys) belt, Quadrant II, commencing at −83 Ma (Olivet et al., 1982).

adjustment (bulge migration) may be compared with the record resulting from intensive surface and subsurface mineral exploration.

6.5.1 Asthenosphere Metastability

After the close of the Pan-African orogenic cycle at about −450 Ma (Cahen et al. 1984a), identified by Kennedy (1964), Africa became almost anorogenic. Until the mid-Mesozoic the continent represented an aggregation of stable cratons, affected by comparatively rare eruptive episodes.

The eruptives and xenoliths have attracted attention for their singular petrology and deep-seated origin. Bailey (1982, 1992), Eggler (1987, 1989), Wyllie (1987, 1989, 1995), Haggerty (1989), and others attribute distinctive characteristics to processes of slow, basal-lithosphere accumulation of volatiles from the upper mantle, resulting in deep-seated metasomatism and spasmodic eruptive episodes. At this time, as described by Scotese and McKerrow (1990), Scotese and Barrett (1990), figure 6.3, Gondwana was moving across the South Pole. Latitude-dependent gravity and pressure changes (figure 6.1) are at that latitude almost nil.

At depth in the metastable state described, mineral facies are in pressure/temperature-controlled equilibrium, subject to local destabilization by the gradual accumulation of abyssal materials, including volatiles. In this connection, Anderson (1982) has pointed out that under superficially unchanging conditions a thick-lithosphere region such as this must be subject to slow accumulation of heat, hence growing instability, due to the effect of thermal blanketing.

The location of the solidus curve has been the subject of active investigation (Brey and Green 1977; Eggler 1989; Mitchell 1986; Wyllie 1989, 1990, 1995) in the field of kimberlite genesis and generation of the Gondwana eruptives (figure 6.7). Analyses display major differences, but almost all call upon phase relations that incorporate topology similar to that shown in figure 6.7. Surfaces defining phase and state transitions and the solubility of volatiles intersect the geothermal surface at a determinate depth, in many cases at a grazing angle or tangentially.

The "stable" condition was one of only slowly progressing metasomatism. At this time (the Paleozoic), only sparse eruptions and release of volatiles characterized the South American, African, and Australian magmatism. Experimental petrology (Wyllie [1987, 1989, 1995]; see also Eggler [1987]) points to unloading, hence cascading or explosive, effects to be expected on occasions of locally passing beyond equilibrium. A disequilibrium condition can be brought about either by an incremental change in confining pressure when in a marginal state, or, as heretofore postulated, by temperature increase as during diapirism.

6.5.2 Incidence of Eruption

Between 200 and 50 Ma BP (Cahen et al. 1984a,b; Skinner 1989; Nixon and Condliffe 1989), southern Africa became the site of more than 1200 kimberlite eruptions (figure 6.8). Bailey (1992) draws attention to the fact that besides being simultaneously "platewide," activity was site-specific (the same sites were subject repeatedly to similar magmatism).

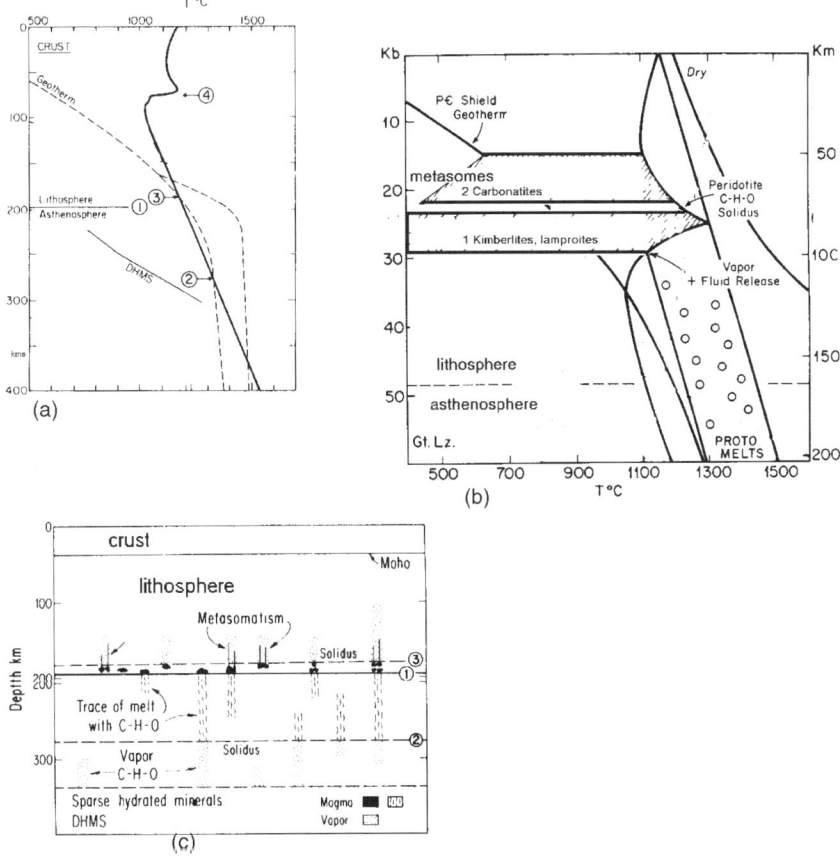

Fig. 6.7. Models of processes responsible for generation of deep-seated Gondwana eruptives. (a) Situation based on laboratory investigations by Wyllie (1995). Within lithosphere and asthenosphere solidus intersects isothermal surfaces at levels 2 and 3. Dense hydrous magnesian silicates (DHMS) are believed to be the residence in the mantle of volatiles, including water. The "ledge" in solidus is associated with upward impediment to flow, causing accumulation of volatiles-rich magmas. (b) Haggerty's model of "metasomes," potentially unstable aggregations of volatile-rich magma, which ultimately produce carbonatitic or kimberlitic eruptives, dependent on their accumulation level. Most researchers point to the critical difference between the "dry" solidus of peridotite, and that intersecting the expected geotherm for magmas containing a small fraction of H_2O or CO_2. (After Haggerty [1989].) (c) Production of intermittent lamproitic eruptions, as a result of the accumulation of volatiles beneath ancient cratonic lithosphere. The critical points 1, 2, and 3 refer to intersections, as in **a**, of the solidus and expected sub-cratonic geotherm. (After Wyllie [1995].)

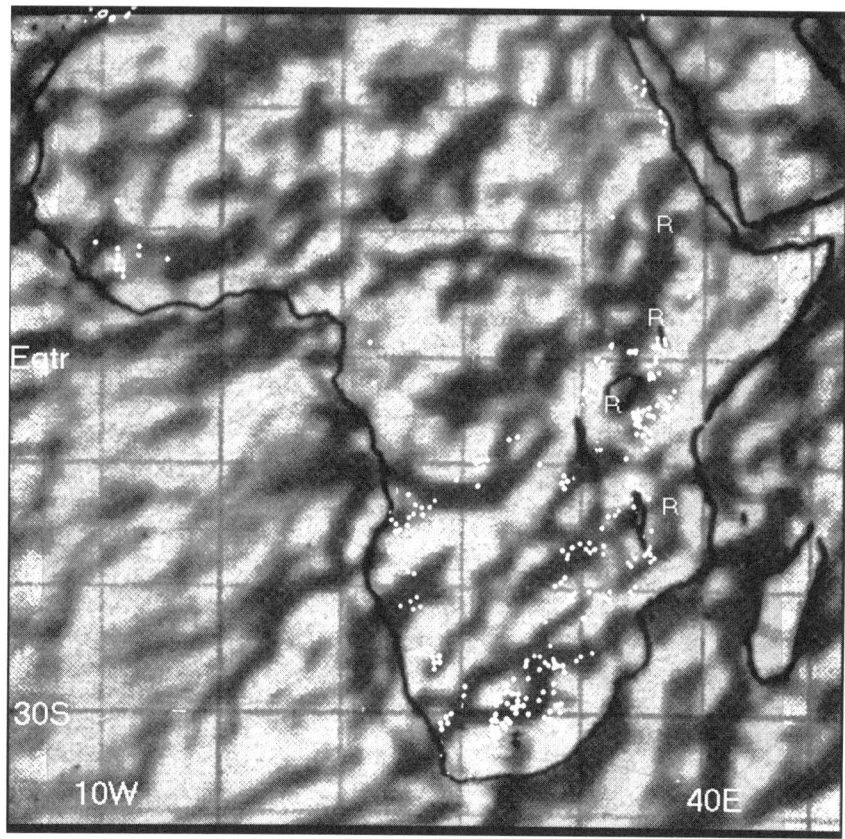

Fig. 6.8. Locations (white dots) of upper Mesozoic kimberlite and carbonatite eruptions, after Cahen et al. (1984a), set against gravimetric image (Rapp 1988) displaying gross structural elements. R . . . R, expression of Tertiary rift system, developing after Africa became stable with respect to rotation and latitude.

In this context, it may be significant that the latter Mesozoic was the interval during which the affected area encountered the zone of most rapid latitude-dependent pressure change and areal extension (figure 6.9), centered at 45° S, figure 6.1. The *lineal* extension is minute, being in proportion to the change in radius; for example, ≈1.6 m/km in experiencing half of the pole-to-Equator radius change. The extension which can be sustained without failure by cold surface rock is generally considered to be nil. The *areal* extension (*supra*) is comparable with the increment represented by the Mesozoic injection of volcanic pipes. The subsurface on which this part of Gondwana rested contained volatile-rich magma accumulations *at the point of eruption*. The lithostatic confining pressure was reduced without reduction of the temperature.

In contrast, latitude change in both high and low latitude zones produces only minor pressure adjustment. Scotese and Barrett (1990) have described the passage

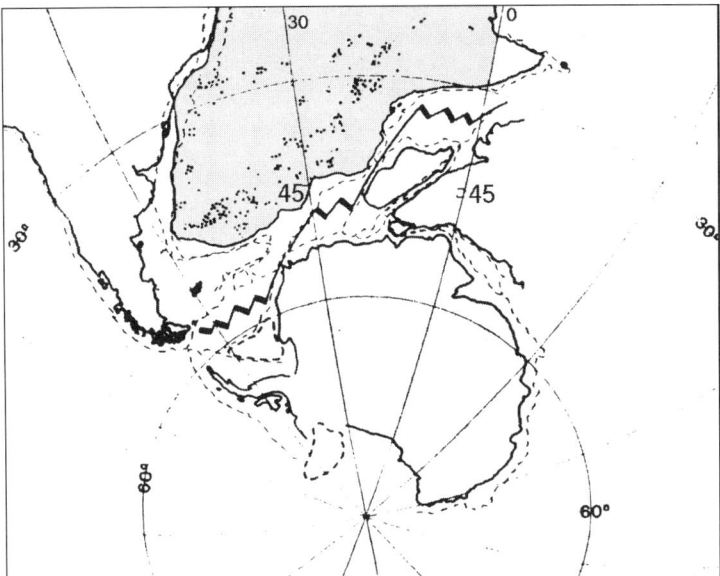

Fig. 6.9. Paleogeography of accelerated Mesozoic eruption of kimberlites and carbonatites (dots). At this time, Africa was traversing the latitude of maximum pressure-decrement and distension (figure 6.1). (Africa contemporary location: Lawver, Sclater, and Meinke [1985]; eruption localities: Cahen et al. [1984b].)

of Gondwana over the south pole. Employing the seafloor magnetic record, Lawver, Sclater, and Meinke (1985) have found that subsequent to −64 Ma, latitude change diminished or stopped. Subsequently, Africa/South America underwent counterclockwise rotation into its present orientation (cf. figure 5.7), followed by the E-W opening of the Atlantic and later the initiation of the Tertiary E-W rift system. The relation of dated eruption episodes to latitude is shown in figure 6.10a. In figure 6.10b an attempt is made to relate these to conditions in the asthenosphere. To a remarkable extent the consequent epeirogenic conditions accord with those deduced on separate grounds by Dawson (1970):

> "Most kimberlites in South Africa are of Cretaceous age and are believed to have been intruded during a period of strong uplift of the continent with attendant downwarping or faulting around the periphery as a result of deepening of the contiguous ocean basins."

Uplift, otherwise enigmatic, appears to have been coincident with the ubiquitous, simultaneous addition of material to the upper asthenosphere and lower lithosphere required to comply with figure adjustment.

Odling (1995) has succeeded in reproducing upper-mantle metasomatic processes in the laboratory. Complicated lithosphere-asthenosphere interactions are

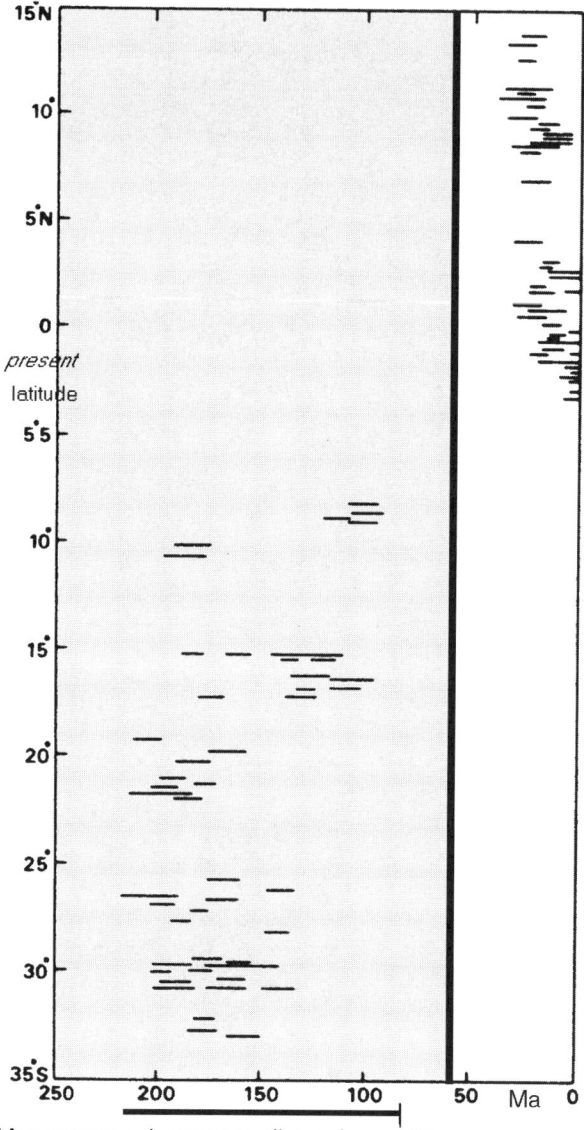

Fig. 6.10. (a) Age vs. latitude and paleo-orientation (figure 6.9) of belts of post-Paleozoic African deep-seated eruptives, following Cahen et al. (1984). Tertiary (post-60 Ma) eruptives include varied rock types, but all are associated with unidirectional extension (E-W rifting), developed under constant orientation of the continent. Tertiary eruptions developed first in the north part of Africa, which first reached low latitudes, and are there most intense. Based on hotspot evidence, (Burke and Wilson 1972), it has been suggested independently that Africa has become stationary with respect to rotation and northward motion during the latter Tertiary.

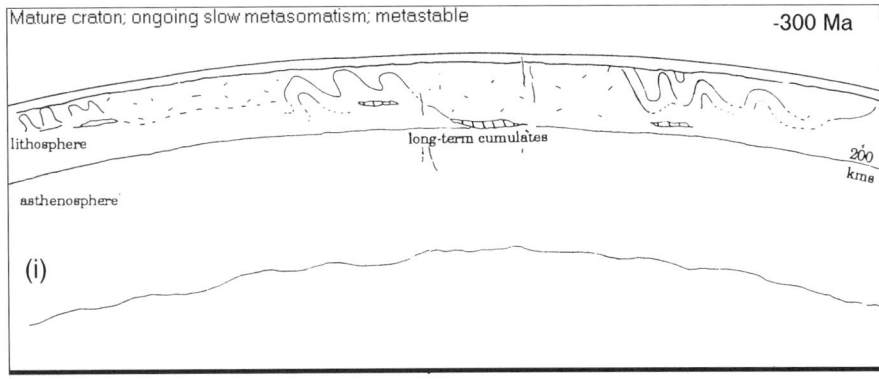

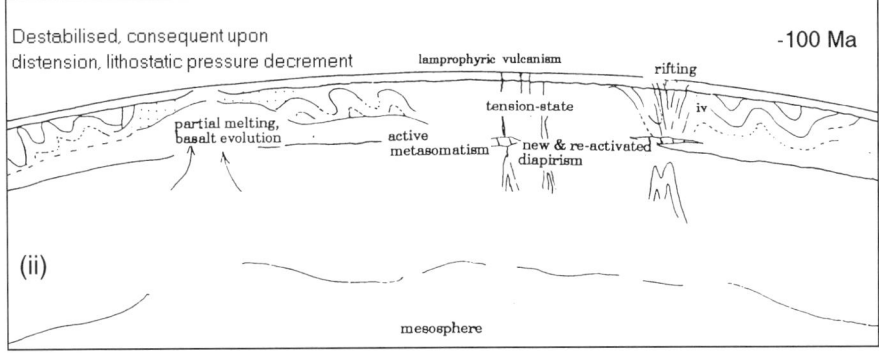

Fig 6.10. (b) Possible event sequence as a result of mature craton experiencing TPW, causing areal extension (distension), together with decrement in gravity, hence in confining-pressure. (i) Long-term metastable state: slow accumulation of volatile-rich material at base of lithosphere results in sparse eruptions, as in Paleozoic Africa (ii) Accelerated occurrence of lamprophyric eruption, in consequence of decreased confining-pressure but maintained magmatic pressure, occurring during latter Mesozoic. Continuing figure adjustment can only be effected via increasingly voluminous addition from the mesosphere. Apparently, the only way in which the volumetric requirements of full-scale adjustment could be met is by way of a repatterning of the convection in the mantle, of which the formation of the Indian Ocean is a manifestation. Reconstruction of the type here shown is vitiated by the absence of known counterpart activity in Australia. The extent to which the environment at depth of the Australian and African archons is similar is unknown. In this regard see Bailey (1992), Atkinson and Smith (1995), Helmstaedt (1995), and Janse (1995).

hard to reproduce in the laboratory or by computer modelling, so that the results of the African "experiment" exist only in the form of the geological record. The simultaneous occurrence of latitudinal plate motion and vulcanism cannot be classed as more than an association, because it is difficult to quantify the pressure-decrement necessary to destabilise, or expedite destabilisation, of such objects as metasomes on the verge of catastrophic eruption.

The geological record seems therefore to offer these alternatives:

1. It is fortuitous that plate-wide Mesozoic eruptive activity took place where and when it did.
2. The eruptive maximum was caused by destabilization of an already unstable assemblage, by its passage through the zone of maximum extension and decrement in confining-pressure.
3. A process still arcane caused simultaneous occurrence of widely-separated similar events.

6.5.3 Tectonics of Africa: Outcome

Pressure changes affect lithosphere plates whether change in latitude represents TPW or independent plate motion. However, in the case of TPW the pressure adjustment affects both the plate and newly emplaced underlying mantle, being the primary agent in its emplacement. If Gondwana's motion incorporated TPW, the absence of a negative gravity anomaly of several thousand mGl requires that destabilization and massive displacement has occurred, keeping pace with the requirements of figure adjustment. With respect to periods in which TPW may have approached the high rates currently suggested (Andrews 1985; Sager and Bleil 1987; Besse and Courtillot 1991; Van der Voo 1994), the question arises as to how material could be transferred so rapidly as to maintain equilibrium figure.

It becomes conspicuous that the widespread abyssal vulcanism characterizing Africa during the Mesozoic was accompanied by the effusion of flood basalts. Contemporaneously with the on-land surge of vulcanism, peri-African rifting became active. This led to the disruption of Gondwana and ongoing initiation of the South Atlantic and Indian Oceans. Storey (1995) has pointed to the probable role of mantle plumes in the process of break up. The on-land eruptives such as kimberlites resemble effects to be expected from pressure decrement. Volumetrically, however, these represent a minute fraction of the 10 or 20 km of material required to maintain figure-equilibrium. Based on earlier discussion the rate required for TPW amounting to 5 cm/yr, some 5 km^3/yr, is within bounds of the flow represented by worldwide convection, but no other processes. Convection as measured by the development of the Indian and South Atlantic Oceans would satisfy the rate requirement. If TPW reaches the high rate suggested by the paleomagnetic record, even for limited periods, repatterning of the convection may be the most efficient and perhaps the only way in which the required figure adjustment can be effected. Once reorganization has resulted in uplimbs forming new oceans, the pattern may be expected to endure, short of further disturbance of planetary scale

Storey (1995, fig. 5) has noted that the formation of the Gondwana igneous provinces was synchronous with convergence at the Pacific margin of Antarctica. It may possibly be significant that in the quadrant antipodal to Gondwana, the East Pacific Rise upwelling, which subsequently has dominated the tectonics of the eastern Pacific, was initiated during its disruption. Whether or not by coincidence, with the commencement of extension in the Gondwana quadrant convergence and re-assimilation was initiated in the adjoining quadrant II, in the form of the Tethys belt of subduction. Commencement of the Mediterranean Tethyan convergence has been placed at −83 Ma (Olivet et al. 1982).

During the period in which Africa has ceased to move northward, the fragmentation of Gondwana has progressed by means of continued opening of the Atlantic, and development of the east Africa rift system. Unfortunately, as originally suggested by Wilson (1970), the north-east part of the EPR has been overridden and obscured, by the westward progression during the Tertiary of the North American continent. The west limb of the rapidly spreading East Pacific Rise has in this interval extended across the Pacific.

6.6 Axis Stabilization (Dynamic Inhibition of TPW)

Physical relationships (Sabadini and Yuen 1989; Ricard, Spada, and Sabadini 1993) suggest that, unless the viscosity of the lower mantle is higher than is generally believed, TPW should frequently exceed 1°/Ma. Such a rate has been postulated to prevail episodically in the past (Van der Voo [1994]). The paleomagnetic data suggest that TPW is erratic but normally much slower. Were viscosity so high as to strongly dephase TPW, the situation would be marked by a persistent wobble.

Stevenson (1993a) has enquired specifically as to why TPW appears to be as slow as observed and points to possible effects of the Earth's rotation on convection (1993b). At present (Goldreich and Toomre 1969; Tanimoto 1989), the axis of figure is nearly coincident with the rotation axis. Observations by the International Latitude Service point to contemporary motion of about 10 cm/yr towards 80°W (Dickman 1979). The rate refers to a time of rapid deglaciation. Williams (1993) has pointed to evidence that, prior to the Late Ordovician (430 Ma BP), external coupling was large enough to have caused significant changes in the obliquity (reorientation of X_1). Subsequently, the obliquity has remained at about the present value, 23.5°. Laskar, Joutel, and Boudin (1993) find evidence that the Moon in orbit has tended to stabilize the obliquity. This would protect the Earth from gross changes in spin-axis direction and seasonality, found to have been experienced by other planets; for related discussion, see section 5.5.4.

Lambeck (1979; see also Dickman [1979]) has pointed out that isostatic compensation, in smoothing the effect of mass anomalies, minimises displacement of the axes of figure. A difficulty is that isostatic adjustment is not instantaneous, but lags mass shift. Furthermore, the large geoidal highs marking subducted cold slab constitute long-lasting departures from hydrostatic equilibrium.

At least three mechanisms might act to impede TPW, by restricting geographic motion of the axes of figure:

176 Tectonic Consequences of the Earth's Rotation

6.6.1 Climatic Pole Entrapment

Munk and Revelle (1952) and Gold (1955) have suggested that the accumulation of on-land polar ice may tend to stabilize the axis of greatest moment about which the Earth rotates. Incipient wander would cause the rotation pole to move toward the location at which most ice is melting, there causing increased accumulation and inhibiting further wander.

6.6.2 Bipolar Organization of Convection

For more than two decades, Pavoni (1969, 1981, 1985, 1993) (figure 6.11) has pointed to the existence of a bipolar pattern of the Earth's tectonics. The geotectonic pattern identified by Pavoni is symmetric about an axis intersecting the Equator at 10°E and 170°W. The satellite-derived geoid has since demonstrated that the axis identified by Pavoni is the location of a principal axis of moment. The existence of Pavoni's bipolar system has been lent further support by seismologic mapping of velocity anomalies in the mantle. Low-velocity, presumably warm, material (Morelli and Dziewonski 1986; Forte, Dziewonski, and Woodward 1993; Matyska, Moser, and Yuen 1994; Su, Woodward, and Dziewonski, 1994) underlies Pavoni's mid-Pacific and Africa axes. S-wave data indicate that complementary, antipodal, fast anomalies underlie the western Pacific and South America (Tanimoto 1990). Seismologically mapped topography at the core-mantle boundary underlies these localities (Forte, Dziewonski, and Woodward 1993).

Matyska, Moser, and Yuen (1994) and Matyska (1995) have incorporated radiative heat transfer in models of the convection in the mantle. The coefficient employed is nonlinear and temperature-dependent, with values to some extent arbitrary and subject to experimental verification. Over a wide range of the ratio $k = \beta_r/\beta_l$ of radiative to lattice thermal conductivity, it is notable that computations result in a convection pattern of order two, comparable with the distribution of hot material beneath the Pavoni poles. As Matyska, Moser, and Yuen point out, if models of this class are valid the development of a convection pattern of order two, with "megaplumes" on the Equator, would not only accord with the structure identified by Pavoni, but would act as a determinant of polar wander. While not itself a product of rotation, by determining a principal axis the pattern of Matyska, Moser, and Yuen would position the Earth so that the "convection poles" are on the Equator. Instead of the discrete masses of Goldreich and Toomre (1969) moving randomly, their movement would be that of the convection poles.

The question has been asked by Uyeda (in Bostrom 1990) whether such events as the development of the Indian Ocean upwelling are the results of TPW or its cause. The investigations of Spada (1993; see also Matyska, Moser, and Yuen [1994]) suggest the former is the cause-effect relationship.

6.6.3 Tidal Effects

An alternative or additional factor impeding TPW acts if convection cannot as in this chapter be modeled in isolation, but is affected by the rotation of the solunar

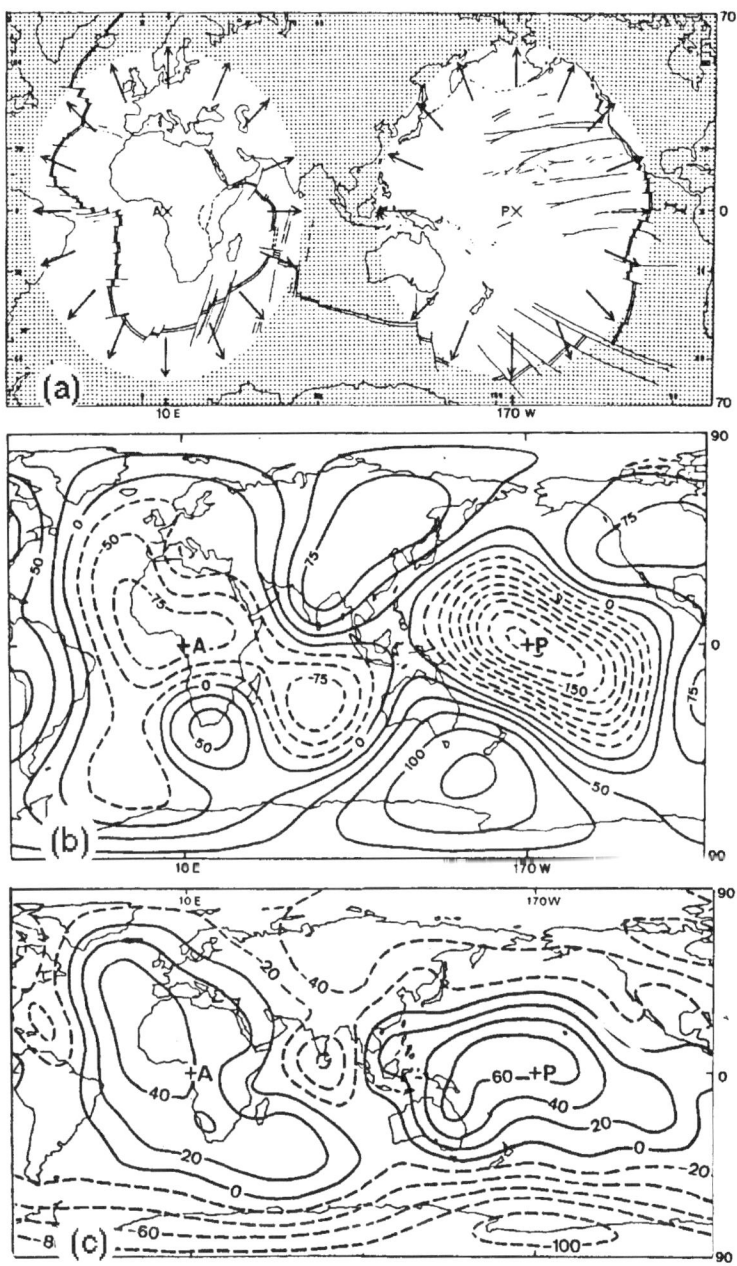

Fig. 6.11. (a) Bipolarity in global tectonics, after Pavoni (1969). The double lines represent seafloor spreading. (b) Lower-mantle equivalent geoid, after Dziewonski (1984), based on seismologic tomography (c) Geoid after subtracting effect of subducted slabs, from Hager (1984). (Reproduced from EOS, vol. 66, no. 25.)

178 Tectonic Consequences of the Earth's Rotation

gravity field relative to the Earth (Bostrom [1994]). Stevenson (1993b) has concluded that tidal heating may have a latitudinal effect on convective flow due to its effect on the viscosity. If the excess velocity of plates in low latitudes (Gordon and Jurdy 1986) is due to tidal vorticity induction, the low-latitude maximum in tectonism systematically contributes to the nonhydrostatic bulge, impeding TPW.

Denis (1989) has discussed the precise value of the excess flattening. The observed value is about 0.5% larger than that for hydrostatic stratification. Inasmuch as we are uncertain as to the cause of the large excess, unobserved changes in its value contribute uncertainty in estimates of the inertial moment (Bursa 1990), hence in the rotation.

6.7 Status: Models of TPW and Figure Adjustment

Prior models of TPW (Vening Meinesz 1947, 1958; Courtillot and Besse 1987; Besse and Cortillot 1991) have considered the tectonic effects of reconfiguration. Hanada (1988) has formulated figure and gravity change in relation to polar wander.

Writing in 1947, prior to the acceptance of mantle convection, Vening Meinesz had good reason, based on his observations of gravity adjacent to tectonic arcs, to suppose that the mantle acts as a fluid in large-scale flow. His model of shear forces (figure 6.12) which might be expected in the crust as a result of TPW assumes a frangible crust or lithosphere overlying a fluid interior. Vening Meinesz (1947) had examined the tectonic effects to be expected of gradual decrease in the rotational flattening. Recently reviewed by Denis and Varga (1990), these had frequently been postulated in the past. The effects of reduction of the flattening under slow despin are insignificant in comparison with motion of the entire bulge. At the time, polar wander was considered less unreasonable than continental drift. Strength of the oceanic crust was believed to prevent continental displacement.

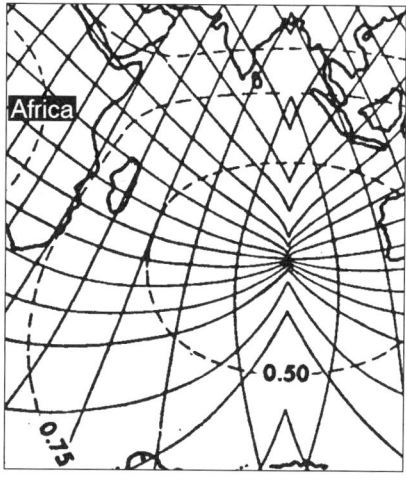

Fig. 6.12. Crustal wrench pattern to be expected as a result of reconfiguration due to motion of the South Pole from 35°S 90°E to its present position, as computed by Vening Meinesz (1947).The dashed contours describe the proportion of stress relative to that expected to cause failure.

Vening Meinesz's numerical experiment assumes pole shift from 35°S 90°E to its present position. His choice was based on observing the rectilinear form of continental margins, in particular on his observation of gravity and continental coastlines in the North Atlantic (Vening Meinesz 1944, 1947). He appears to not have examined the effect of gravity change, therefore pressure change, in effecting the large-scale "isostatic adjustment" necessary in the event of TPW.

Courtillot and Besse (1987), and Besse and Courtillot (1991), figure 6.13, have sought a comprehensive connection between TPW, the active layer D'' at the base of the mantle, continental breakup, and the reversal frequency of the geomagnetic field. A mimimum in the TPW rate during the Late Jurassic to Middle Cretaceous period coincides with a minimum in the rate of continental breakup and with the frequency of reversals. A causative mechanism might be that the emission of warm material from the core-mantle boundary (CMB) increases the rate of mantle convection and TPW. As part of the same regime, cooler material released into the core leads to magnetic reversals. Liu (1974) and Pan (1985) have examined continental breakup and induced mantle flow under TPW using alternative assumptions.

The more limited model here discussed (table 6.1) examines the effects of destabilization of the Gondwana cratonic lithosphere under latitude change, and the mechanism of reconfiguration in the event of TPW. Some intracontinental effects that might be expected are summarized in figure 6.10b, being at best strongly speculative. If TPW reaches a rate commensurate with the paleomagnetic record,

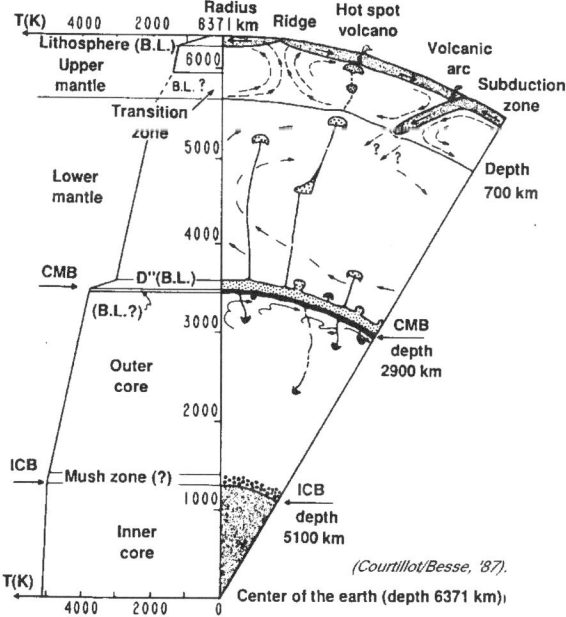

Fig. 6.13. Model describing possible association of TPW, plume formation, and frequency of geomagnetic field reversal, as envisaged (text) by Courtillot and Besse (1987).

180 Tectonic Consequences of the Earth's Rotation

Table 6.1. Causative Mechanism, Assumptions, Consequences, in Models of TPW

Causation	"Rheology" assumption	Evidence/Tectonic consequence	Reference
Mass shifts	Frangible lithosphere, fluid interior	Global fracture system	Vening Meinesz (1947, 1958)
CMB instability → intermittent convection → TPW	Two-layer convective mantle (fig. 6.13)	Geomagnetic field reversals; TPW	Courtillot and Besse (1987, 1991)
Displacement of principal axes under geol. mass shifts	Metastable asthenosphere, convective lower mantle	Asthenosphere destabilization; convection repatterning	Present model

figure adjustment (bulge migration) can only be effected through reorganization of the convection in the mantle, as in development of the Indian Ocean.

A peculiarity is that short of a stabilising mechanism, there is no obvious reason why the value of the nonhydrostatic moment of the intermediate axis should not approach then exceed that of the initial principal axis, effecting their exchange (Bostrom 1990). As noted by Goldreich and Toomre (1969), in terms of geological mass shifts the difference is small between their respective values. An exchange apparently would introduce a period of profound instability, leading to global reorganisation of plate motion. An episode of this type has lately been postulated (Kirschvink, Ripperdan, and Evans 1997) to explain aspects of the Pre-Cambrian geological record. TPW of such rapidity, if real, must monopolize mantle convection to permit preservation of figure. With reference to the question posed by Uyeda (Bostrom 1990), Kirschvink et al. consider that change in the pattern of mantle convection initiated the TPW episode, rather than the inverse. Physically, as they form parts of one process, the structure of convection and the geographic orientation of the space-fixed equatorial bulge are inseparable; they are bound continually to conserve angular momentum.

The outbreak of east-west rifting, upon north-east Africa becoming stationary in low latitudes during the Tertiary, accords with the action of vorticity induction evident in low latitudes elsewhere. The global flow lines inferred by Doglioni (1990 fig. 1) and Ricard, Doglioni, and Sabadini (1991), figure 5.4b, resemble those to be expected in the case of tidal action if the rotational axis was formerly in the vicinity of northern Greenland. In general, TPW and tidal distortion cannot operate independently and must be interactive (see for example Vermeersen et al. 1994). Pursuit of this must await further data.

A summary of the TPW situation is that with the acceptance of convection in the Earth's mantle there has grown increasing awareness of the likelihood of a wandering of the Earth vis-à-vis the axis of rotation, and that the possibility of TPW cannot safely be dismissed. Its best evidence may be that TPW offers an explanation of the concatenation of tectonic events in Mesozoic Africa, otherwise enigmatic. Runcorn's (1957) estimation of the situation forecast to a remarkable extent the discovery of data pointing to TPW, having the characteristic of being normally slow but subject to fast ($\approx 1°$/Ma) episodes. Stabilising factors may exist, for long intervals inhibiting TPW, causing it to be strongly episodic.

7

The Observation Problem

7.1 Overview

Material earlier reviewed indicates that energy deposited in the Earth by tides is not insignificant and may not be dissipated entirely as heat. The supply is ample to influence tectonic displacement. It may be suspected that secular tidal forces account for a number of geological phenomena, such as an asymmetric pattern of the convection in the mantle.

Major difficulties arise. How can these processes be detected and measured? The difficulties are fourfold.

First, manifestations of the solid tide exist as departures from an undisturbed "ground state." The quantities to be monitored, as gravity variations or surface displacement, are typically 10^{-9} of the ground state, challenging the development of instrumentation with which to observe them.

Second, against what standard or model is the departure to be measured? Theoretical reference models have developed through stages, based successively on: an elastic earth; the same with fluid core; and the existence of imperfect elasticity. Routinely, models have assumed a fictitious geostationary potential. Caused to vary with time, this produces oscillatory tidal bulges. The assumption contrasts with the actual, rotating potential, and its response, the progressive wave tides. In benefit of exploring the interior, after excluding the fundamental tide is it logical to ask of observation: what is the response of the Earth to *known* forces?

Third, the Earth is heterogeneous down to the smallest scale. Tides in the world ocean, so far inseparable, affect those observed on dry land. The effect extends to the most remote interior of continents. Tomography has shown that heterogeneity extends throughout the mantle. Major volumes, in incipient phase transition under subduction, are susceptible to dissipation as yet unmeasured. Then at what localities are tidal observations "representative"?

Fourth, with respect to representative measurement, to what extent can observation series of a few decades take into account the effect of long-term processes such as glacial rebound—themselves brief in terms of continental motion?

It might seem that the difficulties are insurmountable. In what follows, current developments are reviewed that may overcome them. Actual and potential advances include comprehensive observation sets obtainable via satellite vehicles, and not less important, computational means of separating the significant signal. The introduction of external-reference observations such as GPS and VLBI offers the possibility of advance beyond any made previously. In particular these may provide validation or otherwise of long ongoing gravimetric observations by the Centre International des Marées Terrestres at Brussels (Melchior and Ducarme 1991). Aspects of these have been inexplicable measured against models composed of a "non-geological" Earth and nonexistent regime in the potential.

Although unable to separate the oceanic from the solid Earth component, the Topex/Poseidon dataset offers welcome means of narrowing the oceanic correction in dry-land gravity observations. Neither the bodily tide nor the marine dissipation is sufficiently well known to separate these using the total (astronomically observed) dissipation. Lambeck's observation (1988) remains valid: "The mechanism by which tidal energy is dissipated remains inadequately quantified if not actually poorly understood. Is it by bottom friction in a few shallow seas, as has traditionally been believed? Or is it by a process that is more uniformly distributed in the oceans as is indicated by the ocean tide models themselves?".

The possibility exists that observation of the internal displacement, as by means of a tiltmeter network forming a minor adjunct to GEOSCOPE, may correlate with seismicity incidence of the long-term type identified by Mogi (1974, 1979), Kanamori (1978), Romanowicz (1993), Du (1993, 1994), and Satyabala and Gupta (1996). It would also seem valuable to extend studies by the USGS of the correlation of eruption and tidal maxima as at Hawaii, inasmuch as the correlation represents tidal modulation of the convection as much as it does "triggering."

In what follows, the sequence aimed at is (1), to review standards or models; (2), to identify the data required to validate or invalidate models; and (3), to review available measurement techniques.

7.2 Standards (Base-Line Models)

7.2.1 The Love Numbers

Since their conception, Love's characteristic numbers h and k have been used to specify the ratio between the observed and a theoretical value of the tidal deformation. Antecedent formulations by Kelvin examined the deformation of a simpler earth.

Love (1909, section 3) formulated the surface-displacement and change in potential at a point in response to forces representing the gradient of a disturbing potential W_n of degree n. As re-emphasized in his *Treatise* (Love 1927, section 180) the orientation of W is defined with respect to axes, for instance the principal axes of moment, fixed in the Earth. The deformation created represents the development of a geographically stationary, oscillatory bulge, in accordance with an external potential caused to vary time-wise in a frame stationary relative to the

Earth (prescribed so as to rotate with it), figure 7.1a,(i). Successors employing a stationary, topologically similar forcing function have introduced relaxation processes, obtaining time-dependent (complex) Love numbers.

Following the initiatives of Kelvin (1863b, 1886), Love (1909) compared the tides in a global ocean overlying an entirely rigid earth with those observed in the Earth itself. To obtain insight as to the actual rigidity he compared values of his characteristic numbers h and k with those resulting from combinations of shells of various rigidities and fluidity. The latter were suggested by published horizontal-pendulum observations (Hecker 1907); the velocity of shear waves as already known (Wiechert 1907); and the extension of the period of the free nutation (Chandler wobble) by elastic deformation (Herglotz 1905). Love (1909 sec. 14) concluded "I think it may be regarded as certain that there is not within a depth of 1400 km a continuous layer of molten matter, separating the inner portion of Earth's body from the outer portion, and behaving as a fluid in respect of forces of the type of tide-generating forces." In respect to "fluidity," the generating forces he created were oscillatory (Love 1927, eq. [38]), his mechanism excluding secular stress. Observation techniques available at the time precluded measurement of earth-tide phase displacement.

Reviewing previous discussion, it has become conspicuous by reason of dissipation that appears unexpectedly large, that in respect to the principal tides the displacement is not in fact as described by Love. In his construction (perhaps prepared as a first approximation), the Earth is not subjected to "forces of the type of the tide-generating attractions of the Sun and Moon" in the form in which these actually exist. Rather than being geostationary, the external potential rotates relative to Earth-fixed axes (figures 7.1a, (ii); 7.1b). Although the potential is a point-denominated scalar, its gradient, a vector quantity, rotates accordingly. The displacement of interior particles is hence not expressible in terms of a self-cancelling, oscillatory, motion of the surface (Love 1927, eq. [38]). M_2 and S_2 exist as waves, manifest as the unbroken E $\rightarrow$ W migration of permanent bulges. The associated stress ellipsoid progresses by rotation, without reversal. Residual strain is cumulative and of the kind denoted by Means (1976) as non-coaxial progressive simple shear (see also Sommerfeld 1950).

Love was well conscious (Love 1927, sec. 176) as was Kelvin, that at great depth the tidal "ground state" must include major hydrostatic compression under gravity. With respect to pre-stress, it is now apparent that the interior is not merely hydrostatically stressed, but metastable at all depths through being continuously stressed to the point of "failure" under convection (refer for instance to figure 5.12c, based on the reconstruction by Vigny, Ricard, and Froidevaux [1991] of the structure of the convection in the mantle). The torques resulting from tidal wave progression are superimposed on elemental torques, per unit volume equally minute, that are responsible for flow under buoyancy gradients. The dissipation is hence likely to be effected not via oscillatory shearing (per se, self-cancelling), but via processes of secular vortex motion (rotational flow).

Recognizing limitations inherent in "the statical Love tidal problem," Krysinski (1992) has examined the effect of rotational deformation. He takes up the case of a fluid planet, incorporating the centrifugal potential and addressing the cases

184   Tectonic Consequences of the Earth's Rotation

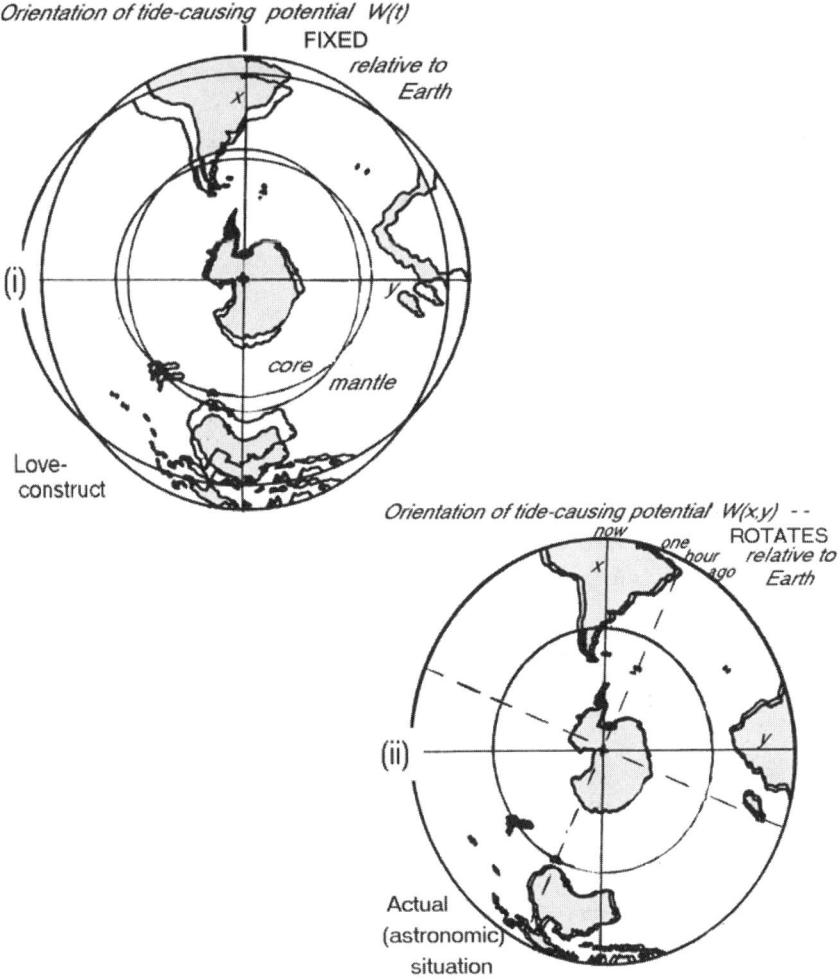

Fig. 7.1. (a) Comparison of the displacement in M_2, S_2 under **(i)** the temporal fluctuation in the potential of AEH Love, and (ii) the actual astronomic regime. South-Polar projection onto Equator plane; x, y are Equatorial and Earth-fixed.

In (i), variation in potential $W_n(t)$, as from a fictional variation in Earth-Moon distance, results in tides producing oscillatory spheroidal deformation, stationary relative to the Earth. In (ii), the Earth continually rotates relative to the perturbing potential $W_n(x,y)$. As a result, bulges are always present and progress around the Earth as waves. Strain ellipsoid progresses by rotation, never reversing. In the course of a complete revolution translations and shears cancel, but rotations (of elements relative to the Earth itself) do not. Failing perfect elasticity, this results in cumulative distortion (vortex flow), exaggerated in drawing, that is responsible for the dissipation. For clarity, one-hour-ago wave position not superimposed.

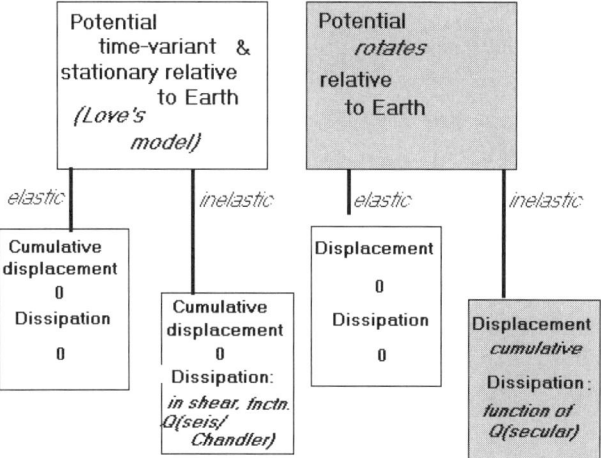

Fig. 7.1. (b) Systems (actual is shaded): the solid-Earth tide, cumulative internal displacement, and dissipation.

(Left) Love's formulation, extended in later work to the incorporation of anelastic processes. His model is based on a time-varying potential static in orientation relative to the Earth. Cumulative displacement is nil (self-canceling). Faute de mieux, dissipation estimates have been based on oscillatory shear using values of the elastic quality factor $Q_{12\,hrs}$ for the periodicity band between seismic processes and the Chandler wobble (14 months).

(Right) The Earth revolves within the gravity field of the Moon and Sun (figure 7.1a,ii). Displacement is cumulative over time. Dissipation is function of secular $Q_{\to 0}$, pertinent to distortion having unlimited period (zero frequency).

of synchronous and asynchronous rotation. So far as known, he has not extended his investigation so as to incorporate the effects of viscosity, hence dissipation, inherent in the terrestrial case.

7.2.2 Reference Models

Reference models employing a Love assumption include those of Jeffreys (1929); Molodenskiy (1976a,b; 1984); Munk and MacDonald (1960a); Kaula (1964); Zschau (1978); Zschau and Wang (1986); Dehant (1987); and Wahr and Bergen (1986). It will be apparent that these exclude a priori the secular (infinite-period) term in visco-elastic processes of cumulative displacement and dissipation.

A particularly thorough examination of the potential of observations such as VLBI (section 7.4.2) to constrain mantle anelasticity, core coupling and other factors has been conducted by Wahr and Bergen (1986). The authors write the deformation in terms of their equation 2:

$$\phi_T(t,\theta,\lambda) = \mathrm{Re}\left[\sum_{m=0}^{2} c_m(t) Y_2^m(\theta,\lambda)\right]$$

where θ, λ are colatitude and eastward longitude. The Y_2^m are second order spherical harmonics. $c_m(t)$ modulating the real part Re of the system represents a sum of periodic terms closely spaced about m cycles/dy. In an ellipsoidal Earth, the $m = 1$ (diurnal) terms produce nutation about an Equatorial axis. The $m = 0$, $m = 1$, $m = 2$ terms are responsible for tidal deformation. Those in $m = 0$ lead to a shape change, hence change in the inertial moment and rotation.

The bounds of this model are as follows. By encompassing the frequency range $0 > m > 2$, Wahr and Bergen permit examination of effects of the precessional torque and nearly-diurnal core resonance, which interacts significantly with the daily tide. The forces conjured by the gradient in a potential rotating with the Earth (their sec. 2) result as described in normal modes of deformation, thus only of a stationary bulge. Formulation as eigenvibrations excludes the internal force-couples of infinite period associated with the migration of the bulge. As earlier apparent, in actuality the latter does not "form," but is always present, progressing in one direction as a wave.

While recognizing the uncertainty in dissipation the authors (Wahr and Bergen 1986, sec. 6) formulate this in terms of the elastic quality factor Q in the frequency range delimited by seismic waves and the Chandler wobble. Dissipation under bulk viscosity (Anderson 1980) is excluded. Producing displacements that are longitudinally symmetrical, the formulation employed excludes dissipation under the zero-frequency primary or fundamental tidal distortion, that is represented by the east-west progression of the semidiurnal wave. In the model, spin-down is attributable solely to dissipation in shear (whence a phase lag and retarding torque), restricted to the seismologic/Chandler frequency range. Solely the minute Darwinian retarding torque acts in the direction of the migration of the M_2, S_2 waves.

It may be noted that with the exception of an assumed, uniform high Q, the limitations described are not applicable to the effects of vertically applied, oscillatory, marine tidal loading.

With reference to computation, models constructed of superimposed "eigenvibrations," employing spherical harmonic components, have been found an economical alternative to ray theory in seismic exploration of the Earth's interior (Dratler et al. 1971; Gilbert and Dziewonski, 1975; Woodhouse, 1988; Um and Dahlen, 1992; Zhao and Dahlen, 1995; Varga and Grafarend, 1996). Nonetheless, the nonreversing bodily wave tides M_2, S_2 cannot be described in terms of oscillatory, symmetric functions, e.g. Xiao and Hsu (1989), but only in terms incorporating the dimension rotation. The introduction of the centrifugal potential, hence ellipticity, and inertial forces under rotation (Smith, 1974; Wahr, 1981; Wahr and Bergen, 1986; Dehant, 1987; Li and Hsu, 1989; Zhang and Kuo, 1989; Dehant, 1993) is not equivalent to rotation of the primary, forcing potential. An attempt is made in figure 7.1a (ii) to illustrate distortion under the regime of bulge progression. Constraining equations must incorporate the dimensions identified by Helmholtz (1857), equation (4.2), specifying rotational as distinct from the solely irrotational flow possible under a stationary potential. In the Earth the distortion includes as its major component a differential displacement which being secular (zero-frequency) and vortical is associated with a Q unrelated to $Q_{seis/Chandler}$.

A history of tidal models is provided by Dehant (1991), commencing with those assuming a spherical elastic Earth, up through models incorporating flattening and various defects in elasticity. All are symmetrical in longitude, based on the Love-number formalism, employing a time-varying, static potential, as distinct from the astronomic potential, rotating relative to the Earth. Dehant notes the extension by Zschau and Wang (1985) of estimates of tidal-Q, hence Love numbers, to unlimited periods, based upon the Gibbs free energy (section 3.5). The forms so modeled represent the development over periods without limit of a stationary spheroidal tidal bulge.

In the field, the observed gravimetric factor δ correctly describes the instantaneous value of the deformation and its complex part, the dissipation. However these are the result not of oscillatory symmetric (spheroidal) deformation in the mode formalized by Love, but of secular (zero-frequency) distortion processes induced by the continuous migration of the bulge. It would be unexpected if "Love numbers" so measured coincide with values computed on the assumption of an oscillatory deformation and seismological Q. The associated dissipation factor should perhaps be termed $(Q^{-1})_{vorticity}$. The distortion under M_2 is not a guide to the structure of the Earth's interior in the fashion assumed in modeling oscillatory spheroidal deformation.

7.2.3 Models of the Marine Dissipation

Dissipation in the seas must account for a major fraction of the tidal phase lag, hence the torque responsible for the astronomically observed despin.

The marine dissipation has been estimated in the following ways:

1. Via estimation of the dissipation in shallow seas in terms of bottom velocity
2. Via estimation of the flux of tidal energy from the open ocean into shallow-sea areas thought to be responsible for the dissipation
3. Via hydrodynamic modeling of dissipation fitted to the record of tide gauges
4. Via estimation of the residue after subtraction of an estimated solid-Earth component from altimetric observation of the total dissipation, over the non-continental area of the Earth

The technique last mentioned incorporates *(a)*, models assuming that almost all marine dissipation takes place in shallow seas; and alternatively *(b)*, those finding that a major fraction takes place in baroclinic internal waves associated with deep-sea bottom topography.

Abbreviating the history of these estimates, the sequence is that Jeffreys (1921) and Heiskanen (1921) concluded that almost all the dissipation required to explain changes in the length of day could be attributed to losses in shallow seas. The limitation of this estimate was that water velocities were geographically almost unmeasured. Compounding this uncertainty, dissipation/unit area is dependent on the square or higher power of the velocity. Estimates of the coefficient of the quadratic dissipation factor varied widely.

In seeking the locale of tidal dissipation, Jeffreys (1952, sect. 8.08) concluded that "It actually becomes rather difficult to see how so much energy gets into the shallow seas [from the open ocean] to be dissipated."

The difficulty still is conspicuous (Cartwright and Ray 1989, 1991). Addressing this problem, Miller (1966) computed the dissipation as the flux of energy from the deep sea into shallow-water dissipation regions. His function

$$<p(t)u(t)> = {}^1\!/\!_2 \rho g a u_0 \cos\phi$$

(where $<, >$, time average: p, pressure difference; u, water velocity; ρ, density; a, tide height; ϕ, phase difference) possesses a major advantage, in that it is dependent on only the first power of the velocity, now more fully charted, and does not require arbitrary imposition of the coefficient of a quadratic term. A *perceived* disadvantage of Miller's estimate has arisen: it yields only between 1.4×10^{19} and 1.7×10^{19} erg/s marine dissipation, between one half and two thirds of the 2.7×10^{19} erg/sec astrometric total required, perforce sought in shallow seas.

An independent approach has been to compute the hydrodynamic dissipation ab initio, based on the three-dimensional configuration of the oceans and the external potential. The computation is tied or fitted to the record of tide gauges (Schwiderski 1979, 1980, 1985a; LeProvost et al. 1994). The resulting estimates vary in proportion to the dissipation function assumed in shallow seas. All estimates incorporate a solid-Earth component based on the Love-number assumption (high-Q, entirely oscillatory deformation). In that Hendershott (1972) and Zahel (1978) have shown that a large proportion of the ocean tide is a function of solid-Earth seafloor "heaving," errors in the solid component must strongly affect the marine tides themselves. Departure from equilibrium is amplified in the fashion experienced in carrying water in a tray.

It might be expected that the large and valuable Topex/Poseidon data set would permit resolution of the marine-dissipation problem. Andersen (1995), (see our section 7.4.5), has displayed the difference between tidal solutions resulting from the altimetric observations. As all incorporate similar assumptions (that the loading factor is linear, high-Q, and similar everywhere; and that the solid-Earth component is a function of oscillatory, high-Q deformation), difference maps obscure the principal uncertainties.

Compounding the difficulty in estimation of the marine dissipation, Sjoberg and Stigebrandt (1992) identify mixing processes via internal tides (a deep-water process) as a potential major energy sink. Ray and Mitchum (1996) conclude that substantial dissipation may be associated with the internal disturbance caused by the Hawaii Ridge. Morozov (1995; see also Baines [1982]) has measured internal waves resulting from the interaction of the semidiurnal marine tides and the topography of the world ocean floor, by deploying temperature-recording moored buoys.

Ridges such as the Mascarene in the tropical Indian Ocean, and in particular the lengthy meridional Mid-Atlantic Ridge (figure 7.2), may in sum account for one quarter (11×10^{11}W) of the total dissipation, identified by Cartwright (1977). Other ridges may be greatly dissipative but are as yet not as closely investigated. Morozov surmises that the reason the deep-water ridges are so dissipative in comparison with the continental slopes is that the currents of the barotropic tidal wave tend to be parallel to the shore, so that only a small part of the mass flux crosses the slope. In contrast, the ridges tend to be perpendicular to the tidal currents. Their tendency to be meridional in orientation in low latitudes may be genetic.

Ironically, the sum of shallow-sea estimates of dissipation and that attributed to internal waves now is greater than permitted by lunar motion. Munk (1997) has reduced Morozov's estimate of internal dissipation to 0.2 TW, that is, in the direction of reconciliation, on the basis of more recent measurements. Kagan and Sundermann (1996) have reviewed what is known of the entire tidal energy budget, taking into account dissipation, evidence as to tides in the geological past, and the evolution of the Earth-Moon system.

Based on the concentration of tidal work input near the Equator displayed by satellite data (figure 3.2), Cartwright and Ray (1990) remark "There is no relationship between the zonal areas where work is done on the tides by generating forces and the location of the various shallow seas, known for decades as the main dissipating areas." Ray (1994) comments that attribution of almost all tidal dissipation to the shallow seas has become "geophysical folklore," based on slender evidence. In review, the situation remains not far from that perceived by Munk and MacDonald (1960a) and MacDonald (1964): "The oceanographic data suggest, but do not establish conclusively, that about half to two-thirds of the dissipation is within shallow seas."

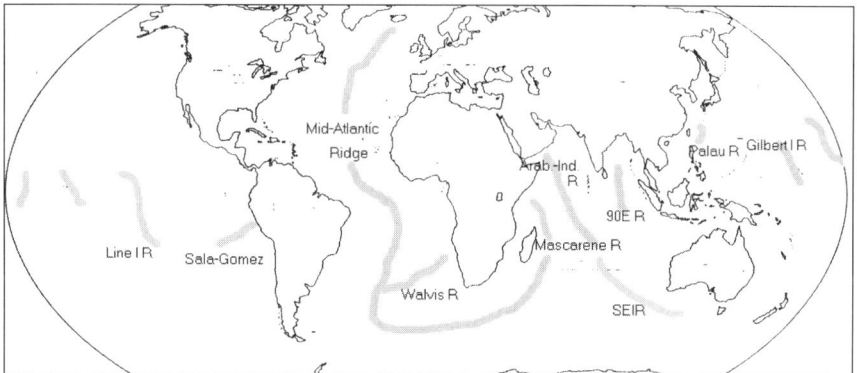

Fig. 7.2. Deep-sea ridges responsible for formation of internal waves, resulting in marine tidal dissipation. Those most dissipatory are named. Segments of the Mid-Atlantic Ridge are believed responsible in sum for 4×10^{11}W, concentrated near the Equator. (After Morozov 1995.)

7.3 Validation

In an attempt to identify uncertainties that might be set up for observation, tables 7.1 and 7.2 compare assumptions on which estimates of the dissipation have been based:

In table 7.2 the top row refers to the vector component denominated $[R,\rho]$, (figure 7.3b) by Melchior, as the model value or datum against which have been measured global gravimetric observations, and his tidal-loading correction component, $[L,\lambda]$.

7.4 Observation Techniques

The following attempts to identify avenues of advance whilst taking into account constraints imposed by astronomy, geophysics, and not least the geologic record.

7.4.1 Gravity Observations

The development of gravimetric observation of the earth tides has been recorded in the pages of the Bull. d'Inf. des Marées Terrestres (BIM), issued at Brussels since 1956, and summarised in IUGG Chronicle 288 (IUGG 1988). Observations based on instrumental sensitivity of $10^{-9}g$ or better now amount to several hundred. A culled data bank (Melchior 1994a,b) exists, listing 211 stations in which the error within numerous variables, including notably oceanic effects, is believed smaller than the signal. The majority of observations has been made using spring gravimeters, developed originally for geophysical prospecting. Essentially, these act as means of comparing gravity from point to point.

Although sufficiently sensitive, spring gravimeters are subject to uncertainties in calibration. Furthermore in a system comprising a long-period sensor with feedback electronics, a formidable problem exists in determining phase lag within a small fraction of a degree. The introduction of the "super-conducting gravity meter" (Prothero and Goodkind 1972) and ballistic absolute gravimeter (Niebauer, Klopping, and Faller 1995) has circumvented these difficulties. The calibration of superconducting gravimeters can itself be effected by the accession of known masses (Achilli et al., 1995). Shuhei et al. (1997) have compared the output of superconducting and ballistic gravimeters. Side-by-side operation of spring-actuated and new instruments appears to have established that fractional rather than

Table 7.1. Inclusion of the Secular Term: Models Based on Temporal versus those Based on Orientational Variation (Rotation) of the Tide-Causing Potential.

Secular term	Geostationary; temporal variation (Darwin/Love assumption)	Under rotation (actual)
Displacement (a vorticity)	0	✔
Dissipation (in vorticity)	0	✔

Table 7.2. Estimations of the Tidal Dissipation.

Marine contribution Shallow seas	Bodily-tide contribution (vector component $[R,\circ]$) Open ocean	Due temporal variation in potential (Love assumption)	Load-tide (component $[L,\lambda]$) Due rotational (actual) variation in potential	(Assumption: uniform passive-Earth)	Reference
P	I	I	0	I	1
P	S	S	0	I	2
P	S	S	0	I	3
–	–	I	0	I	4
P	–	I	0	I	5
–	–	I	0	I	6
–	–	I	0	I	7
–	–	I	0	I	8

P: principal contribution; S, significant; I, insignificant; 0, nil; –, not addressed.
References: 1, Jeffreys (1929); 2, Munk and MacDonald (1960a); 3, MacDonald (1964); 4, Molodenskiy (1976a,b); 5, Zschau (1978); 6, Zschau and Wang (1986); 7, Wahr and Bergen (1986); 8, Dehant (1991).

fundamental adjustments need be made to observations obtained using spring instruments (Melchior 1994a,b).

The observation problem is illustrated by data such as those shown in figure 7.3a. To minimize ocean effects potentially as large as the signal sought, it is desirable in on-land observations to set up in the vicinity of an amphidrome, albeit uncertainly known. With respect to tectonics, the observations illustrated were made in a region where subduction is intense. The vector components of greatest uncertainty are identified in figure 7.3b. [R,ρ] and [L,λ], representing respectively the reference-model and a correction for oceanic effects, are based on the following.

[R,ρ] is computed, employing the assumption that dissipation takes place in an inert, uniform Earth. In respect to those models not perfectly elastic, ($Q_{elastic} \to \infty$), the elastic quality factor is equated with that pertinent to a purely oscillatory spheroidal deformation, $Q_{seis} \approx 200$. The zero-frequency term associated with the primary tidal distortion, that under the lateral progression of the solid-Earth bulge, is omitted. The oceanic term [L,λ] correctly assumes a purely oscillatory forcing, but also, less correctly, the existence of an inert Earth in which Q is high and geographically uniform (section 7.4.5).

In surveys such as that illustrated an anomaly appears (table 7.3), in that the amplitude and phase in M_2 is several *times* larger and more strongly variable relative to the global average, than expected. Side-by-side gravimeter calibration has confirmed that the result would be the same were observations to be repeated using absolute instruments. The validity of the results is furthermore corroborated by the value of the difference between the separated cos X and sin X components. Worldwide, many "anomalous" regions have become evident (Melchior and De Becker 1983; Melchior and Ducarme 1991; Melchior 1995a,b). Davis et al. (1987) have shown that a discrepancy in tidal tilt observations in Hawaii might be explained by locally assuming one half the average oceanic lithospheric rigidity.

Extraneous to the above, it may be hoped that introduction of an orbital gradiometer (Rummel 1996; Tapley, Melbourne, and Reigber 1996; Jekeli and Shum 1998) will provide independent tidal information.

In table 7.3, the values quoted have in common that they represent departure from a basic, theoretical model or datum, of the bodily tide, namely the vector component entitled [R,ρ] by Melchior (1989), see figure 7.3b. [R,ρ] assumes oscillatory, geostationary deformation of an elastic Earth. Melchior (1989) notes that values having such anomalous characteristics as those mapped might be understood if (R,ρ) represents other than a correctly modeled vector component. The geographic variation has been tentatively attributed to regional variations in heat flow (Melchior and Ducarme, 1991; Yanshin et al. 1986; Melchior, 1995a). The correlation has been challenged by Rydelek, Zurn, and Hinderer (1991), but the values vindicated (Melchior 1994a, 1995a) by culling stations in which a generous estimate of the uncertainty exceeds a fraction of the observed variation.

An alternative, geological explanation of the variation might be ventured by attributing to the anomalous regions, e.g., southeast Asia, Australasia, a significantly reduced Q_{bulk} under the permanent state of phase instability there prevalent (section 4.4). Subduction regions such as the great western Pacific zone of inflow

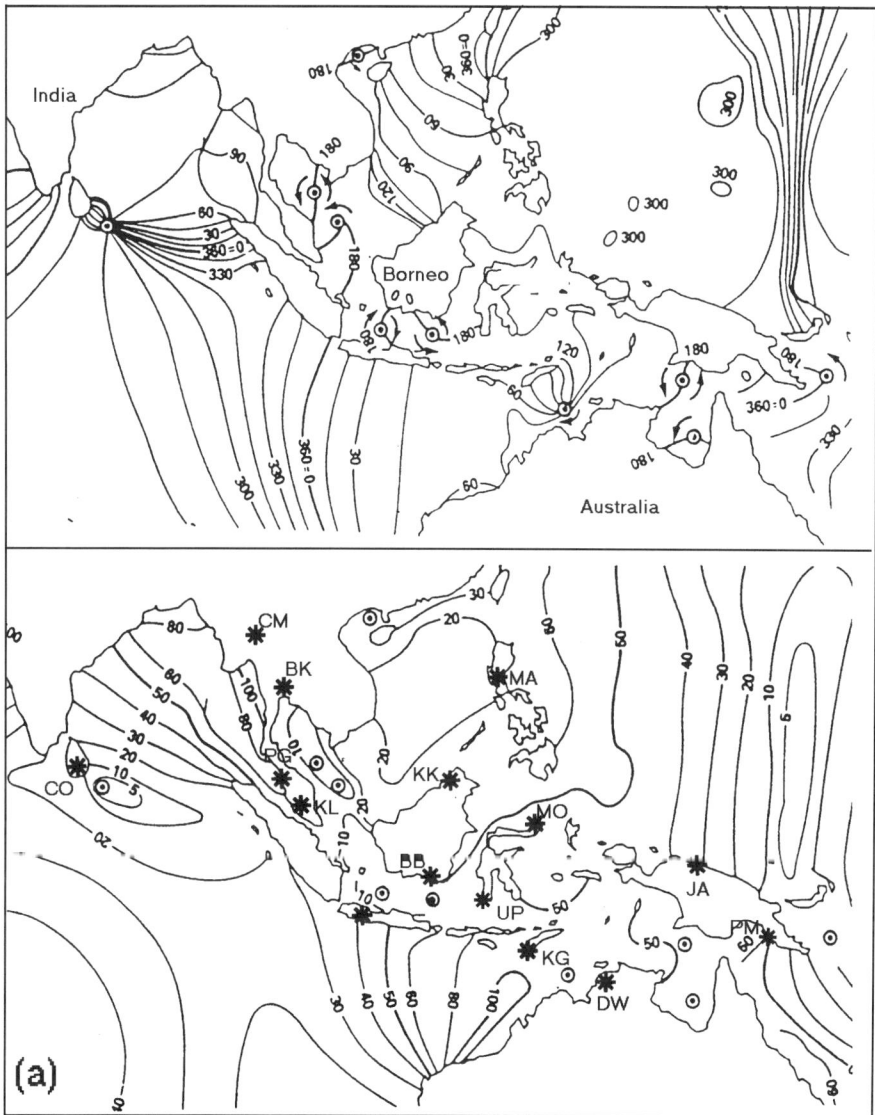

Fig. 7.3. (a) Gravity stations (asterisks) in southeast Asia, to illustrate observational difficulties. From place to place the apparent variation is rapid, because the region is subject both to oceanic and tectonic effects. The possibility of instrumental miscalibration has been minimized by comparison with the diurnal tide O_1. The oceanic contribution to the signal is major, and varies strongly with locality. To minimize the contribution of this component, stations have been located in the vicinity of amphidromic points, shown in the cotidal map (upper) [after Schwiderski (1979, 1980)]. Schwiderski's points differ slightly from those computed by Le Provost et al. (1994). Both computations are based on hydrodynamic dissipation assuming a high-Q, uniform, seafloor load factor. A lower, and geographically variable, Q is suggested by recent observations by Matsumoto et al. (1998). (Map data: Melchior, Francis, and Ducarme [1995].)

194 Tectonic Consequences of the Earth's Rotation

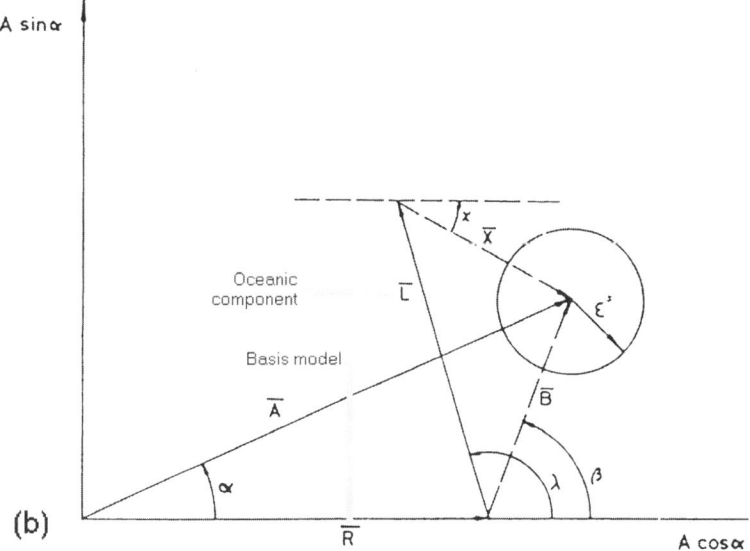

Fig. 7.3. (b) Oceanic and reference-model vector components of the gravimetric bodily tide, as observed. Both components are estimated, modeling an inert, uniform, hi-Q earth. Reduction of an observation entails correction of the observed value [A,α] for the estimated contribution [R,ρ] of the model, yielding [B,β]; a further correction [L,λ] is applied for oceanic effects (attraction and loading), leaving [X,χ]. ε describes circle of uncertainty due to unidentifiable noise. (After Melchior [1989].)

Table 7.3. Gravimetric Observations, Bodily M_2: Global Mean as Against those in a Tectonically Active Region (Sundaland[1]).

	Global average[2]	S'land, average[2] employing Schwiderski[3] marine loading	S'land, average[2] employing Grenoble[4] marine loading
Xcos χ (μgal)	+.002 ± .0445	1.76	1.68

[1]On the basis of deep seismicity and gravity relief "Sundaland" (figure 7.3a) is believed to be the site of subduction. This entails downflow, secular pressure increase, and mineral phase compression.
[2]ICET Databank DB92 (Melchior 1994a,b, 1995a; Melchior, Francis, and Ducarme 1996). Melchior, Francis, and Ducarme (1995, table 6) details values for stations comprising the average shown in figure 7.3a.
[3]Schwiderski (1979, 1980, 1985a). Based on oscillatory bodily tide and Q ≈ 200, representative of Chandler/seismologic damping. To reconcile the lunar recession with the computed oceanic dissipation, Schwiderski found it necessary to add an arbitrary extraneous phase lag, 0.4°. Tentatively, the terrestrial deceleration was attributed to a pressure increment acting on east-sloping continental slopes.
[4]Le Prevost et al. (1994). The authors employ similar bodily tide, but loading assumption is limited to hi-Q, isotropic Earth. Reconciliation with astronomic data is attainable by selection of shallow seas friction coefficient.

of cool lithosphere are not invariably the site of decreased heat efflux (Pollack, Hurter, and Johnson [1993]; Ida [1983]), figure 5.9. Additionally, it might be postulated that the astronomically required phase lag attributed by Schwiderski to pressure against the continental flanks, is in major part a result of dissipation under the omitted secular component of the dissipation. The observation problem as to oceanic effects is taken up in section 7.4.5.

Summarizing, the models tabulated have in common that they assume (1) Love's deformation construct, excluding a priori the fundamental (rotational) term in the tidal potential, and (2) the existence of an inert and homogeneous earth.

7.4.2 VLBI Observations

Means of referring tidal observations to an inertial frame are becoming available, in the form of space-based geodesy. The basis of a celestial system was outlined by Gold (1967); see also Broten et al. (1967) and demonstrated by Thomas et al. (1976), figure 7.4a; see also Robertson (1987).

Radio-frequency signals reach the Earth from astronomic objects so distant that the wave front is plane. VLBI (Very Long Baseline Interferometry) permits monitoring of length changes and changes in orientation of the arc separating ter-

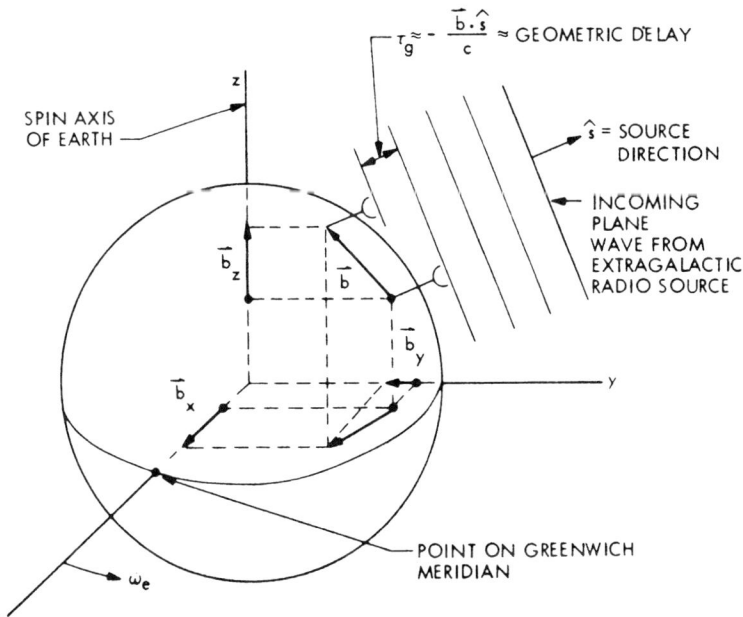

Fig. 7.4. (a) Prototype configuration of antennae permitting construction of geodetic VLB profile (after Thomas et al. [1976]). The geometric delay τ_g in signal arrival was approximated to a function of the baseline vector **b** and the arrival direction **s**, permitting establishment of changes in the Earth-fixed parameters describing baseline length and orientation and in Earth rotation.

restrial observation stations in terms of the arrival time of signals. The logistical and instrumental requirement is considerable (Lambeck 1988; Ma et al. 1993; Ryan et al. 1986; Ryan et al. 1993), entailing precisely synchronised recording of signals at widely separated large antennas, and the physical transport of massive taped data sets. This is followed by central signal processing, to identify the correlation and difference in arrival time.

The results must be corrected for numerous error sources represented by temporal and local variation in the atmosphere, troposphere, and ionosphere (Elgered et al. 1991; vanDam and Herring 1994). VLBI is believed to have the potential to determine the relative position of radio telescopes with a precision better than 1 part in 10^9 of the inter-antenna distance (Herring, Davis, and Shapiro 1990). A global receiver network (figure 7.4b) incorporating refinements has been established, permitting cooperative observation of intercontinental VLBI profiles.To an extent, results are validated by accord with geologically "known" plate displacement. Observation series ongoing over more than a decade now provide independent data as to extension, elevation change, and the elastic quality factor Q at various frequencies (e.g., Herring and Dong, 1994). Giunchi, Spada, and Sabadini (1996) have demonstrated that the accuracy of VLBI is sufficient to pick up effects of lateral viscosity variations and post-glacial rebound.

Ryan et al. (1986) have employed VLBI to constrain the value of the Love numbers, and Ryan et al. (1993) have observed the relative motion of lithosphere plates. Profiles of base-length change have been monitored in some cases over periods exceeding ten years. Several profiles traverse regions such as the Atlantic, believed on geological grounds to be experiencing ongoing extension (figure 7.5 a–c). Profiles have been optimally corrected for the following effects: diurnal spin; polar motion and precession; ionospheric refraction via dual-frequency observation; tropospheric moisture content via separate measurement of this parameter; general

Fig. 7.4. (b) VLBI receiver stations. A "session" consists of simultaneous microwave recording by 2 or more stations. The centimetric signal is heterodyned against a lower frequency fiduciary to reduce volume, and the taped record transported to a correlation center; this process may be replaced by a broad-band communication link.

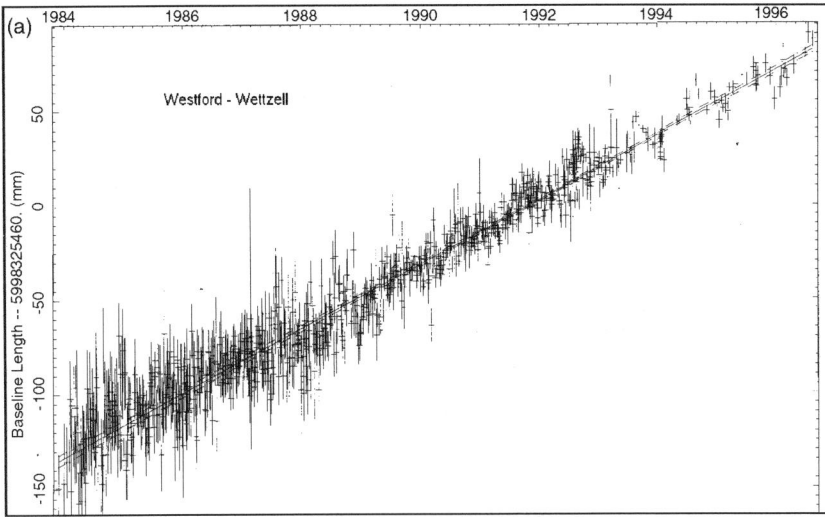

Fig. 7.5. (a) Length change in VLB profile optimally corrected, extending from Westford (North America) to Wettzell in Europe, years 1984–1996. Length, 5998 km. The extension rate (solid line) is 17.0 ± 0.1 mm/yr (1-sigma). Refinement in corrections has caused the probable error gradually to diminish. The profile traverses the Mid-Atlantic Ridge zone of seafloor spreading believed to be associated with increasing distance between North America and Eurasia lithosphere plates. (By courtesy of Goddard Space Flight Center [NASA], per Dr. J. W. Ryan.)

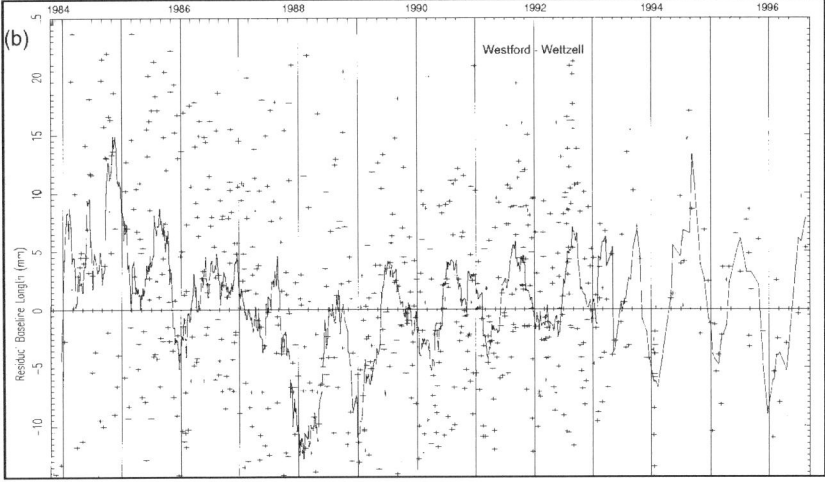

Fig. 7.5. (b) Transatlantic (Westford-Wettzell) VLBI profile as in (a), but secular component removed by application of 90-day box-car filter. After "known" corrections have been applied, the signal contains quasi-yearly components visible, for example, y 1988–1995, and longer-period modulation, minimum at 1988.

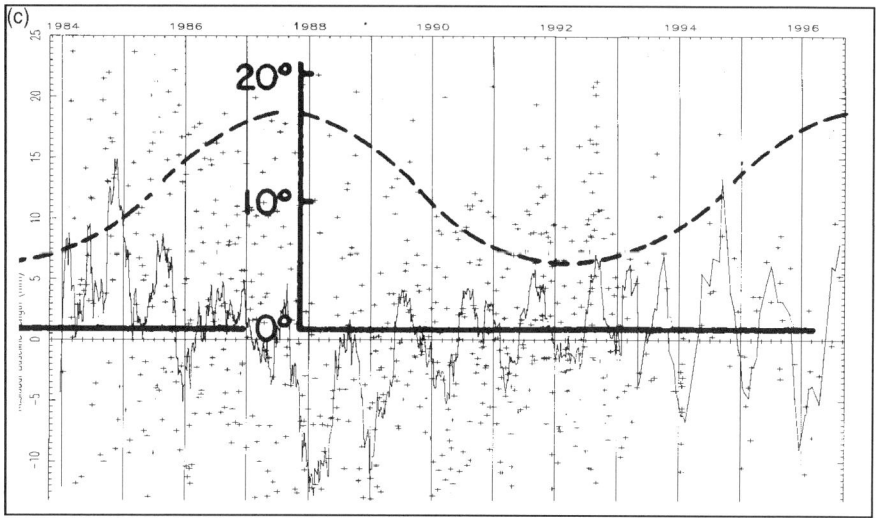

Fig. 7.5. (c) Westford-Wettzell profile. Perhaps coincidentally, the multiyear modulation is in antiphase with variation in the 8.5-yr cycle representing Earth-Moon distance change and maximum in lunar declination (dashed curve). The degrees represent latitude of neutral zone, separating depressed region from the equator-ward tidal bulge, believed by Hamilton (1973) also to be represented in global vulcanicity.

and special relativity, having an effect on clock rates under acceleration and changing potential; clock errors; the solid tides, using orthodox estimation; and ocean and atmospheric pressure loading. Observation sessions (sampling interval) are too infrequent to observe semidiurnal, diurnal and fortnightly tidal effects if present; note, however, Schuh and Schwegmann (1998).

After filtration to remove the component of secular extension, a quasi-yearly modulation component (figure 7.5b) remains unexplained. Separately, an annual variation in gravity has been observed by Hinderer and Legros (1989) and De Mayer and Ducarme (1989). On the basis of an observation series too short to be statistically significant, it is perhaps notable that variation in the transatlantic Westford-Wettzell profile varies with perigee/apogee and lunar declination. The identical unidentified modulation affects the transatlantic profiles Onsala-Westford and Richmond-Wettzell. In respect to the Kohee (Hawaii)-Millcreek (Alaska) profile, bridging a zone of crustal convergence, the modulation is present, but in the inverse phase. The Earth's center of mass is believed adequately located.

Tidal corrections applied so far have assumed deformation under high-Q oscillatory stresses. If the unexplained effects are tidal, they constitute departure from orthodox models. Unless perfect elasticity exists, the secular, rotational component of displacement contributed by the M_2, S_2 waves (clockwise in the northern hemisphere), is partially annulled by the countervailing induction contributed by the fortnightly M_f, modulated by the lunar declination (see also section 7.5). The declination affects both the latitude at which the semidiurnal bulges pass east-to-west along latitude small-circles, and the angle at which the lunar fortnightly bulges pass in the contrary sense. A continuation of the invaluable international VLBI observations is desirable, to permit separation of long-term aliasing combinations.

7.4.3 Global Positioning System (GPS)

GPS like VLBI refers surface points to external objects, in this case to a constellation of artificial satellites, figure 7.6. The latter constitute sources in orbits that make not less than six visible to a receiver at any time. The location of a ground station is specified as the intersection of not less than four spheres. Source output can be recorded in several modes, figure 7.7a. Delikaraoglou and Steeves (1985) have discussed trade-offs in speed and accuracy employing combinations of multiple ground stations and satellites.

It has been possible to miniaturize GPS receivers, unlike VLBI stations, to the point of being portable and economically available. Theoretically, altimetric control is possible to the sub-centimeter level. Using their "Fiducial GPS Technique," Ashkenazi et al. (1993) have shown that GPS may be used to monitor changes in mean sea level down to millimeters (a process permitting long-term averaging, not directly applicable to observation of the "high-frequency" tides). A geocentric height

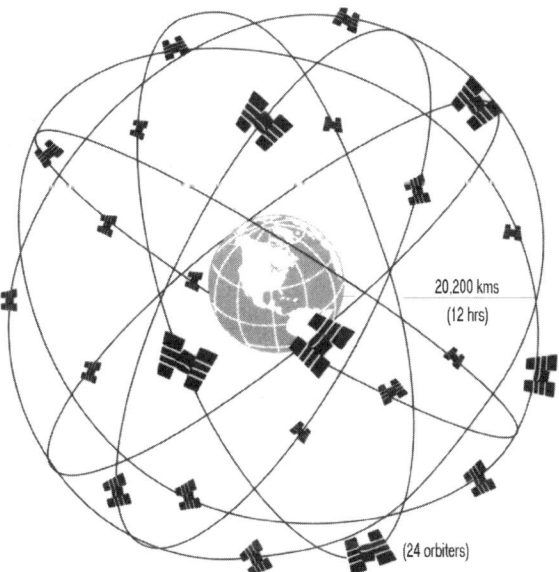

Fig. 7.6. GPS 24-satellite constellation. Not less than six transmitters are observable at a surface point almost 100% of the time. Ranging codes are continuously transmitted at two L-band frequencies. Optimal accuracy is achieved via selected transmitter/receiver combinations, see figure 7.7a. (NASA schematic.)

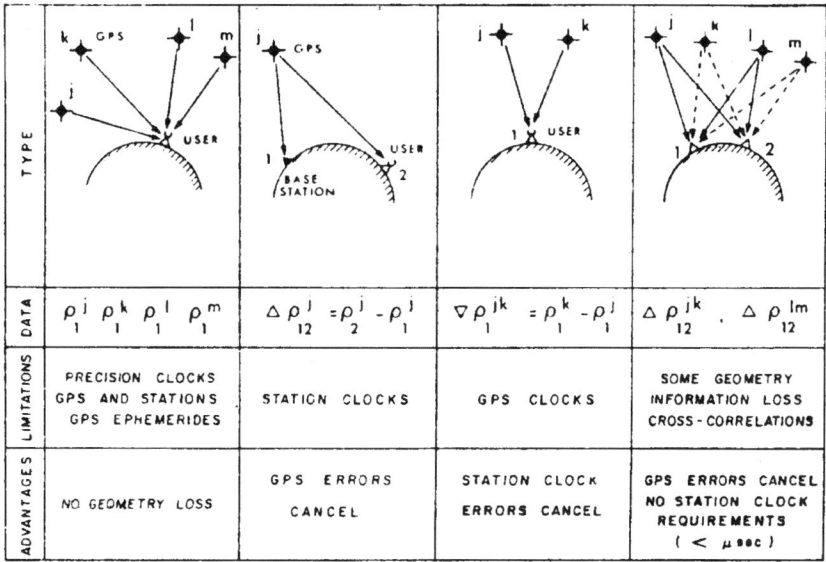

Fig. 7.7. (a) Advantages vs. limitations of various GPS transmitter-receiver configurations, incorporating differential signals and multiple transmitters j, ..., m, at the cost of slowness of measurement. Sub-centimeter accuracy is attainable via simultaneous analysis of dual-band carrier-phase data received from all orbiters by a global receiver network. (Schematic by Delikaraoglou and Steeves 1985; see also Vanicek et al. 1984.)

change of 3 mm corresponds approximately to a gravity increment of 1 microgal. It might therefore be hoped that a network of GPS stations would permit independent backup of gravimetric stations, which are expensive to install and maintain.

Degrading accuracy, GPS signals are affected by arrivals via multiple paths and such factors as the vapor content of the atmosphere (Ware and Rocken, 1986), which must be measured separately. Multipath and environment problems have focused attention on antenna form, placement and stability (Johnson and Agnew 1995; Mao, Harrison, and Dixon 1996; Mader, 1996; Niell 1996). The corrections include terms due to atmospheric pressure variations (vanDam, Blewitt, and Heflin 1994); these terms may be site-dependent, due perhaps to inhomogeneity (vanDam 1997). Noise forms are accentuated in the case of low-elevation arrivals, which may therefore not be usable. A major consideration is that the orbital period of the GPS transmitters is close to that of the semidiurnal tides, making it desirable (Malla, Wu, and Lichten 1993) to incorporate data provided by low-level orbiters having a 90-min period.

It seems to emerge, that to separate the altimetric GPS signal to the necessary precision requires antenna-installation plus corrections comparable in effort to those required in tidal gravity observations. Nevertheless the noise in gravimetry and GPS observation overlaps mainly with respect to oceanic effects, making GPS on-land altimetry a valuable adjunct to gravimetry. The bodily phase lag, several times the gravimetric lag, may be more readily measured; but gravimetry is more sensitive to geocentric height change. Side by side output of a super-conducting gravimeter and fully corrected GPS receiver is being recorded at a site in Colorado (Phillipsen 1996).

7.4.4 Detection of Absolute Motion

The motion of lithosphere plates relative to the mantle ("absolute motion") is a vexed but important question. Tectonicists (O'Connell, Gable, and Hager 1991; Ricard, Doglioni, and Sabadini 1991; Lithgow-Bertelloni et al. 1993) have concluded that in sum the lithosphere rotates relative to the mantle at a rate between 1.9 and 6.0 cm/yr. Controls are few, other than somewhat precarious reference to the trace of mantle plumes (section 5.4). Such an approach takes advantage of the displacement which has accumulated over geological time periods. Detection of topical (day-to-day) or decadal displacement encounters the difficulty of measuring displacement of the order 2 cm/y (5×10^{-3} cm/d), and especially, of identifying fiduciary points at depth within the Earth.

VLBI and GPS stations define their position relative to each other. Net-lithosphere-rotation implies undistorted longitudinal motion of the lithosphere as a whole, undetectable in terms of the motion of surface objects (for instance, GPS stations).

Distortion of the form previously illustrated in the Equatorial plane (figure 3.11) is represented at surface by displacement in the form shown in figure 7.7b. In the course of a decade, such displacement (if it exists) must make itself apparent as zonal displacement of GPS stations. Those already installed comprise the foundation of a global network (Johnson and Agnew 1995; Bevis et al. 1995).

If zonal motion eventually is observed, it should be recalled that in itself it does not describe "absolute" motion. If tidal in origin, high-latitude zones are displaced eastward simultaneously with displacement westward of the equatorial zone (figure 7.7b). This reflects the fact that the induced distortion is internal in origin, produced by the continual reconfiguration of the Earth in the direction of equilibrium, in the field of itself and the Moon (figure 2.3c). The action is not such as directly to brake the rotation. The latter is accomplished by the feeble, secondary retarding torque, product of the dissipation-caused phase lag.

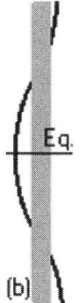

Fig. 7.7. (b) Zonal displacement of surface points (black line) relative to an initial north-south alignment, within gray band. In terms of motion relative to the Earth as a whole, low-latitude surface points are displaced westward, those in higher latitudes eastward.

Observation of the zonal motion, if this exists, will be greatly complicated by the independent motion of lithosphere plates. It might be hoped that such data will corroborate, or otherwise, the finding by Gordon and Jurdy (1986) that during Tertiary time plate motion has been fastest in low latitudes.

7.4.5 Quantification of Ocean Loading

It will be evident from previous discussion that presently the limiting factor in observation of the bodily tide is uncertainty as to the oceanic correction. Longman (1963) early pointed out the importance of the load factor and developed Green's function making estimation possible assuming Gutenberg's model of the elasticity. Farrell (1972, 1973) extended the calculation to take into account not only the elastic deformation, but the potential of the water. In many regions, the latter is half as large as the primary lunar potential. Zschau (1978) has pointed to the effect of the underlying asthenosphere on tidal loading. To reduce the computational load Ray and Sanchez (1989) have investigated the advantage of employing a high-degree harmonic expansion in lieu of integration using the Green's function written by Longman (1963) and Farrell (1972). The authors compare the results of computation routes, proving their equivalent outcome, but assume an orthodox model of the dissipation.

The significance of the load tide, even assuming a seismologic, uniformly high Q as if the Earth were inert, is apparent in their atlas of co-amplitude in M_2, figure 7.8a. The advent of the Topex/Poseidon altimetric survey, having great geographic detail, renders it essential not only to economize computation extending to high harmonics, but to consider whether a radially-symmetric, isotropic and high-Q computation, partly contradicted by observation, is sufficiently accurate; further to this, see Matsumoto et al. 1997; Matsumoto et al. 1998.

7.4.5.1 The Schwiderski Accounting.
A form of tidal loading has been invoked by Schwiderski (1979, 1980, 1985a,b) in analysis of the marine dissipation constrained by tide-gauge records and the lunar motion. His technique minimises the question of the marine dissipative mechanism by means of accounting for the dissipation as a whole, delimited by the lunar orbital change. The step requires attributing a priori the whole of the tidal couple to marine dissipation, the solid Earth being accounted elastic.

The conservation equation (1985a, eq. 25) can be stated

$$d(W°g)/dt = d(W°p + W°f + W°e)/dt$$

in which W is work, g, p, f, and e denote the total and that due to bottom pressure, bottom friction, and eddy dissipation. The superscripts refer to the ocean. Schwiderski equates the total with the sum of the dissipation due to bottom pressure and friction, eddy dissipation being trivial. By reason of the bodily tide assumption, dissipation in this, likewise, is trivial.

The assumption would be justified in terms of using an orthodox Love-type model based on a high-Q earth under stationary oscillatory deformation. As noted, the Love model does not mimic the solid tides, which consist of the lateral motion

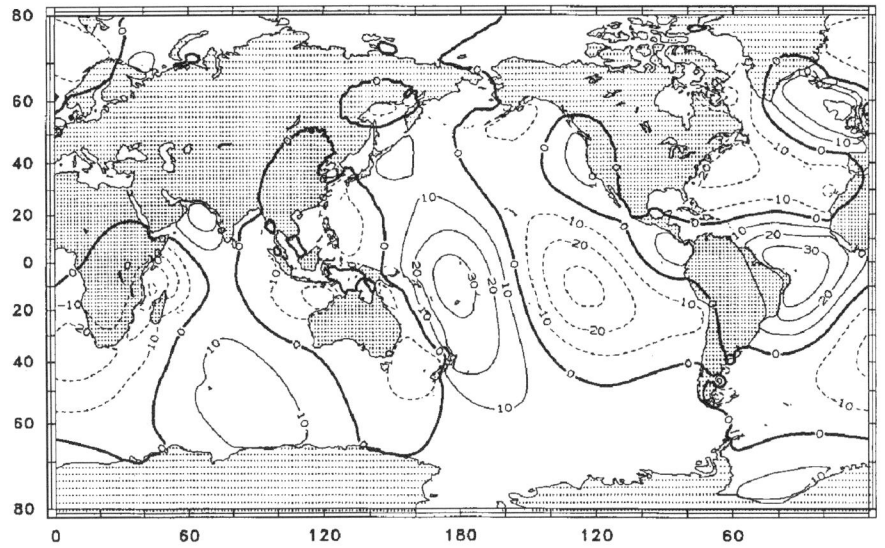

Fig. 7.8. (a) Co-amplitude of M_2 load tide, in millimeters, computed by Ray and Sanchez (1989). Based on "rule-of-thumb" relation that the load tide is 7% of the ocean tide. The writers show that minor variants of this rule lead to insignificant change. By assuming high-Q (seismologic) dissipation, orthodox models of the load tide permit only insignificant complex component.

of permanent bulges. While it is acceptable to model Laplacian oceanic tides, blocked by the continental margins, using harmonic vertical loading, it is incorrect to do so with respect to the solid-Earth wave tides. In terms of his accounting Schwiderski finds that the marine tidal mass, taking account of its phase, tends actually to accelerate the Earth. To overcome this and account for the deceleration, Schwiderski calls on braking by asymmetric pressure loading of the deformed ocean floor. Somewhat akin to this, Knopoff and Leeds (1972) postulated regionally-variable effects of the retarding torque upon lithosphere segments.

Schwiderski's accounting makes conspicuous a shortfall identified by him as 0.4° in the solid-earth phase lag. He himself finds this "exclusively due to to the postulated earth tide model with real and constant Love numbers and, hence, zero phase lag in the tidal response." It is now logical to postulate that the missing component is the omitted secular term, responsible for distortion.

7.4.5.2 Comparison, Computed vs. Altimetric Data. Worldwide gravimetric observation of the tides, revealing a variation maximum in tectonically active regions (Melchior and De Becker 1982; Melchior, Ducarme, and DeBecker 1986; Melchior and Ducarme 1991; Melchior 1995a) suggests that the assumption of mechanical isotropy may be inadequate, in particular in respect to the applicable Q. Following Hendershott (1972), Zahel (1978) has pointed to the effect of solid-Earth deformation on the ocean tides themselves.

Based on prolonged VLBI observations, Sovers (1994) has examined vertical ocean-loading amplitudes. He finds that ocean-load values are considerably larger than those calculated from models ordinarily computed (on a linear, homogeneous high-Q basis). With VLBI representing in his reckoning a global average, it would be desirable to separate loading values on a spot basis in such regions as Australasia, where gravimetric observations reduced on the supposition of a high-Q earth model are known to produce anomalous residuals. Numerous valiant attempts have been made to separate the solid-Earth from the water-depth part of the Topex dataset. Ray, Eanes, and Chao (1996) have based estimation of the solid dissipation upon equating the astrometric phase-lag and total Topex phase-lag, with respect to which the solid-Earth and marine component are both uncontrolled.

Estimations that include an arbitrary assumption of the dissipation in the continental one-third of the Earth fit laser-satellite observations of the lag in the total tide. A hydrodynamic tidal solution independent of altimetric data has been constructed by Le Provost et al. (1994). Within their equation 1, the coefficient K of the squared term in the water velocity, denominating bottom friction, has been set variously to 3×10^{-3} and 2.5×10^{-3}. A comparison by Andersen, Woodworth, and Flather (1995) maps the difference between the results and satellite-based solutions describing the marine tides. A categorical disparity (figure 7.8b,c) emerges between sets that are solely hydrodynamically computed, and those that additionally incorporate direct observation (Topex data).

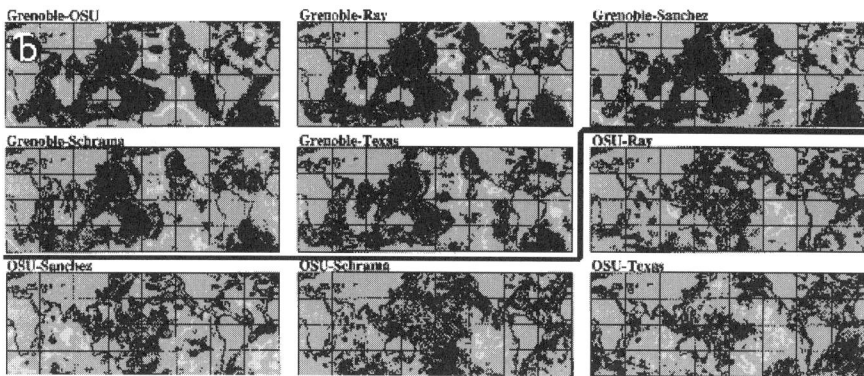

Fig. 7.8. (b) Difference-maps, displaying distinction between hydrodynamic tidal solutions absent Topex constraints, which *omit* complex component of load tide ζ_1^* by reason of being based on high-Q model (those maps above and left of black line); and those *including* the complex component, by reason of being based on observation (Topex altimetric data). The images above and left display the difference between the Grenoble hydrodynamic solution and the solutions named, based on Topex. The remaining images show the difference between solutions each containing the complex component; their difference omits the latter. After Andersen, Woodworth, and Flather (1995); refer to color original, displaying six additional Topex-based difference-images. The components of the difference-images are from: Grenoble: Le Provost et al. (1994); Ray (1994); Sanchez and Pavlis (1995); Schrama and Ray (1994); Texas: Eanes (1994); OSU: Egbert, Bennett, and Foreman (1994).

Grenoble minus Schrama

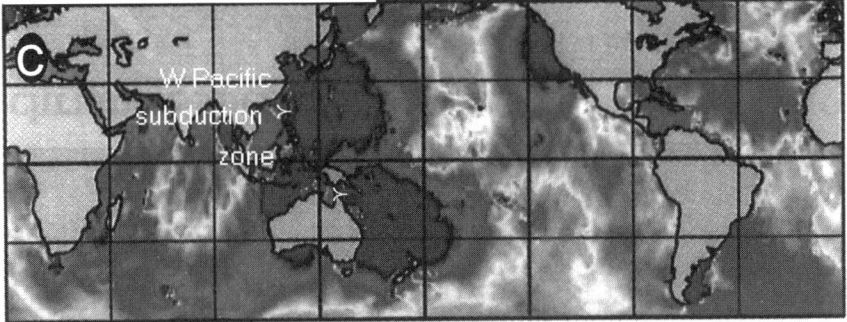

OSU minus Sanchez

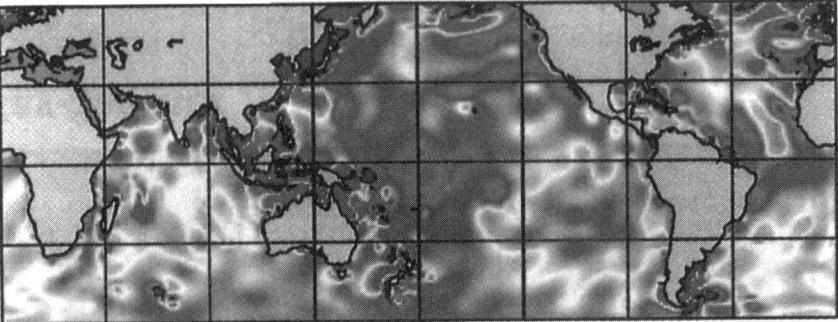

Fig. 7.8. (c) Comparison of (top) typical difference-image containing complex component of load tide with (below) difference-image *not* containing the complex component, because both constituents contain this equally. The complex component is coincident with the large-scale western Pacific zone of subduction (top), hence of secular pressure increase and phase instability.

Let ζ_g (Cartwright and Ray, 1989; Ray and Sanchez, 1989) represent the directly satellite-measured altimetric tide.

As pointed out by Ray and Sanchez ζ_g consists of the sum of the bodily tide ζ_b, the gauge-relative ocean tide ζ_o, and a load tide ζ_l. An incomplete aim of Topex has been to separate the sum of the latter two terms. A valuable incidental benefit would be to provide access to the residual, the bodily tide.

The load tide has to date been calculated in orthodox fashion (*supra*) on the supposition of a high-Q uniform earth. Unlike the case of the bodily tide (not blocked by the continental margin), it is legitimate to model the effect of the oceanic tide by employing a stationary potential and load. The tectonic situation (chapter 5) and the disparity between purely hydrological and observation-bounded tides shown by Andersen et al. suggests that ζ_l may be complex, the imaginary part representing dissipation under such effects as loss in modulated phase compression in the Pacific zones of subduction.

Both the Grenoble purely-hydrodynamic solution and the Topex-including solutions include an equivalent computed correction for the real part, ζ_1. In contrast, only the Topex observations have picked up the imaginary part, ζ_1^*, by direct observation.

The altimetric tide

$$\zeta_a(\text{Grenoble}) = \zeta_0 + \zeta_1;$$

whereas

$$\zeta_a(\text{Topex}) = \zeta_0 + \zeta_1 + \zeta_1^*,$$

so that the systematic difference displayed in Andersen, Woodworth, and Flather's "Grenoble minus OSU," and the other Grenoble-containing difference-maps (figure 7.8b,c), brings into relief the imaginary part ζ_1^* of the load tide.

It seems evident that its magnitude is large. Furthermore, its coherent features coincide with the western Pacific zone of subduction, pressure increase and incipient phase transition. The quadrature component (Andersen, Woodworth, and Flather's Plate 2; their Grenoble-containing difference-maps) displays features coinciding with the rapidly-spreading East Pacific Rise zone of upwelling, pressure decrease and phase- and state-transition. ζ_1^* is not precisely imaged by the difference map, because shallow-water dissipation in the Grenoble solution is calculated on the basis of a dissipation factor, as described by Schwiderski "at the disposal of the analyst."

Options in improving control of the fundamental ocean-loading effect include positioning subsequent gravimetric and GPS stations as profiles extending from a stable region into one of known subduction. It would furthermore be interesting to compare the gravimetric phase with bottom-pressure via use of manometers such as have been successfully employed in the Atlantic (Cartwright, Spencer, and Vassie 1987; Cartwright et al. 1988; in this respect see also Davis and Becker 1994; Davis et al. 1995). Matsumoto et al. (1998) have commenced separation of the ocean-loading component in some detail in the vicinity of Japan.

The control options mentioned are without exception expensive and time-consuming. Nevertheless, the position might be taken that unless and until a satisfactory ocean-loading factor is established, it will be difficult to make headway in expensive but invaluable measurements of the solid tide. Separately, few geologists would be confident that the outer Earth is homogeneous with respect to a twice-daily displacement of tens of centimeters, and a pressure pulse exceeding that in the advance in the convection. From the geophysical point of view (section 4.4) the situation is that, to an unknown extent, dissipation in zones of vertical flow includes the effect of water loading and the pressure pulse on the advance in the convection.

As a stopgap short of measurement, it would be possible pragmatically to correct excellent, extant gravimetric data for the regionally varying imaginary part of the load tide, displayed in the difference maps of Andersen et al. To the extent that shallow-water dissipation coefficients have been fitted to the deficit in the retarding torque, this correction would seem legitimate.

7.4.6 Geophysical Tiltmeter

Almost all geophysical instruments detecting the tidal disturbance produce a time-varying scalar signal. In the output of a gravimeter or strain meter, or potentially of GPS stations, dissipation is signalled as a phase displacement attributable to any loss process, for instance either oscillatory shear or bulk viscosity. As exceptions the tiltmeter and defunct torsion balance belong to a class of detector that yields real-time vectorial information. A multiaxial laser strain-meter (Bostrom and Vali 1969) yields similar information but requires a cavern site, itself a preclusive noise source.

Melchior (1983) has traced the development of the tiltmeter in the form of a horizontal pendulum from its conception by Hengler in the 1830s. Its fostering by the Observatoire at Brussels has shown that this type of instrument is readily capable of detecting earth-tidal tilts. However, the output is grossly distorted by site mechanical heterogeneities, such as cavern-shape and surounding geological structure. Harrison (1978) has examined this form of noise, and demonstrated almost preclusive difficulty in its eradication by computation. It has become apparent that the site difficulty can be overcome by emplacement in deep (≥ 500 ms) boreholes.

The area, axial ratio, azimuth, and trace direction of the tiltmeter ellipse (figure 7.9a) are a function not only of the amount of the dissipation but of its mode. Thus it is to be expected and is found, that at mid-latitudes in the Northern Hemisphere the tiltmeter ellipse is aligned NW-SE and written clockwise. Its form signals the

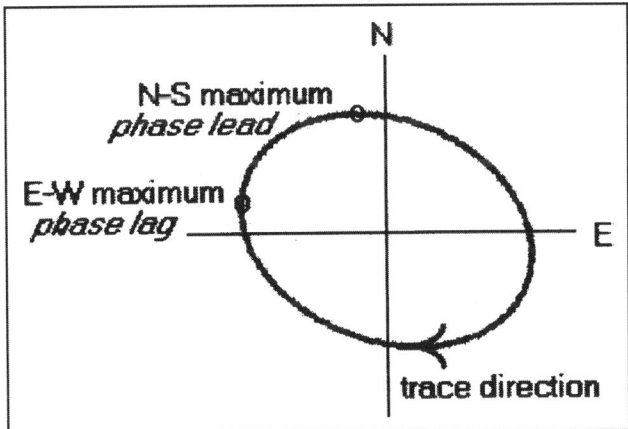

Fig. 7.9. (a) Trace of vertical pendulum or equivalent tiltmeter responding to M_2 at mid-latitude, Northern Hemisphere, in ideal site. The ellipse tilted in azimuth is the resultant of the E-W tilt component, in phase lag with respect to the E-W axis, and the N-S component, in phase advance relative to the N-S axis. The trace direction (sense), azimuthal tilt, and area of the ellipse are intrinsic in the case of an earth imperfectly elastic, rotating with respect to the tidal potential. (Drawing based on Melchior [1983].)

occurrence not only of dissipation, but that due to westward progression of the semidiurnal tidal bulge; namely, displacement as circulation induction "somewhere in the Earth."

Large-scale engineering and military requirements have brought about a rapid development of tiltmeters (Agnew, 1986), many capable of recording the tidal distortion. Electronic amplification makes it possible to record the displacement of simple-pendulums only a few centimeters in length. Designs also include observation of the displacement of a graphite proof body under frictionless suspension in a magnetic field; and electronic observation of a leveling-bubble. Agnew has explored the trade-off between such factors as emplacement and operational expense; long-term reliability; and sensitivity.

Industry has shown the feasibility of continuously recording tilt in deep boreholes. Thus, experiments have been conducted under the auspices of the Department of Energy and Sandia National Laboratories (Branagan et al. 1996), designed to observe clinometric changes as a result of hydraulic fracturing of a gas reservoir at Mesa Verde, Co. Six biaxial inclinometers operated for an interval of months under elevated temperature conditions while located in boreholes ranging in depth from 1302 to 1527 m. The logistical difficulties in borehole instrument emplacement and operation are considerable but are routinely overcome in the global petroleum industry. Bore-hole inclinometers are capable of recording the tides (figure 7.9b) and adaptable to in situ calibration The output from outlying deep-seated instruments is transmissible via satellite-based telemetry, to extant central recording stations.

It might be worth considering the deployment of a sparse borehole-emplaced tiltmeter network as modest supplement to the World-Wide Standard Seismograph

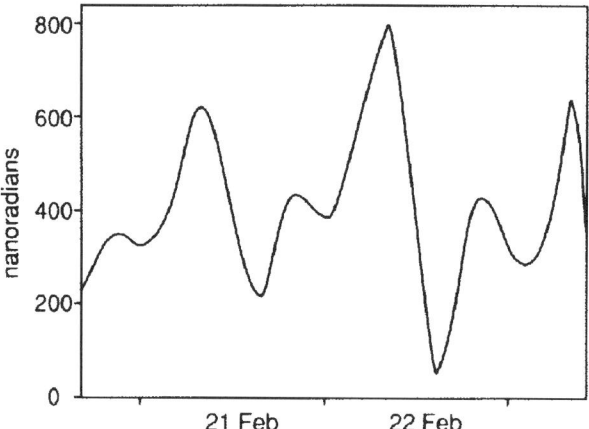

Fig. 7.9. (b) Earth-tide signal from a borehole tiltmeter located near the California coast. In this instrument platinum electrodes sense resistance changes caused by motion of a bubble in a conductive fluid. (From image supplied by G. Holzhausen [1997]).

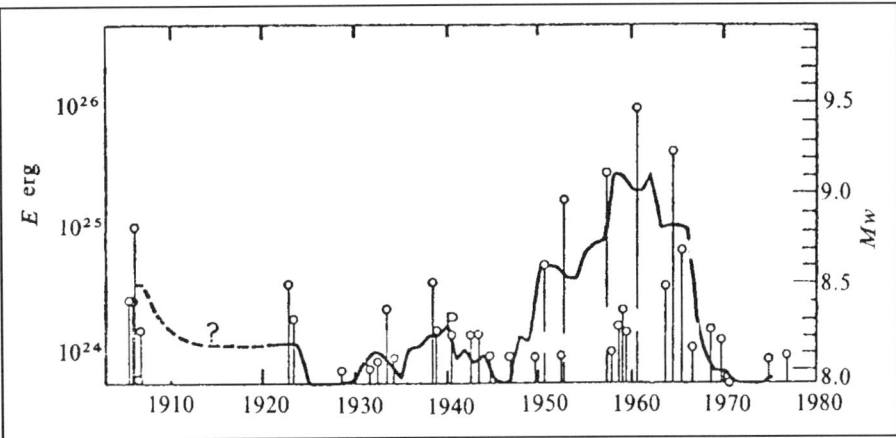

Fig. 7.10. Global release of energy in seismicity since introduction of calibratable seismograms, after Kanamori (1978). Almost all seismic energy is released in the form of major events, believed recorded without exception. The graph represents the five-year running average of energy release per year. Kanamori identifies the individual events and points to the long-term hundredfold variation in release rate.

Network, its successors, and the GEOSCOPE network (Montagner and Romanowicz 1994). In benefit of separating the zero-frequency component of tilt, double integration of long-period seismograms cannot replicate the output of deep-seated bore-hole tiltmeters, due to high level of surface noise.

It would appear that tiltmeters located in a borehole in each of the seven stable mid-latitude cratons would provide insight as to the long-term variation in global vorticity induction. The tidal signal is separable from secular (tectonic) tilt, typically a hundredfold smaller. Separation of the vorticity component of the signal would be facilitated in that there would be required the detection of output change, not the absolute value of tilt (the latter entailing correction for deep geological structure).

It might be expected that meteorological effects such as rainfall and barometric loading, troublesome in tilt observation, would not introduce the phase signature representing vortical motion. It would be of interest to compare observed vorticity induction with the well-documented but unexplained long-term variation in global seismicity perceived by Mogi (1974, 1979), Kanamori (1978; our figure 7.10, below), Pacheco and Sykes (1992), Romanowicz, (1993), Du (1993, 1994), and Satyabala and Gupta (1996). Romanowicz speculates as to whether observed switches from poloidal to toroidal motion are of external or rotational origin. Unlike that of the ultra-long period seismograph network, the tiltmeter cumulative record might correlate with the association reported by the USGS between tides and volcanicity (below).

Ingenious use by investigators beginning with Lodge (1893, 1897), Michelson (1904), and Sagnac (1913) of the absolute velocity of light travelling in a closed path has led to its use in ring lasers to observe Earth rotation. Anderson, Bilger, and Stedman (1994) and Rautenberg et al. (1997) are using a closed path to observe areal strain, changes in the normal vector to the ring laser plane, and the vorticity resident in tidal distortion at a site in New Zealand. The limitation in this important technique lies as it formerly did with tiltmeters, in the necessity for borehole emplacement in order to escape surface inhomogeneities.

7.5 The Volcanicity Association

The USGS and other organizations have for some decades reported a correlation between tidal phases and some volcanic phenomena (Mauk and Johnston 1973; Hamilton 1973; Michael and Christoffel 1976; Dzurisin 1980; McNutt and Beavan 1981; Rydelek, Davis, and Koyanagi 1988). As an irreversible, entropic process, eruption "triggering" represents tidal modulation of one of the multifarious processes that together comprise mantle convection.

Mauk and Johnston (figure 7.11a,b) have displayed correlation between eruptive activity and the fortnightly gravimetric component of the tide. As a group andesitic centers have been found to erupt principally at the time of tidal maximum. Basaltic centers display an additional peak in probability at tidal minimum. At the time of analysis, the identification of various components in mantle convection was less fully explored than now. A case might be made that injection and effusion at the crest of the global seafloor ridge system constitute the principal present-day volcanism. If this is the case, information as to its tidal modulation (or absence of

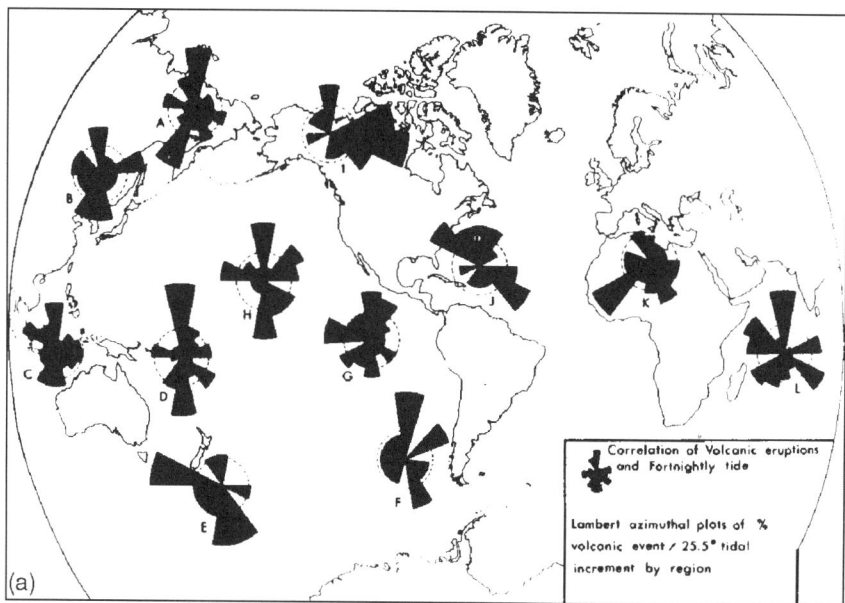

Fig. 7.11. (a) Correlation of eruptions and phase of the Mf tide. The latter is represented by the gravimetric (symmetric) component. (From Mauk and Johston [1973]). The authors subdivide the distribution into andesitic and basaltic regions (figure 7.11b). Their data have been modified to only a minor extent by Dzurisin (1980) and McNutt and Beavan (1981).

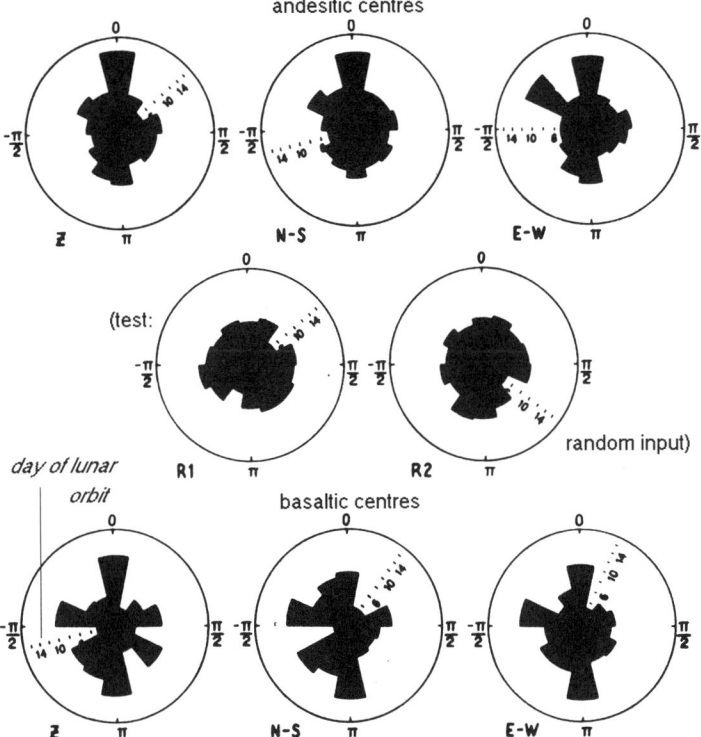

Fig. 7.11. (b) Histograms for andesitic (top row) and basaltic (bottom row) eruptive centers. In contrast to andesitic centers, basaltic centers display a secondary peak at tidal minimum. *Center row:* For comparison with tidal histograms, two distributions produced by a random number generator. Total number of eruptions, 633. Zero phase is plotted at tidal maximum. (From Mauk and Johnston [1973].)

this) may be present in the continuous recording of submarine noise conducted for military purposes. It will be recalled (section 4.4.5) that Klein (1976) has demonstrated a correlation between tide phase and seismicity at the crest of the Mid-Atlantic Ridge.

The lunar orbital declination with respect to the equatorial plane varies in an 18.6-yr cycle between the values of 28° and 18°. Shirokov (1983) (see also Du [1994]) believes that volcanicity in the high-latitude Kamchatka subduction region is modulated by the long-term declination cycle, being maximum at maximum declination (figures 7.12). Superimposed on the declinational is the apogee-perigee cycle in the course of which the Earth-Moon distance varies, a quantity that disproportionately varies the magnitude of the tides. The resultant is the 8.85-yr anomalistic-declinational cycle. Slichter (1963) analogized tidal dissipation with the electrical dissipation in a brass ring rotating about a diameter in a magnetic field, without identifying the mode of the tidal dissipation. Only angular rotation components

212   Tectonic Consequences of the Earth's Rotation

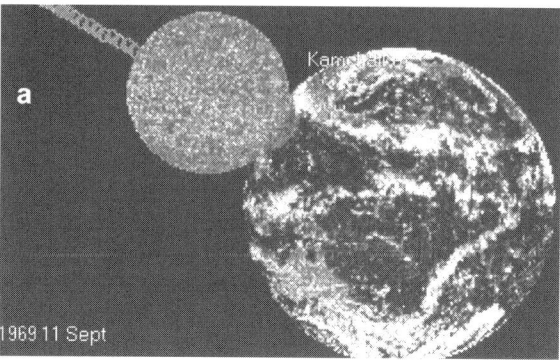

Fig. 7.12. (a) Declination of lunar orbit near maximum (11 Sept. 1969). The track of the bulge crest, due to the rotation of the Earth being faster than the orbit of the Moon about the Earth, is almost east → west, as small-circle at latitude 28°. Inscribed circles: orbital trace.

normal to the field direction incur dissipation in the ring. It may be significant that a varying portion of tidal vorticity induction is about axes inclined to the Earth's rotation axis (due to the varying lunar declination and the obliquity of the ecliptic). Imposition of rotation upon Tomaschek's graph (1957 fig. 1) of stress variation with declination makes plain the complicated geometry of TVI.

Speculation may center around the trace of the tidal bulges at times of high lunar declination (figures 7.12). Physically, the monthly rotation of the Moon about the Earth represents the passage of a bulge in sense inverse to that of the semidiurnal bulges. Its track is the resultant of motion of the M_f tidal bulge and that of the much faster semidiurnal bulges. Is it possible that the Earth responds separately to the components of the potential, in the same fashion that it seems to respond separately to modes and frequencies excited by the passage of the semidiurnal waves? The conspicuous volcanic response to the lunar-orbital (fortnightly) tides picked

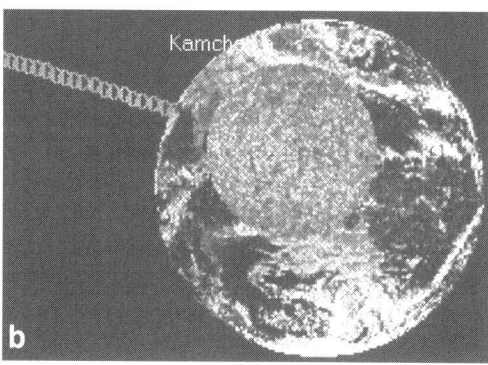

Fig. 7.12. (b) Declination of lunar orbit near declination mimimum. The track of the bulge-crest is at latitude 18°, more nearly coincident with the east-to-west solar bulge passage, in Equator plane at this time (fall equinox). Images prepared using program "Redshift" (Maris 1994).

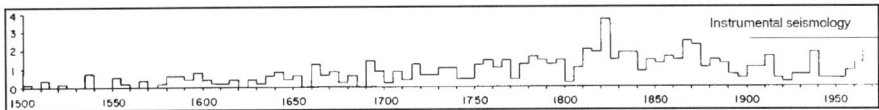

Fig. 7.13. (a) Five-century record of global volcanicity, prepared by Hamilton (1973) on the basis of Lamb's analysis of dust precipitation. Record is incomplete for far southern hemisphere. Inset for comparison: period, since about 1900, in which major earthquakes have been instrumentally recorded (cf. figure 7.10)

up by volcanogists suggests that this is the case, but the situation must be complicated by the distribution and, in particular, by the characteristic period of viscous components of the mantle.

There exists a bewildering number of difference and sum frequencies corresponding to aliasing of superimposed modes, few of which we are in a position to separate. If dissipation is effected by passage of the M_f bulge (as distinct from a modeled stationary imposition), geographical inhomogeneity in the tidal response may be expected to excite a nutation. Based on VLBI observations, Schuh (1989) finds a polar variation having a phase lag of 1 day, and larger influence of the Earth tide on polar motion than was previously estimated.

A particularly tempting data set from which to separate the long-period tidal harmonic components would seem to exist in the form of the global record of undersea volcanic noise, amassed in the course of military monitoring; relative to this, see the connection between magmatic heat and El Niño proposed by Shaw and Moore (1988). Is it possible that the enigmatic triggering of El Niño correlates

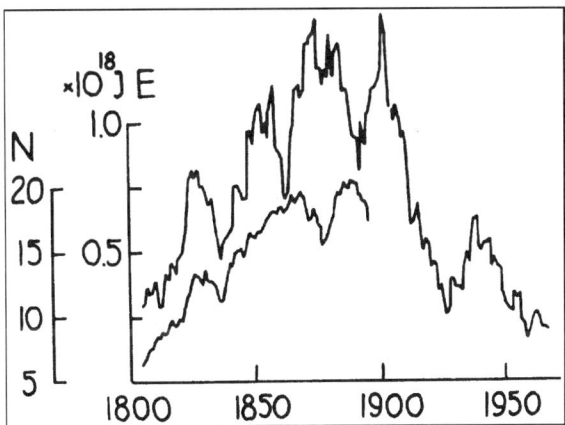

Fig. 7.13. (b) Released seismic energy (E) vs. annual number of volcanic eruptions since 1800. Constructed from historic records by Kalinin and Kiselev (1980); curves were subjected to an 11-yr sliding mean.

with changes in vorticity induction, detectable for instance by tilt-meter observations, and with the EPR seismicity identified by Walker (1988, 1995, 1999)?

Because of their short history instrumental observations cannot separate rotational effects which may be fundamental. Hamilton (1973) has used the five-century record of global volcanic dust precipitation prepared by Lamb (1970), figure 7.13a to examine its fluctuations. Hamilton points out the possible correlation of the 1821 volcanic maximum with tidal factors. Kalinin and Kiselev (1980), figure 7.13b, compare the historical records of volcanicity and seismicity.

7.6 Orbital Interaction; Multiple Outcomes (Indeterminacy)

> Figuring the nature of the Times deceas'd:
> The which observ'd, a man may prophecie
> With a neere ayme, of the maine chance of things,
> As yet not come to Life, which in their Seedes
> And weake beginnings lye entreasured:
> Such things become the Hatch and Brood of Time
> *Henry IV (2)*

Lately, the Earth's rotation and orbital elements have been found basic in the development of "ice ages," thus polar loading and deformation affecting the inertial moment. Precession and cyclical changes in the eccentricity and the obliquity have been shown to correlate with age-dated seafloor sediments.

Based on the sedimentary record of paleotemperatures Emiliani (1955, 1957), Emiliani and Geiss (1957) and Hays, Imbrie, and Shackleton (1976) claimed that orbital variations are "the pace-maker of ice ages." Building on the correlation between climate change and orbital variations suggested in pioneering studies by Milankovitch (1941a,b), Kukla (1975) concluded that changes in the "cryosphere" are related to insolation changes in the interior of North America and Eurasia, an effect that is maximum in the early autumn. Martinson et al. (1987) have developed a high-resolution correlation of age dates and orbital elements extending from the present to 300 kyr BP. Berggren et al. (1995) have established a correlation extending over the Late Neogene of astrochronological components, the seafloor sedimentary record, and revised "magnetochronology." The latter represents the seafloor record of discontinuous and aperiodic magnetic reversals. As the authors point out, the higher frequency astronomical components (figure 7.14) such as the precession present the possibility of continuous dating, as distinct from opportunistic use of magnetic reversals with interpolation.

Hilgren et al. (1995) have extended the correlation into the Miocene across a sedimentary hiatus represented by dessication of the Mediterranean basin. Their use of sapropel clustering permits enhanced recognition of precession extrema. Their calibration incorporates the 65°N insolation curve and present-day values of the dynamical ellipticity and tidal dissipation. As they emphasize, the accuracy of

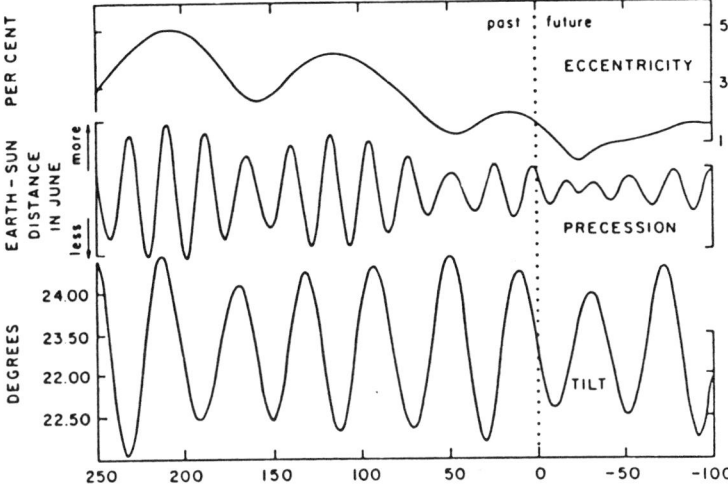

Fig. 7.14. Rotational and orbital factors modulating insolation, climate and polar glacial loading. Although appearing minor, acting on the highly nonlinear global climatic system these have been shown to result in the major fluctuations identified by Milankovitch (1941a,b). (Reproduced from Berger, [1988].)

astronomical extrapolations is uncertain because of uncertainties in the tidal dissipation and the ellipticity. Tidal effects in the form of the couple acting on the equatorial bulge have been made clear by Laskar, Joutel, and Boudin (1993), figure 7.15. Ivanov (1985) has related some long-term tides to the rotational precession.

Extrapolation back into the past has involved such assumptions as that spindown has been approximately constant. In point of fact, glacial episodes may result in drawdown of world sea-level to the edge of the continental shelves. Besides deep-sea sedimentation effects, it might be supposed that this results in reduction in the frictional marine dissipation, believed to be major in the shelf seas. It is considered (Berrgren et al. 1995) that astrochronological events have in general been detected despite variations in the sedimentary record.

Drawdown during the Neogene ice ages has resulted in the intermittent appearance of the Bering land bridge, creating a barrier to circulation. Large tidal changes might be expected at earlier dates, resulting for example from the Paleocene development of the Panamanian isthmus (Ziegler, Scotese, and Barrett 1982). During earlier geological history, the marine tidal torques may have differed from those at present (Krohn and Sundermann 1982). The major lunar contribution to the obliquity (figure 7.15), acting on the equatorial bulge, might be expected to vary with the amplitude of the marine tide. Change in the length of day (spindown) is not obviously related to the length of the year (but see Brosche and Wunsch [1990]).

Vening Meinesz (1947) and Denis and Varga (1988) have examined possible

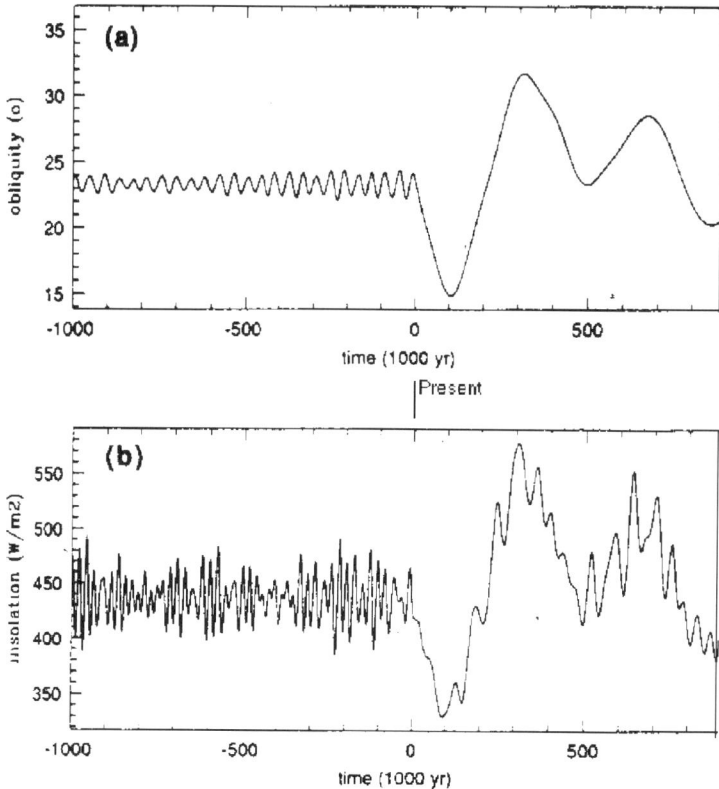

Fig. 7.15. Lunar influence on the obliquity and insolation (a determinant of "ice ages," ice-loading of the polar regions, and marine drawdown). The computation displays (a) the evolution of obliquity; (b), insolation as from 20 Ma BP. Extrapolation forward from the present results from Laskar, Joutel, and Boudin (1993) experimentally "suppressing" the Moon and its tides. (By courtesy of the authors and "Astronomy and Astrophysics" [ibid].)

tectonic effects of the reduction in flattening under the tidal spin-down. Williams (1994) has pointed to the early Paleozoic Gondwana tectonothermal event as a possible consequence of core resonance in passage through a critical day-length, about 22.2 hr. The occurrence of catastrophic tides, suggested if the lunar regression has long been as fast as it is at present, has to date not been perceived in the geological record (Munk, 1968). Williams (1997) has used the record of tidal rhythmites (Williams, 1989), previously extending to −620 Ma, to place bounds on the long-term rate of lunar recession, indicating lack of a catastrophic close proximity of the Moon since at least −3Ga. Extrapolating back to the Archean, it might be suspected that the lunar distance is to an extent self-regulating, in that a highly non-linear tidal response to close approach must strongly govern the lunar couple, hence orbital expansion. As originally shown by G. H. Darwin (1879a), the couple goes as the inverse sixth power of the distance.

Darwin's examination of tides in viscous spheroids (1879a,b and 1908) revealed proliferating interactions, for example the effect of solar on lunar tides. Rather than the solar system "clockwork" perceived by Newton, Laplace, and Lagrange, there has become evident a system of worlds each affected by endogenous internal dynamics—each interacting, additionally, with planetary siblings. Over geological time the rotation and orbital periods of Earth, Venus and Mercury, for example, have become related, due to interactions not yet elucidated.

Chaos theory delimits our ability "To look into the Seeds of Time, and say which grains will grow, and which will not." Unlike Macbeth's witches Laskar (1989, 1994) has enlisted a Liapunov formulation, to evaluate the long-term stability of the solar system. Because resonances result eventually in exponential interaction, the inner planets are governed by a regime of chaos, entirely resisting prediction. Numerically, a change in assumed initial planetary locations of less than one Planck's length leads to unlimited difference in perturbation within a period of geological significance (characteristically, tens to hundreds of millions of years). A significant probability exists of profound past and future changes in the orbital and rotational elements, hence climate, of Earth, Mars, Venus, and Mercury. For instance, Mercury is prone to develop eventually an "escape" eccentricity, unity, due to ongoing interaction with Venus. It has been speculated (Bostrom 1997) that the Hesperian flooding of Mars resulted from a tidal encounter, of which no record can remain in present orbital elements.

As noted earlier, in view of a possible correlation between vorticity induction and the geomagnetic field (Bostrom 1998c), it might be valuable to examine the correlation, if any, between reversals and changes in the lunar declination and terrestrial obliquity.

It has become increasingly clear that in respect to periods important to the geologist, the orbital and rotational evolution of the inner planets including the Earth is not only indeterminable, but in the absolute sense (Prigogine 1995) indeterminate. A similar restriction applies to models, discussed below, of flow within the evolutionary and highly non-linear mantle; a given initial state, *no matter how minutely specified*, leads within geological time to innumerable, entirely dissimilar "solutions." The restriction is not one of computability, including the potential of quantum computation.

7.7 Numerical Models

With respect to analytic solutions of fluid-dynamic models, the situation has not advanced dramatically since reviews by Birkhoff (1960) and Lugt (1979); solutions to problems far simpler than the regime in the mantle are scarce. Numerically, Ferziger and Peric (1996; see also Leonard [1997]) point to restrictive limits in computational fluid dynamics. Despite this, in skilled hands present computational capacity has made it possible to reach short-term realistic solutions via stepped numerical integration (Hager and Clayton 1989; Forte and Peltier 1991; Forte,

Woodward, and Dziewonski 1994; Tackley et al. 1993); see, for example, figure 4.3.

Rayleigh's masterly analysis of stability change in a fluid heated from below (Rayleigh 1916) elucidated the point at which a system "falls away from unstable equilibrium ... into steady motion." His analysis made it clear that a switch in stability states must occur, first at values of his parameter σ^{-1} (his eq. 20) specifying large cell size. Large (low order) cell size is the first to be reached after attainment of his critical number. It is less obvious from Rayleigh's analysis that the greatest cell size (smallest σ) results in integration of the driving force, buoyancy under gravity, over the greatest spatial dimensions. Rayleigh perceived that the flow in other than simple cases would have to be determined by numerical computation, at that time beyond reach.

Neither the experiments of Kelvin (1881) and Benard (1900), nor the analyses of Rayleigh and successors employing Euler's formulation fluid flow, could bring into relief a phenomenon now emergent as fundamental (Prigogine 1997): the proclivity of flow within an evolutionary dissipative system to be self-organizing. In convection, this results in the formation of "cells" optimizing the dissipation path.

Although revealing, numerical modelling has to date employed classical methodology, for lack of a better approach by progressing from initial, geophysically observed, mass anomalies. Torques induced by passage of the tidal waves (section 3.6.3) are of the same order of magnitude and dimensionally identical to those generated by the density variations driving buoyancy convection. The phase lag recorded by the Observatoire at Brussels (section 3.9; figure 3.14) suggests that the viscous dissipation may be comparable with that in the convection. Flow under the induction must like the convection be subject to self-organization, almost certainly through commingling with it, taking the form of modifying cell geometry.

It might therefore be worth examining the effect of a vorticity contribution to numerical solutions. In a rotating inhomogenous earth, it would seem that models of convection must incorporate the effect of axial drift (polar wander) and concomitant figure-adjustment. The energetics of loss in the convection vs. the tidal distortion suggest that it may not be possible to simulate the flow in the mantle, producing such effects as the grossly excess flattening and dominance of surface-west flow cells, *devoid* of the tidal component.

7.8 Status of Field

In the field of tectonics and Earth rotation, the tidal aspect is paramount. Due to remarkable effort on the part of the "tidalists," the gravimetric-measurement problem has in large part been overcome. At the present stage instrumental detection of an altimetric or gravimetric signal, whilst essential, is no longer the limiting factor.

With reference to identifying the seat of tidal dissipation (Plate 1, frontispiece), priorities lie (1) in the construction of a reference model both astronomically and geologically plausible; and (2), in obtaining data quantifying the oceanic contribution. In re the latter, the investigations of Matsumoto et al. (1998) show unprecedented promise.

Lambeck's (1988) stricture still holds: it is not possible to evaluate solid-Earth dissipation by subtracting an oceanic contribution from the well-measured astronomic total. More than one skilled attempt has seemed finally to delimit the elusive "solid" factor. Zschau (1978) estimated dissipation in a spheroid assuming timevariation of a stationary potential. NASA/Goddard (Ray, Eanes, and Chao 1996) compares the total Topex-observed dissipation with the astrometric total. The "solid" contribution on both sides of the equation is based upon the orthodox model, excluding the fundamental tides (the M_2, S_2 waves). Until a means can be found of separating the seafloor from the marine contribution, as many solutions to equating Topex with the astrometric data exist as the unlimited ways in which it is possible to allot the apportionment. By the same token, it is unfortunately not possible to term the Topex-derived marine tides "accurate" until the apportionment problem is overcome. Most notably, the data amassed by project WOCE (World Ocean Circulation Experiment) (Chapman 1998) now provide the resolution and accuracy necessary to re-evaluate Miller's (1969) estimate of the marine dissipation.

Astronomically, with admiration of the effort they have inspired, it may be necessary to pass ab initio beyond fictional Darwin/Love models, in favor of acknowledging the rotation of the potential and the real action of the tides. Theory of his time attributing dissipation mainly to slippage of the Earth's crust over a less viscous layer, the caution exhibited by Eddington (1923) still is well merited: "I am tempted to add further speculation arising out of the location of the dissipation"; and "all may not be so certain as our own one-sided view seemed to indicate."

In respect to observation, only scattered sampling of the continental one-third of the Earth has been achieved. We have little idea, other than assuming on the basis of unreal models that dissipation in these regions is trivial, of the action of the Earth tides in Australia, invested by plate-boundaries; in South America, overlying the Andean subduction; in North America, overriding the EPR; in Africa including the Rift upwelling; and in the lengthy Tethys belt, encompassing both the Mediterranean and Asian convergences. The investigations of researchers such as Schmeling and Bussod (1996) and Bailey (1998) suggest that extremely lowviscosity material may be present within parts of the continental lithosphere.

It may be hoped that deployment of the orbital gradiometer, narrow beam radar, or orbital lidar will permit coverage of the unsurveyed expanses. In particular, inter-satellite range observation (Jekeli and Shum 1998) may provide uniform ocean/continent coverage, of the kind strikingly demonstrated by Rapp's (1988) high-level projection of the gravity gradient. McKenzie (1995) has pointed out that thanks to orbital radar, in several respects Venus's topography is more accurately known than Earth's. We await the development of a device as potent as the Navy's altimetric radar, which has illuminated the oceanic Earth. As tidal observation requires measurement of elevation change rather than absolute value, it might conceivably be possible to mobilize radar interferometry of the sort demonstrated recently by Carnec, Massonet, and King (1996; see also Vincent and Rundle [1998]), to extend tidal observation to the continents. It is technically attractive to take advantage of the stability of the ancient shields to monitor vorticity induction, using borehole tiltmeters already available.

7.9 Scope

The chapters of this volume are no more than cuts through a matrix, multidimensional and of unlimited extent. Since antiquity humans have perceived the Moon as enigmatic, and at times even baleful. The alarmingly proficient sorcerer, Prospero, "could control the Moon, make flows and ebbs," and "make shake the strongbas'd promontory." Can it be that Prospero's lunar orb has in fact contorted the Earth—in fashion infinitely more profound than conceived by Shakespeare?

Despite the peculiarities in plate motion to which Runcorn (1957, 1973a,b), Sullivan (1974) and others drew attention decades ago, it has seemed that the only mechanism capable of "driving plate tectonics" has been buoyancy convection, so plainly affecting the mantle. To what other agency may we appeal?

It appears essential to examine the contribution of tidal vorticity induction (TVI) to geotectonics. It is scarcely logical to assume that the Earth, subjected to the endless one-way passage of waves of amplitude several cm, is for some reason immune to residual distortion. TVI is cumulative, dimensionally matched to endogenous convection, and in direction suited to effect its modulation. Convection within the Earth, of any nature, at any scale, at any depth, incorporates branch points (see, for example, Nicolis and Prigogine 1989). At these an *infinitesimal* perturbation determines its direction, such as clockwise or counterclockwise. An increasingly conspicuous phase lag indicates that the tidal distortion, tending to align convection with its own direction, is in fact far from merely infinitesimal. It seems conceivable that absent the lunar member of Kuiper's closely bound doubleplanet, Earth's physiography would bear little resemblance to today's. Then to what extent can we measure this effect?

Lacking the Moon, Earth would lack also a mobile lithosphere—thus jointly the subduction standing between us and conditions on Venus, and the upwelling that generates procreant submarine vents. Does the extreme age of the Earth-Moon partnership (Boss and Peale 1986; Newsom and Taylor 1989; O'Neill 1991; Halliday et al., 1996), and growing evidence (e.g., Lowe 1994) that subduction dates from the planet's early days, indicate the existence of sheltered conditions since primordial times? Akin to this, see also Laskar, Joutel, and Robutel (1993).

Biologists may consider whether devoid of shelter of such extraordinary duration, there could have developed Earth's rich biota—of late, supporting a form of intelligence.

References

Achilli, V. P. Baldi, G. Casula, M. Errani, S. Focardi, M. Guerzoni, F. Palmonari and G. Raguni. 1995. A calibration system for superconducting gravimeters. *Bull. Geod.* 69: 73–80.
Agnew, D. C. 1986. Strainmeters and tiltmeters. *Rev. Geophys.* 24(3): 579–624.
Ahrens, T. J. 1989. Water storage in the mantle. *Nature* 342: 122–123.
Akaogi, M. E. Ito and A. Navrotsky. 1989. Olivine–modified spinel–spinel transitions in the system Mg_2SiO_4–Fe_2SiO_4; calorimetric measurements, thermochemical calculation, and geophysical application. *J. Geophys. Res.* 94(11): 15,671–15,685.
Alessandrini, B. 1989. The hydrostatic equilibrium figure of the Earth: an iterative approach. *Phys. Earth Plan. Int.* 54: 180–192.
American Association of Petroleum Geologists. 1981. Map 1/10m, *Plate Tectonics of the Circum–Pacific Region: Northeast Quadrant.* K. J. Drummond, Chairman, NE Quadrant; compilation by Addison, W. O. and P. W. Richards. Tulsa, OK: AAPG.
Ampferer, O. 1906. Über das Bewegungsbild von Faltengebirgen. *Jahrb. Geol. Reichsanstalt* 56 (Pts. 3–4): 539–622.
———. 1925. *The origin of continents.* Editorial report as to Amferer's favoring "under-streaming currents" in respect to continental mobility. *Nature* 116: p. 481.
———. 1941. Gedanken ueber das Bewegensbild des atlantischen Raumes. *Sitz. Der Akad. Wissen. Wien, Math. Naturwiss. Klasse. Abt. I: Mineral. Biol. Erdkunde*, 150; 1–2, 19–35.
Ampferer, O. and W. Hammer. 1911. Geologischer Querschnitt durch die Ostalpen vom Allgäu zum Gardasee. *Jahr. Geol. Reichsanstalt.* 61 (Pts. 3–4): 531–710.
Amstutz, A. 1951. Sur l'evolution des structures alpines (notes pour la legende d'une serie de schemas embryotectoniques). *Arch. Sci.* 4: 323–329.
Andersen, O. B. 1995. New ocean tide models for loading computation. *Bull. Inf. Centr. Mar. Terr.* 122: 9256–9264.
Andersen, O. B., P. L. Woodworth and R. A. Flather. 1995. Intercomparison of recent ocean tide models. *J. Geophys. Res.* 100(C12): 25,261–25,282.
Anderson, D. L. 1980. Bulk attenuation in the Earth and the viscosity of the core. *Nature* 285: 204–207.
———. 1982. Hotspots, polar wander, Mesozoic convection and the geoid. *Nature* 297: 391–393.

———. 1987. Thermally induced phase changes, lateral heterogeneity of the mantle, continental roots, and deep slab anomalies. *J. Geophys. Res.* 92(B13): 19,968–19,380.

———. 1989. Where on Earth is the crust? *Phys. Today* (Amer. Inst. Physics) 42(3): 38–46.

———. 1994a. The sublithospheric mantle as the source of continental flood basalts. *Earth Plan. Sci. Lett.* 123: 269–280.

———. 1994b. Superplumes or continents? *Geology* 22: 39–42.

Anderson, D. L. and C. Sammis. 1969. Partial melting in the upper mantle. *Phys. Earth Plan. Int.* 3: 41–50.

Anderson, D. L. and J. B. Minster. 1979. The frequency dependence of Q in the Earth and implications for mantle rheology and Chandler wobble. *Geophys. J. Roy. Astron. Soc.* 58: 431–440.

Anderson, D. L. and J. B. Minster. 1981. The physics of creep and attenuation in the mantle. In: *Anelasticity in the Earth* (F. D. Stacey, M. S. Paterson and A. Nicholas, eds.). Pp. 5–11.

Anderson, D. L. and J. D. Bass. 1986. Transition region of the upper mantle. *Nature* 320: 321–328.

Anderson, D. L., Yu–Shen Zhang and T. Tanimoto. 1992. Plume heads, continental lithosphere, flood basalts and tomography, in: Magmatism and the Causes of Continental Break–up (Eds. B. C. Storey, T. Alabaster and R. J. Pankhurst), *Geol. Soc. Spec. Pub.* no. 68: 99–124.

Anderson, D. L. and J. W. Given. 1982. Absorption band Q model for the Earth. *J. Geophys. Res.* 87(B5): 3893–3904.

Anderson, D. L., and C. Sammis. 1970. Partial melting in the upper mantle. *Phys. Earth Plan. Int.* 3: 41–50.

Anderson, R., H. R. Bilger and G. E. Stedman. 1994. "Sagnac" effect: a century of Earth–rotated interferometers. *Am. J. Phys.* 62: 975–985.

Anderson, J. G., P. Bodin, J. N. Brune, J. Prince, S. K. Singh, R. Quaas and M. Onata. 1986. Strong ground motion from the Michoacan, Mexico, earthquake. *Science* 233: 1043–1049.

Andrews, J. A. 1985. True polar wander: an analysis of Cenozoic and Mesozoic paleomagnetic poles. *J. Geophys. Res.* 90(B9): 7737–7750.

Argus, D. F. and R. G. Gordon. 1991. No–net–rotation model of current plate velocities incorporating plate motion model NUVEL–1. *Geophys. Res. Lett.* 18(11): 2039–2042.

Ashkenazi, V., R. M. Bingley, G. M. Whitmore and T. F. Baker. 1993. Monitoring changes in mean–sea–level to millimeters using GPS. *Geophys. Res. Lett.* 20(18): 1951–1954.

Atkinson, W. J., and C. B. Smith. 1995. Diamond deposits of Australia. *Min. Engin.* 47(8): 733–737.

Atwater, Tanya. 1970. Implications of plate tectonics for the Cenozoic evolution of western North America. *Bull. Geol. Soc. Amer.* 81: 3513–3536.

Atwater, T. J. Sclater, D. Sandwell, J. Severinghaus and M. S. Marlow. 1993. Fracture zone traces across the North Pacific Cretaceous Quiet Zone and their tectonic implications. In: The Mesozoic Pacific: Geology, Tectonics and Volcanism. *Geophys. Monograph. AGU*, 77, 137–150.

Audley–Charles, M. G., P. D. Ballantyne and R. Hall. 1988. Cenozoic rift–drift sequence of Asian fragments from Gondwanaland. *Tectonophysics* 155: 17–330.

Bailey, D. K. 1982. Mantle metasomatism—continuing chemical change within the Earth. *Nature*, 296: 525–530.

———. 1992. Episodic alkaline igneous activity across Africa: implications for the causes of continental break–up, in: *Magmatism and the Causes of Continental Break–up*

(B. C. Storey, T. Alabaster and R. J. Pankhurst, eds.). *Geol. Soc. Spec. Publ.* no. 68: pp. 91–98.

Baines, P. G. 1982. On internal tide generation models. *Deep–Sea Res.* 29(3A): 307–338.

Balmino, G. 1986. Present status and future improvements in measuring the gravity field of the Earth and planets. In: *Space Geodesy and Geodynamics* (A. J. Anderson and A. Cazenave, eds.), London, Academic Press. 19–54.

Barazangi, M. and B. Isacks, 1971. Lateral variations of seismic wave attenuation in the upper mantle above the inclined earthquake zone of the Tonga island arc: deep anomaly in the upper mantle. *J. Geophys. Res.* 76: 8493–8515.

Barrell, J. 1914. The Strength of the Earth's crust. 2: VI: Relation of isostatic movements to a sphere of weakness—the asthenosphere. *J. Geophys. Res.* 22: 537–555.

———. 1915. The strength of the Earth's crust, Pt. VII. *J. Geology*, 23: 27–44.

Batchelor, G. K. 1967. *Introduction to Fluid Mechanics.* Cambridge Univ. Press.

Baudry, N. and L. Kroenke. 1991. Intermediate–wavelength (400–600 km), South Pacific geoidal undulations: their relationship to linear volcanic chains. *Earth Plan. Sci. Lett.* 102: 430–443.

Becker, G. F. 1898. Kant as a natural philosopher. *Amer. J. Sci. Fourth Ser.* vol. v, no. 26: 97–112.

Becquerel, A. H. 1903. Recherches sur une proprieté nouvelle de la matière: activité radiante spontanée ou radioactivité de la matière/par Henri Becquerel. Paris: Fermin–Didot.

Bell, D. R. and G. R. Rossman. 1992. Water in Earth's mantle: the role of nominally anhydrous minerals. *Science*, 255: 1391–1397.

Bénard, H. 1900. Rev. Gen. Sci. xii, pp. 1261, 1309 *(cited by Lord Rayleigh, 1916, q.v.).*

Benioff, H. 1951. Earthquakes and rock creep. *Bull. Seis. Soc. Amer.* 41(1): 31–62.

Bercovici, D. 1993. A simple model of plate generation from mantle flow. *Geophys. J. Intern.* 114: 635–650.

Bercovici, D. and P. Wessel. 1994. A continuous kinematic model of plate-tectonic motions. *Geophys. J. Intern.* 119: 595–610.

Berg, F. and H. Pulpan 1971. Tilts associated with small and medium size earthquakes. *J. Phys. Earth* 19(1): 59–77.

Berg, E. and W. Lutschak. 1973. Crustal tilt fields and propagation velocities associated with earthquakes. *Geophys. J. Roy. Astron. Soc.* 35: 5–29.

Berger, A. 1988. Milankovitch theory and climate. *Rev. Geophys.* 26(4): 624–657.

Berggren, W. A., F. J. Hilgen, C. G. Langereis, D. V. Kent, J. D. Obradovich, I. Raffi, M. E. Raymo and N. J. Shackleton. 1995. Late Neogene chronology: new perspectives in high–resolution stratigraphy. *Geol. Soc. Amer. Bull.* 107(11): 1272–1287.

Besse, J. and V. Courtillot. 1991. Revised and synthetic apparent polar wander paths of the African, Eurasian, North American and Indian plates, and true polar wander since 200 Ma. *J. Geophys. Res.* 96(B3): 4029–4050.

Bevis, M. Y. Bock, P. Fang, T. Herring, J. Stowell and R. Smalley, Jnr. (1997). Blending old and new approaches to regional GPS geodesy. *EOS (Trans. Amer. Geophys. Un).* 78(6): pp 61, 64, 66.

Bina, C. R. and G. Helffrich. 1994. Phase transition Clapeyron slopes and transition zone seismic discontinuity topography. *J. Geophys. Res.* 99(B8): 15,853–15,860.

Bird, R. T., D. F. Naar, R. L. Larson, R. C. Searle and C. R. Scotese. 1998. Plate tectonic reconstruction of the Juan de Fernandez microplate: transformation from internal shear to rigid rotation. *J. Geophys. Res.* 103(B4): 7049–7067.

Birkhoff, G. 1960. Hydrodynamics: A Study in Logic, Fact and Similitude. Princeton U. P. 186 pages. Ch. 1, "Hydrodynamical Paradoxes."

Blackett, P. M. S. 1952. A negative experiment relating to magnetism and the Earth's rotation. *Phil. Trans. Roy. Soc.* A245: 309–370.
Bobryakov, A. P., A. F. Revuzhenko and E. I. Shemyakin. 1991. Tidal deformation of planets: experience in experimental modeling. *Geotectonics*, 25(6): 473–481.
Bock, G. and J. R. Clements. 1982. Attenuation of short–period P, PcP, ScP and pP waves in the Earth's mantle. *J. Geophys. Res.* 87(B5): 3905–3918.
Bolt, B. A. 1985. Constraints from core reflections on mantle Q and density at the core boundary. *Phys. Earth Plan. Int.* 38(1): 1–8.
_____. 1986. Seismic energy release over a broad frequency band. *Pure Appl. Geophys.* 124(4–5): 919–930.
Bolt, B. A. 1991. The precision of density estimation deep in the Earth. *Q. J. Roy. Astron. Soc.* 32(4): 367–388.
Bonatti, E. 1990. Not so hot "hot spots" in the oceanic mantle. *Science* 250: 107–111.
Bonatti, E. C., G. A. Harrison, D. E. Fisher, J. Honnorez, J.–G. Schilling, J. J. Stipp and M. Zentilli. 1977. Easter volcanic chain (Southeast Pacific): a mantle hot line. *J. Geophys. Res.* 82(17): 2457–2478.
Bonatti, E. and C. G. A. Harrison. 1976. Hot lines in the Earth's mantle. *Nature* 263: 402–404.
Bonatz, M. 1978. Tides of the solid Earth from gravimetric measurements. In: *Tidal Friction and the Earth's Rotation* (P. Brosche and J. Sundemann, eds.); Springer-Verlag. Pp. 55–61.
Borch, R. S. and H. W. Green II. 1987. Dependence of creep in olivine on homologous temperature and is implications for flow in the mantle. *Nature*, 330: 345–348.
Boschi, E., A. Morelli and P. Gasperini. 1994. A network of multi-sensor stations for continuous monitoring of ground motion and deformation. *Phys. Earth Plan. Int.* 84(1–4): 289–298.
Boss, A. P. and S. J. Peale, 1986. Dynamical constraints on the origin of the Moon. In: *Origin of the Moon* (Hartmann, W. K., et al., eds.). Lunar and Planetary Institute, Houston. 59–101.
Bostrom, R. C. 1973. Arrangement of convection in the Earth by lunar gravity. *Phil. Trans. Roy. Soc. London* A 274: 397–407.
_____. 1976. Westwanderung and the lunar tidal couple: modulation of convection by bulge stress. *Moon*, 15: 109–117.
_____. 1978a. Motion of the Pacific plate and formation of marginal basins: asymmetric flow induction. *J. Phys. Earth*, 26 Suppl.: S103–S122.
_____. 1978b. Tectonics of the tidal, convective Earth, I: shaping of the convection in the mantle by the passage of the tidal bulge; w. Russian abstract. *Mod. Geol.* 6: 171–183.
_____. 1978c. Tectonics of the tidal, convective Earth, II: pattern of sea floor spreading, deep subduction, and plate motion; w. Russian abstract. *Mod. Geol.* 6: 185–198.
_____. 1981a. Lithosphere creep. *J. Phys. Earth* 29: 145–161.
_____. 1981b. Tidal dissipation: stress transmission in the lithosphere. *Proc. Ninth Intern. Symp. Earth Tides, New York, Aug.* 17–22 1981: 593–605.
_____. 1981c. Formation of the cordillera: westward extension of the Atlantic realm. *Pacific Geology*, 15: 51–64.
_____. 1984. Westward Pacific drift and the tectonics of eastern Asia. *Tectonophysics* 102: 359–376.
_____. 1990. Figure adjustment and tectonics during polar wander: the flowing–apart of Gondwana. *Tectonophysics* 182: 393–402.
_____. 1994. Tidal constructs: excess flattening; TPW inhibition. *EOS (Trans. Amer. Geophys. Un.)*, 75(16, Suppl.): 120.

_____. 1998a. Tectonics of Earth and Venus: existence of waves M_2, S_2. *Ann. Geophys.* *XXIII* Gen. Assemb. Europ. Geophys. Soc. Nice, April, 1998 (abstract).

_____. 1998b. Earth and Venus: Moon, subduction, fixation of CO_2. *EOS (Trans. Amer. Geophys. Un.)*, 79 (Suppl. Western Pacific Meeting, Taiwan, July 21–24), in pr.

_____. 1998c. Excitation of core magnetic fields; tidal vorticity input *EOS (Trans. Amer. Geophys. Un.)*, 79 (Suppl. San Francisco Meeting), in pr.

Bostrom, R. C. and V. Vali. 1968. Strains recorded on a high-magnification interferometric seismograph. *Nature* 220 (5171): 1018–1020.

Bostrom, R. C. and V. Vali. 1969. A seismograph to observe deviatoric strains. *Tr. Engin.* 20(4): 7–12.

Bostrom, R. C., K. K. Saar and D. A. Terry. 1983. Basin formation; the mass anomaly at the West Pacific margin. In: *Geodynamics of the Western Pacific-Indonesian region. AGV Geodynamics Ser.* 11:51–62.

Bott, M. H. P. and D. S. Dean. 1973. Stress diffusion from plate boundaries. *Nature*, 243: 339–341.

Bowman, J. R. 1988. Body wave attenuation in the Tonga subduction zone. *J. Geophys. Res.* 93(B3): 2125–2139.

Branagan, P. T., N. R. Warpinski, B. Engler and R. Wilmer. 1996. Measuring the hydraulic fracture–induced deformation of reservoirs and adjacent rocks employing a deeply buried inclinometer array. *SPE Ann. Techn. Conf. Colo. 6–9 October, 1996.* SPE, Box 833836, Rchardson, TX 75083.

Brey, G. and D. H. Green. 1977. Systematic study of liquidus phase relations in olivine melilitite at high pressure and the role of CO_2 in the Earth's upper mantle. *Contr. Miner. Petrol.* 55: 217–230.

Briden, J. C. and I. G. Gass. 1974. Plate movement and continental magmatism. *Nature*, 248: 650–653.

Brosche, P. and J. Wunsch. 1990. The solar torque–a leak for the angular momentum of the Earth-Moon system. In: *Earth's Rotation from Eons to Days* (Eds. P. Brosche and J. Sundermann); Springer, Berlin, pp 135–145.

Broten, N. W., T. H. Legg, J. L. Locke, C. W. McLeish, R. S. Richards, R. M. Chisholm, H. P. Gush, J. L. Yen and J. A. Galt. 1967. Long base–line interferometry, a new technique. *Science*, 156: 1592–1593.

Brown, E. W. 1925. Tidal oscillations in Halemaumau, the lava pit of Kilauea. *Am. J. Sci. Fifth Ser. vol. IX (whole number CCIX):* 95 –112.

Brown, J. M. and T. J. Shankland. 1981. Thermodynamic parameters in the Earth as determined from seismic profiles. *Geophys. J. Roy. Astron. Soc.* 66: 579–596.

Brune, J. N. 1991. Seismic source dynamics, radiation and stress. *Rev. Geophys.* 1991. Cntrbtns. in Seismology, U. S. National Report, IUGG, Vienna. 688–699.

Bullen, K. E. 1936. The variation of density and the ellipticities of strata of equal density within the Earth. *Mon. Not. Roy. Astron. Soc. Geophys. Suppl.* 3: 395.

Bunge, H.-P. and M. Richards. 1996. The origin of large–scale structure in mantle convection: effects of plate motions and viscosity stratification. *Geophys. Res. Lett.* 23(2): 298–299.

Burke, K. and J. T. Wilson. 1972. Is the African plate stationary? *Nature*, 239: 387–389.

Burke, K., W. S. F. Kidd and J. T. Wilson. 1973a. Plumes and concentric plume traces of the Eurasian plate. *Nature Phys. Sci.* 241: 128–129.

Burke, K., W. S. F. Kidd and J. T. Wilson. 1973b. Relative and latitudinal motion of Atlantic hot spots. *Nature* 245: 133–137.

Burke, K. and J. F. Dewey. 1974. Two plates in Africa during the Cretaceous? *Nature* 249: 313–316.

Burke, K. C. and J. T Wilson. 1976. Hot spots on the Earth's surface. *Sci. Amer.* August 1976, pp 46–57.

Bursa, M. 1990. The variation in J_2 and in the moments of inertia: satellite results and consequences for the angular momentum budget of the Earth-Moon-Sun system. In: *Earth's Rotation from Eons to Days* (Eds. P. Brosche and J. Sundermann); Springer, Berlin. Pp 52–57.

Buster, P. B. H. Hager and T. H. Jordan. 1995. Mantle convection experiments with evolving plates. *Geophys. Res. Lett.* 22(16): 2223–2226.

Cadek, O. and Y. Ricard. 1992. Toroidal/poloidal partitioning and global lithospheric rotation during Cenozoic time. *Earth Plan. Sci. Lett.* 109: 621–632.

Cahen, L. N., J. Snelling, J. Delhal and J. R. Vail. 1984a. Review and discussion of tectono-thermal events in Africa. Ch. 23 in: *The Geochronology and Evolution of Africa*. Clarendon Press. pp 512, 420–441.

_____. 1984b. Phanerozoic anorogenic igneous activity in Africa. Ch. 22 in: *The Geochronology and Evolution of Africa*. Clarendon Press. Pp 512, 375–419.

Cajori, F. 1962. *A History of Physics*. New York: Dover. 58–59.

Campbell, I. H. and S. R. Taylor. 1983. No water, no granites–no oceans, no continents. *Geophys. Res. Lett.* 10 (11): 1061–1064.

Caputo, M. 1965. The minimum strength of the Earth. *J. Geophys. Res.* 70: 55–963.

Carey, S. W. 1955. The orocline concept in *Geotectonics*. Pap. Proc. R. Soc. Tasmania, 89: 255–288.

_____. 1970. Australia, New Guinea and Melanesia in the current revolution in concepts of the evolution of the Earth. *Search*, 1: 78–189.

_____. 1958. A tectonic approach to continental drift. In: Continental Drift—A Symposium. Univ. of Tasmania, 177–355.

Carlowicz, M. 1995. New map of seafloor mirrors surface. *EOS (Trans. Amer. Geophys. Un.)* 76(44): 441–442.

Campbell, I. H. and S. R. Taylor. 1983. No water, no granites–no oceans, no continents. *Geophys. Res. Lett.* 10(11): 1061–1064.

Carnec, C., D. Massonet and C. King. 1996. Two examples of the use of SAR interferometry on displacement fields of small extent. *Geophys. Res. Lett.* 23(24): 3579–3582.

Carter, W. E., D. S. Robertson and J. R. Mackay. 1985. Geodetic radio interferometric surveying: applications and results. *J. Geophys. Res.* 90: 4577–4587.

Cartwright, D. E. and R. D. Ray. 1989. New estimates of oceanic tidal energy dissipation from satellite altimetry. *Geophys. Res. Lett.* 16(1): 73–76.

_____. 1990. Ocean tides from satellite altimetry. *J. Geophys. Res.* 95(C3): 3069–3090.

Cartwright, D. E., and R. D. Ray. 1991. Energetics of global ocean tides from Geosat altimetry. *J. Geophys. Res.* 96(C9): 16,897–16,912.

Cartwright, D. E., R. Spencer and J. M. Vassie. 1987. Pressure variations on the Atlantic Equator. *J. Geophys. Res.* 92(C1): 725–741.

Cartwright, D. E., R. Spencer, J. M. Vassie and P. L. Woodworth. 1988. The tides of the Atlantic Ocean. *Phil. Trans. Roy. Soc. London* A324: 513–563.

Cartwright, D. F. 1977. Oceanic tides. *Reps. Progr. Phy.* 40: 665–708.

Cavendish, H. 1798. *Phil. Trans. Roy. Soc. London* 88: 469.

Cazenave, A. S. Houry, B. Lago and K. Dominh, 1992. GEOSat–derived geoid anomalies at medium wavelength. *J. Geophys. Res.* 97: 7081–7096.

Cazenave, A. B. Parsons and P. Calcagno. 1995. Geoid lineations of 1000 km wavelength over the central Pacific. *Geophys. Res. Lett.* 22(2): 97–100.

Chadwick, J. 1962. *Ernest Rutherford: Collected Papers*. Vol. 1, New Zealand, Cambridge, Montreal. Cambridge: Cambridge Univ. Press.

Chandrasekhar, S. 1961. *Hydrodynamic and Hydromagnetic Stability.* Oxford: Clarendon Pr. Ch. 3.

———. Eddington: the most distinguished astrophysicist of his time. Ch. 6 in: *Truth and Beauty; Aesthetics and Motivation in Science.* Chicago: Univ. Of Chicago Press. Pp. 92–109.

Chapman, S. 1948. The main geomagnetic field. *Nature* 161: 462–464.

Chapman, M. G. and R. L. Kirk. 1996. A migratory mantle plume on Venus: implications for Earth? *J. Geophys. Res.* 101(B7): 15,953–15,967.

Chastel, Y. B., P. R. Dawson, H. R. Wenk and K. Bennett. 1993. Anisotropic convection with implications for the upper mantle. *J. Geophys. Res.* 98(B10): 17,757–17,771.

Cheney, E. S. 1996. Sequence stratigraphy and plate tectonic signficance of the Transvaal succession of southern Africa and its equivalent in Western Australia. *Precambrian Research* 79: 3–24.

Chinnery, M. A. 1961. The deformation of the ground around surface faults. *Bull. Seismol. Soc. Am.* 51: 355–372.

Choy, G. L. and J. L. Boatwright. 1995. Global patterns of radiated seismic energy and apparent atress. *J. Geophys. Res.* 100(B9): 18,205–18,228.

Christodoulides, D. C., D. E. Smith, R. G. Williamson and S. M. Klosko. 1988. Observed tidal braking in the Earth/Moon/Sun system *J. Geophys. Res.* 93(B6): 6216–6236.

Cohen, I. B. 1978. Isaac Newton's Papers and Letters on Natural Philosophy and related documents. Cambridge, MA: Harvard Univ. Press. 540 pp.

Coleman, P. J. and G. H. Packham. 1976. The Melanesian Borderlands and India–Pacific plates boundary. *Earth–Sci. Revs.* 12: 197–233.

Cormier, M. H. and K. C. Macdonald. 1994. East Pacific Rise 18 degrees–19 degrees S; asymmetric spreading and ridge reorientation by ultrafast migration axial discontinuities. *J. Geophys. Res.* 99(1): 543–564.

Corrieu, V., Y. Ricard and C. Froidevaux. 1994. Converting mantle topography into mass anomalies to predict the Earth's radial viscosity. *Phys. Earth Plan. Int.* 84: 3–13.

Courtillot, V. and J. Besse. 1987. Magnetic field reversals, polar wander, and core mantle coupling. *Science* 237: 1140–1147.

Cowan, D. S. and C. J. Potter, 1991. Transect B-B′, Juan de Fuca Spreading Ridge to Montana Thrust Belt. In: *Tectonic Section Displays; Centenial Continent–Ocean Displays, Geological Society of America. Sheet 1–Pacific Alaska Region* (R. C. Speed, Principal Compiler). Boulder, CO: Geological Society of America.

Cowie, P. A., C. H. Scholz, M. Edwards and A. Malinverno. 1993. Fault strain and seismic coupling on mid–ocean ridges. *J. Geophys. Res.* 98(B10): 17,911–17,920.

Creager, K. C. and T. H. Jordan. 1986. Slab penetration into the lower mantle beneath the Mariana and other island arcs of the northwest Pacific. *J. Geophys. Res.* 91: 3573–3589.

Creer, K. M. 1964. A reconstruction of the continents for the upper Palæozoic from palæomagnetic data. *Nature*, 203: 1115–1120.

———. 1965. Paleomagnetic data from the Gondwanic continents. *Phil. Trans. Roy. Soc. London* A258: 27–40.

———. 1970. A review of paleomagnetism. *Earth–Sci. Rev.* 6: 369–466.

Creer, K., M. E. Irving and S. K. Runcorn. 1957. Geophysical interpretation of paleomagnetic directions from Great Britain. *Phil. Trans. Roy. Soc. London* A250: 144–156.

Crook, K. A. W. 1978. Stage maps to illustrate the development of the Southwest Pacific, 90 m.y. to present; a consequence of Earth rotation? *Bull. Austral. Soc. Explr. Geophys.* 9(3): 152–156.

_____. 1980. The origin of west Pacific–type geosynclines by zonal spreading. *Tectonophysics* 63: 235–259.

Crook, K. A. W., B. Taylor, N. F. Exon and R. W. Johnson. 1987. Woodlark triple–junction structure and tectonics: arc volcanoes on an incoming oceanic plate. *EOS (Trans. Amer. Geophys. Un.)*, 68(14): 1445.

Curie, M. 1904. *Recherches sur les substances radioactives* (Thesis, Faculte des Sciences de Paris) Paris: Gauthier–Villars. pp. 155.

Curray, J. R. and Tissa Munasinghe. 1992. Reply (In re Kent et al. 1992): *Geology*, Oct. 1992, 958–959.

Daly, R. A. 1938. *Architecture of the Earth.* Appleton–Century, N. Y.

_____. 1951. Relevant facts and inferences from field geology, in: *Internal Constitution of the Earth* (B. Gutenberg, ed. New York: Dover), pp. 23–49.

Danes, Z. F. 1973. Mainstream mantle convection: a geologic analysis of plate motion: discussion. *Bull. Amer. Assoc. Petrol. Geol.* 57: 410–411.

Darwin, Sir G. H. 1876. On the influence of geological changes on the Earth's axis of rotation. *London Edin. Dubl. Phil. Magaz. J. Sci.* 51: 328–333.

_____. 1878. On the bodily tides of viscous and semi-elastic spheroids, and on the ocean tides on a yielding nucleus. *Phil. Trans. Roy. Soc. London* 170: 1–25.

_____. 1879a. On the precession of a viscous spheroid, and on the remote history of the Earth. *Phil. Trans. Roy. Soc.* 170: 447–530.

_____. 1879b. Problems connected with the tides of a viscous spheroid. *Phil. Trans. Roy. Soc. London* 170: 539–593.

_____. 1908. Attempted evaluation of the rigidity of the Earth from the tides of long period. In: *Scientific Papers by Sir George Howard Darwin*, Vol. I. Cambridge Univ. Press: 340–346.

_____. 1908. Dynamical theory of the tides. In: *Scientific Papers by Sir George Howard Darwin*, Vol. I. Cambridge Univ. Press. Pp. 347–365.

Davies, G. F. 1984. Geophysical and isotopic constraints on mantle convection. *J. Geophys. Res.* 89(B7): 6017–6040.

_____. 1988. Ocean bathymetry and mantle convection. 1, Large–scale flow and hotspots. *J. Geophys. Res.* 93: 10,467–10,480.

Davis, E. E. and K. Becker. 1994. Formation temperatures and pressures in a sedimented rift hydrothermal system: 10 months of CORK observations, holes 857D and 858G. *Proc. Oc. Drilling Program: Sci. Results* 139: 649–666.

Davis, E. E., K. Becker, K. Wang and B. Carson. 1995. Long–term observations of pressure and temperature in hole 892B, Cascadia accretionary prism. *Proc. Oc. Drilling Program: Sci. Results* 146 (Pt. 1): 299–311.

Davis, J. L., T. A. Herring and I. I. Shapiro. 1991. Effects of atmospheric modeling errors on determination of baseline vectors from very long baseline interferometry. *J. Geophys. Res.* 96(B1): 643–650.

Davis, P. M., P. A. Rydelek, D. C. Agnew and A. T. Okamura. 1987. Observation of tidal tilt on Kilauea volcano, Hawaii. *Geophys. J. Roy. Astron. Soc.* 90: 233–244.

Dawson, J. B. 1970. The structural setting of African kimberlite magmatism. In: *African Magmatism and Tectonics* (T. N. Clifford and I. G. Gass, Eds.). Darien, Conn.: Hafner. 462 pp. ch. 15, 321–333.

Day, A. A. and S. K. Runcorn. 1955. Polar wandering. *Nature*, 176: 422–426.

De Mayer, F., and B. Ducarme. 1989. (unititled abstract) *Proc. 11th Intern. Symp. Earth Tides, Helsinki.* Stuttgart: E. Schweizerbart'sche Verlag. p. 30.

De Sitter, W. 1927. On the secular accelerations and the fluctuations of the longitudes of the moon, the sun, Mercury and Venus. *Bull. Astron. Inst. Netherlands.* 4: 21–38.

Dehant, V. 1985. Body tides for an elliptical rotating Earth with an inelastic mantle. *Proc. 10th Int. Symp. on Earth Tides, Madrid* pp. 367–377.

———. 1986. Tidal parameters for an inelastic Earth. *Phys. Earth Plan. Int.* 49: 97–116.

———. 1987a. Tidal parameters for an inelastic Earth. *Phys. Earth Plan. Int.* 49: 97–116.

———. 1987b. Integration of the gravitational motion equations for an elliptical uniformly rotating Earth with an inelastic mantle. *Phys. Earth Plan. Int.* 49: 242–258.

———. 1990. On the nutations of a more realistic Earth model. *Geophys. J. Intern.* 100: 477–483.

———. 1991. Review of the Earth tidal models and contribution of Earth tides in geodynamics. *J. Geophys. Res.* 96(B12): 20,235–20,240.

———. 1992. Conclusions drawn during the meeting of the Working Group on Theoretical Tidal Model. *Bull. Inf. Mar. Terr.* 115: 8419–8422.

———. 1993. Recommendations of the WG on "Theoretical Tidal Model." *Bull. Inf. Cent. Mar. Terr.* 30 Sept. 1993, p. 8716.

———. 1995. Concerning the definition of Love numbers. *Bull. Inf. Mar. Terr.* 121: 9027–9028.

Dehant, V. and J. Zschau. 1989. The effect of mantle inelasticity on tidal gravity: a comparison between the spherical and elliptical Earth model. *Geophys. J.* 97: 549–555.

Dehant, V. B. Ducarme, G. Jentzsch, J. Kaariainen, G. Y. Li, S. M. Molodensky, S. Okubo, J. M. Wahr, X. Qin-Wen and J. Zschau, 1991. Report of the working group on Theoretical Tidal Model. *Proc. 11th Int. Symp. on Earth Tides, Helsinki*, 533–547.

Dehlinger, P. 1971. Tidal friction and plate tectonics. *Symp. Ocean Floor Spreading, IUGG Intern. Meeting, Moscow* Abstr. v. 4, p. 34.

Delikaraoglou, D. and R. R. Steeves. 1985. The impact of VLBI and GPS on geodesy in Canada. *Proc. First Intern. Symp. Precise Positioning with GPS, Rockville, Maryland*; vol. II. (Cnvnr. C. C. Goad). Nat. Geod. Inf. Center, Rockville, MD. 743–752.

DeMets, C. R., G. Gordon, D. F. Argus and S. Stein. 1990. Current plate motions. *Geophys. J. Intern.* 101: 425–478.

Denis, C. 1989. The hydrostatic figure of the Earth. In: Gravity and Low-Frequency Geodynamics (ed. R Teisseyre); Elsevier. Pp. 171–185.

Denis, C., and P. Varga. 1990. Tectonic consequences of the Earth's variable rotation on geological time scales. In: *Earth's Rotation from Eons to Days* (P. Brosche and J. Sundermann, Eds.) Berlin: Springer. Pp. 146–161.

Desonie, D. L., R. A. Duncan, and J. H. Natland. 1993. Temporal and geochemical variability of volcanic products of the Marquesas hotspot. *J. Geophys. Res.* 98(10): 17,649–17,665.

Dewey, J. F., and J. M. Bird. 1972. Mountain belts and the new global tectonics. *J. Geophys. Res.* 75(14): 257–279.

Dewey, J. F. 1980. Episodicity, sequence, and style at convergent plate boundaries. In: The Continental Crust and Its Mineral Deposits, *Geol. Assoc. Can. Sp. Paper* 20 (Ed. D. W. Strangway), 553–573.

Dicke, R. H. 1966. The secular acceleration of the Earth's rotation and cosmology. In: *The Earth–Moon System*, (B. G. Marsden and A. G. W. Cameron, Eds.) New York: Plenum. P. 98.

———. 1969. Average acceleration of the Earth's rotation and the viscosity of the deep mantle. *J. Geophys. Res.* 74(25): 5895–5902.

Dickinson, W. R. 1978. Plate tectonic evolution of north Pacific rim. In: *Geodynamics of the Western Pacific* (S. Uyeda, R. W. Murphy and K. Kobayashi, Eds.). *Adv. in Earth Plan. Sci. 6:* Japan: Centr. Acad. Pub., pp. 1–19.

Dickman, S. R. 1979. Continental drift and true polar wander. *Geophys. J. Roy. Astron. Soc.* 57: 41–50.
Dickman, S. R. and Y. S. Nam. 1998. Constraints on Q at long periods in the Earth's interior. *Geophys. Res. Lett.* 25(2): 211–214.
Dietz, R. S. 1961. Continent and ocean basin evolution by spreading of the sea floor. *Nature.* 190: 854–857.
Doglioni, C. 1990. The global tectonic pattern. *J. Geodynamics.* 12: 21–38.
_____. 1991. Basal lithosphere detachment, eastward mantle flow and Mediterranean geodynamics: a discussion. *J. Geodynamics* 13(11): 47–65.
_____. 1993. Geological evidence for a global tectonic polarity. *J. Geol. Soc.* [London] 150(5): 991–1002.
Doglioni, C., I. Moretti and F. Roure. 1991. Basal lithosphere detachment, eastward mantle flow and Mediterranean geodynamics: a discussion. *J. Geodynamics*, 13(1): 47–65.
Doodson, A. T. 1921. The harmonic development of the tide–generating potential. *Proc. Roy. Soc. Lond.* A100: 305–329.
Dorman, J., M. Ewing and J. Oliver. 1960. Study of shear-velocity distribution in the upper mantle by mantle Rayleigh waves. *Bull. Seis. Soc. Amer.* 50: 87–115.
Dratler, J. W., E. Farrell, B. Block and F. Gilbert. 1971. High–Q overtone modes of the Earth. *Geophys. J. Roy. Astron. Soc.* 23: 399–410.
Du, P. 1993. Modulation effect of the magma tide in the asthenosphere: preliminary exploration for the cause of the 18.6-yr seismic cycle. *Earthq. Res. China* 7(4): 397–401.
_____. 1994. 18.6-year seismic cycle and the preliminary exploration for its cause. *Acta Geophys. Sin.* 37(3): 362–369.
_____. 1996. 18.6 years cycles of large intermediate and deep focus earthquakes in the world's major seismic regions. *Acta Geophys. Sin.* 39(3): 327–335.
Dubrovskiy, V. A. 1985. Tectonic waves. Izv. Akad. Nauk SSSR, *Fizika zemlii* no. 1 1985, 29–34.
Duncan, R. A. 1981. Hotspots in the southern oceans–an absolute frame of reference for motion of the Gondwana continents. *Tectonophysics* 74: 29–42.
_____. 1982. The New England Seamounts and the absolute motion of North America since mid–Cretaceous time. *EOS (Trans. Amer. Geophys. Un.)*, 63: 1103.
_____. 1984. Age progressive volcanism in the New England Seamounts and the opening of the central Atlantic Ocean. *J. Geophys. Res.* 89(B12): 9980–9990.
Duncan, R. A., N. Petersen and R. B. Hargraves. 1972. Mantle plumes, movement of the European plate, and polar wandering. *Nature*, 239: 2–96.
Durek, J. J. and G. Ekstrom. 1995. Evidence of bulk attenuation in the asthenosphere from recordings of the Bolivia earthquake. *Geophys. Res. Lett.* 22(16): 2309–2312.
Dziewonski, A. M. 1984. Mapping the lower mantle. Determination of lateral heterogeneity in P velocity up to degree and order 6. *J. Geophys. Res.* 89: 5929–5952.
_____. 1993. Negative velocity anomalies from mid–ocean ridges to CMB. *EOS (Trans. Amer. Geophys. Un.)*, 74 Suppl. p. 76.
Dziewonski, A. M., B. H. Hager and R. J. O'Connell. 1977. Large-scale heterogeneities in the lower mantle. *J. Geophys. Res.* 82: 239–255.
Dzurisin, D. 1980. Influence of fortnightly earth tides at Kilauea volcano, Hawaii. *Geophs. Res. Lett.* 7(11): 925–928.
Eanes, R. J. 1994. Diurnal and semi-diurnal tides from TOPEX/POSEIDON altimetry. *EOS (Trans. Amer. Geophys. Un.)*, 75(16): 108.
Eddington, Sir Arthur S. 1920. Space, Time and Graviatation. Cambridge Univ. Press.
_____. 1923. The borderland of astronomy and geology. *Nature*, no. 2775, v. 111: 18–21.
_____. 1958. The Nature of the Physical World. Ann Arbor: Univ. of Michigan Pr. 361 p.

Egbert, G. D., A. F. Bennet and M. G. G. Foreman. 1994. TOPEX/POSEIDON tides estimated using a global inverse model. *J. Geophys. Res.* 99(C12): 24,821–24,852.

Eggler, D. H. 1987. Discussion of recent papers on carbonated peridotite, bearing on mantle metasomatism and magmatism: an alternative. *Earth Plan. Sci. Lett.* 82: 398–400.

———. 1989. Kimberlites: how do they form? In: *Kimberlites and Related. Rocks*, vol. 1, p. 1. *Geol. Soc. Austral. Spec. Pub.* 14, (J. Ross, Ed): 489–504.

Ehrenfest, P. and T. Ehrenfest. 1959. *The Foundations of the Statistical Approach in Mechanics*; transl. by M. J. Moravcsik. Cornell Univ. Press. pp. 114.

Ekman, Martin. 1991. A concise history of the theories of tides, precession-nutation and polar motion (from antiquity to 1950). *Bull. Inf. Marees Terrestres* 109, 15 Jan. 1991, 7795–7848.

Elgered, G. J., L. Davis, T. A. Herring and I. I. Shapiro. 1991. Geodesy by radio interfemometry; water vapor radiometry for estimation of the wet delay. *J. Geophys. Res.* 96(B4): 6541–6555.

Elkana, Y. 1974. The Discovery of the Conservation of Energy. Cambridge, MA: Harvard Univ. Press. 175–197.

Elliot, D. H. 1987. Triassic–Early Cretaceous evolution of Antarctica. In: *Geological Evolution of Antarctica*. (M. R. A. Thomson, J. A. Crame and J. W. Thomson, Eds.) Cambridge Univ. Press. 541–548.

Elsasser, W. M. 1969. Convection and stress propagation in the upper mantle. In: *The Application of Modern Physics to the Earth and Planetary Interiors* (S. K. Runcorn Ed.); Wiley-InterScience. 223–246.

Emiliani, C. 1955. Pleistocene temperatures. *J. Geol.* 538–578.

———. 1966. Paleotemperature analysis of Caribbean cores P 6304–8 and P 6304–9 and a generalized temperature curve for the last 425,000 yrs. *J. Geol.* 63: 538–578.

Emiliani, C. and J. Geiss. 1957. On glaciations and their cause. *Geol. Rundsch.* 46: 576–601.

Engebretson, D. C., A. Cox and R. G. Gordon. 1984. Relative motions between oceanic plates of the Pacific basin. *J. Geophys. Res.* 89(B12): 10,291–10,310.

Epp, D. 1984. Possible perturbations to hotspot traces and implications for the origin and structure of the Line Islands. *J. Geophys. Res.* 89(B13): 11,273–11,286.

Fairhead, J. D. 1988. Mesozoic plate tectonic reconstructions of the central South Atlantic Ocean: the role of the West and Central African rift system. *Tectonophysics* 155: 181–191.

Farrell, W. E. 1972. Deformation of the Earth by surface loads. *Rev. Geophys. Sp. Phys.* 10(3): 761–797.

———. 1973. Earth tides, ocean tides and tidal loading. *Phil. Trans. Roy. Soc. London* A274: 253–259.

Ferrel, W. 1864. On the influence of the tides in causing an apparent secular acceleration of the Moon's mean motion. *Proc. Amer. Acad. Arts Sci.* 6: 379–383.

Ferziger, J. H. and M. Peric. 1996. *Computational Methods for Fluid Dynamics*. Springer–Verlag, New York. Pp. 356.

Fisher, O. 1889. *Physics of the Earth's Crust*, London, Macmillan, 2nd Ed. 1889. Cited by W. Sullivan. 1976, q.v.

Fjeldskaar, W. 1994. Viscosity and thickness of the asthenosphere detected from the Scandinavian uplift. *Earth Plan. Sci. Lett.* 126: 399–410.

Fjeldskaar, W. and L. Cathles. 1991. Rheology of mantle and lithosphere inferred from postglacial uplift in Fennoscandia. *NATO ASI Ser. C*, 334: 1–19.

Flanagan, M. P. and D. A. Wiens. 1994. Radial upper mantle attenuation structure of inactive back arc basins from differential shear wave measurements. *J. Geophys. Res.* 99(B8): 15,469–15,485.

Florsch, N., F. Chambat, J. Hinderer and H. Legros. 1994. A simple method to retrieve the complex eigenfrequency of the Earth's nearly diurnal free wobble; application to the Strasbourg superconducting gravimeter data. *Geophys. J. Roy. Astron. Soc.* 116: 53–63.

Forsyth, D. and S. Uyeda. 1975. On the relative importance of the driving forces of plate motion. *Geophys. J. Roy. Astron. Soc.* 43: 163–200.

Forte, A. M., A. M. Dziewonski and R. L. Woodward. 1993. Aspherical structure of the mantle, tectonic plate motions, nonhydrostatic geoid and topography of the core-mantle boundary. In: *Dynamics of Earth's Deep Interior and Earth Rotation. AGU Monograph* 72: 135–163.

Forte, A. M. and W. R. Peltier. 1991a. Viscous flow models of global geophysical observables. I, forward problems. *J. Geophys. Res.* 96: 20,131–20,159.

Forte, A. M. and W. R. Peltier. 1991b. Mantle convection and core-mantle boundary topography: explanations and implications. *Tectonophysics* 187(1–3): 91–116.

Forte, A. M., R. L. Woodward and A. M. Dziewonski. 1994. Joint inversions of seismic and geodynamic data for models of three–dimensional mantle heterogeneity. *J. Geophys. Res.* 99(B11): 21,857–21,877.

Forte, A. M., W. R. Peltier and A. M. Dziewonski. 1991. Inferences of mantle viscosity from tectonic plate velocities. *Geophys. Res. Lett.* 18(9): 1747–1750.

Forte, A. M., W. R. Peltier, A. M. Dziewonski and R. L. Woodward. 1993. Reply to comment by M. Gurnis. *Geophys. Res. Lett.* 20(15): 1665–1666.

Forte, A. M. and J. X. Mitrovica. 1996. New inferences of mantle viscosity from joint inversion of long wavelength mantle convection and postglacial rebound data. *Geophys. Res. Lett.* 23(10): 1147–1150.

Fotheringham, J. 1920. Secular accelerations of sun and moon as determined from ancient lunar and solar eclipses, occultations, and equinox observations. *Mon. Not. Roy. Astron. Soc.* 80: 578–581.

Fowler, A. C. and S. B. G. O'Brien. 1996. A mechanism for episodic subduction on Venus. *J. Geophys. Res.* 101(E2): 4755–4763.

Francis, G. and P. Massega. 1990. Global charts of ocean tide loading effects. *J. Geophys. Res.* 95 (C7): 11,411–11,424.

Francis, O. 1992. Interactions between earth and oceanic tides. *Bull. Inf. Centr. Mar. Terr.* 112: 8131–8144.

Francis, O. and V. Dehant. 1987. Recomputation of the the Green's functions for tidal loading estimations. *Bull. Inf. Cent. Mar. Terr.* 100: 6962–6986.

Francis, T. D. J. 1968. Seismicity of mid-oceanic ridges and its relation to properties of the upper mantle and crust. *Nature* 220: 899–901.

———. 1974. A new interpretation of the 1963 Fernandina Caldera collapse and its implications for the mid-oceanic ridges. *Geophys. J. Roy. Astron. Soc.* 39: 301–318.

Frohlich, C. 1987. Kiyoo Wadati and early research on deep focus earthquakes: introduction to special section on deep and intermediate focus earthquakes. *J. Geophys. Res.* 92(B13): 3,777–13,788.

———. 1989. The nature of deep–focus earthquakes. *Ann. Rev. Earth Planet. Sci.* 17: 227–254.

Fukao, Y., K. Tanabe and Y. Ogata. 1990. Whole mantle P–wave travel time tomography. *Phys. Earth Plan. Int.* 59: 294.

Furukawa, Y. 1993. Depth of the decoupling interface and thermal structure under arcs. *J. Geophys. Res.* 98(B11): 20,005–20,013.

Gable, C. W., R. J. O'Connell and B. J. Travis. 1991. Convection in three dimensions with surface plates: generation of toroidal flow. *J. Geophys. Res.* 96: 8391–8405.

Garfunkel, K. 1975. Growth, shrinking, and long–term evolution of plates and their implications for the flow pattern in the mantle. *J. Geophys. Res.* 80: 4425–4432.

Garthe, C. 1852. *Foucault's Versuch als direkter Beweis der Achsendrehung der Erde.* Wiesbaden: Martin Sandig oHG. Reprinted 1969 in Unveranderter Neudruck der Ausgabe von 1852, (Germany), Titel–Nummer 2133.

Gasparik, T. 1993. The role of volatiles in the transition zone. *J. Geophys. Res.* 98(B3): 4287–4299.

Gebhart, B. 1962. Effects of viscous dissipation in natural convection. *Fluid Mech.* 14: 225–232.

Giardini, D. and P. Lundgren. 1995. The June 9 Bolivia and March 9 Fiji deep earthquakes of 1994: II. Geodynamic implications. *Geophys. Res. Lett.* 22(16): 2281–2284.

Gilbert, F. and A. M. Dziewonski. 1975. An application of normal mode theory to the retrieval of structural parameters and source mechanisms from seismic spectra. *Phil. Trans. Roy. Soc. London* A278: 187–269.

Giunchi, C., G. Spada and R. Sabadini. 1996. Lateral viscosity variations and post–glacial rebound: effects on present–day VLBI baseline deformations. *Geophys. Res. Lett.* 24(1): 13–16.

Glatzmaier, G. A. and G. Schubert. 1993. Three–dimenional spherical models of layered and whole mantle convection. *J. Geophys. Res.* 98(12): 21,969–21,976.

Gold, T. 1955. Instability of the Earth's axis of rotation. *Nature*, 175: 526–529.

———. 1967. Radio method for the precise measurement of the rotation period of the Earth. *Science*, 157: 302–304.

Goldreich, P. and A. Toomre. 1969. Some remarks on polar wandering. *J. Geophys. Res.* 74: 2555–2567.

Gordon, R. B. 1965. Diffusion creep in the Earth's mantle. *J. Geophys. Res.* 70(10): 2413–2418.

Gordon, R. G. 1983. Late Cretaceois apparent polar wander of the Pacific plate: evidence for a rapid shift of the Pacific hotspots with respect to the spin axis. *Geophys. Res. Lett.* 10: 709–712.

———. 1987. Polar wandering and paleomagnetism. *Ann. Rev. Earth Plan. Sci.* 15: 567 593.

———. 1995. Plate motions, crustal and lithospheric mobility, and paleomagnetism: prospective viewpoint. *J. Geophys. Res.* 100(B12): 24,367–24,392.

Gordon, R. G. and C. D. Cape. 1981. Cenozoic latitudinal shift of the Hawaiian hotspot and its implications for true polar wander. *Earth Plan. Sci. Lett.* 55: 37–47.

Gordon, R. G. and D. M. Jurdy. 1986. Cenozoic plate motions. *J. Geophys. Res.* 91(B12): 12389–12406.

Green, D. H. and R. F. Cooper. 1993. Dilatational anelasticity in partial melts: viscosity, attenuation and velocity dispersion. *J. Geophys. Res.* 98(B11): 19,807–19,817.

Green, D. H., R. R. Cooper and S. Zhang. 1990. Attenuation spectra of olivine basalt partial melts: transformation of Newtonian creep response. *Geophys. Res. Lett.* 17(12): 1997–2100.

Griffin, V. S. 1970. Relevancy of the Dewey-Bird hypothesis of cordilleran type mountain belts and the Wegmann stockwork concept. *J. Geophys. Res.* 75:(35): 7504–7507.

Griggs, D. T. 1939. A theory of mountain building. *Am. J. Sci.* 237: 11–650.

———. 1972. The sinking lithosphere and the focal mechanism of deep earthquakes. In: *The Nature of the Solid Earth* (E. C. Robertson, J. F. Hays and L. Knopoff, Eds.) New York: McGraw–Hill; p. 361–384.

Gripp, A. E. and R. G. Gordon. 1990. Current plate velocities relative to the hotspots incorporating the Nuvel-1 global plate motion model. *Geophys. Res. Lett.* 17(8): 1109–1112.

Groten, E. and J. Brennecke. 1973. Global interaction between earth tides and sea tides. *J. Geophys. Res.* 78: 8519–8526.

Groten, E., H. Leonhardt and S. M. Molodensky. On the period and damping of polar motion. *J. Geophys. Res.* 96(12): 20,241–20,256.

Gudmundsson, O. B., L. N. Kennett and A. Goody. 1994. Broadband observations of upper mantle seismic phases in northern Australia and the attenuation structure in the upper mantle. *Phys. Earth Plan. Int.* 84: 207–226.

Guegen, Y. and J. M. Mercier. 1973. High attenuation and the low–velocity zone. *Phys. Earth Plan. Int.* 7: 39–46.

Gurnis, M. and G. F. Davies. 1986. Mixing in numerical models of mantle convection incorporating plate kinematics. *J. Geophys. Res.* 91(B6): 6375–6395.

Gutenberg, B. 1914. Uber Erdbebenwellen, VIIA. Beobachtungen an Registrierungegn von Fernbeben in Gottingen. *Nachr Gesell. Wiss. Gottingen, Nachrichten Math.–physik.* Kl. 1–52.

———. 1951. *Internal Constitution of the Earth.* New York: Dover. Pp. 208–213.

———. 1926. Untersuchungen zur Frage, bis zu welcher Tiefe die Erde kristallin ist. *Zeitschr. f. Geophysik*, 2: 4–29.

———. 1936. Structure of the Earth's crust and the spreading of the continents. *Geol. Soc. Amer. Bull.* 47: 587–1610.

———. 1948. On the layer of relatively low velocity at a depth of about 80 kms. *Bull. Seis. Soc. Amer.* 38: 21–148.

Gutenberg, B. and C. F. Richter. 1951. Structure of the crust: continents and oceans, in: *Internal Constitution of the Earth.* New York: Dover. Pp. 314–339.

Haas, R. and H. Schuh. 1996. Determination of frequency dependent Love and Shida numbers from VLBI data. *Geophys. Res. Lett.* 23: 1509–1512.

Hager, B. H. 1984. Subducted slabs and the geoid: constraints on mantle geology and flow. *J. Geophys. Res.* 89: 6003–6015.

———. 1991. Mantle viscosity: a comparison of models from postglacial rebound and from the geoid, plate driving forces and advected heat flux. In: *Glacial Isostasy, Sea–Level, and Mantle Rheology* (R. Sabadini et al., eds.). Kluwer. Pp. 493–513.

Hager, B. H. and R. W. Clayton. 1989. Constraints on the structure of mantle convection using seismic observations, flow models and the geoid. In: *Mantle Convection; Plate Tectonics and Global Dynamics.* New York: Gordon and Breach. Ch. 9; 657–763.

Haggerty, S. E. 1989. Mantle metasomes and the kinship between carbonatites and kimberlites. In: *Carbonatites: Genesis and Evolution.* K. Bell, ed. Allen and Unwin. Ch. 21: 546–557.

Hales, A. L. and S. Bloch. 1969. Upper mantle structure: are the low–velocity layers thin? *Nature*, 221: 930–933.

Halley, E. 1697. The true theory of the tides. *Phil. Trans. Roy. Soc. London* no. 226: 45–457; facsimile in Cohen, (1978), q.v.

Halliday, A., M. Rehkamper, L. Der-Chuenand and Y. Wen. 1996. Early evolution of the Earth and Moon: new constraints from Hf-W isotope geochemistry. *Earth Plan. Sci. Lett.* 142(1–2): 75–89.

Hamilton, W. L. 1973. Tidal cycles of volcanic eruptions: fortnightly to 19 yearly periods. *J. Geophys. Res.* 78(17): 3363–3375.

Hanada, H. 1988. Deformation of the viscoelastic earth due to the secular change in the Earth's axis of rotation. *Geophys. J.* I 95: 315–321.

Hara, T., K. Kuge and H. Kawakatsu. 1995. Determination of the isotropic component of the 1994 Bolivia deep earthquake. *Geophys. Res. Lett.* 22(16): 2265–2268.

Harabaglia, P. and C. Doglioni. 1998. Topography and gravity across subduction zones. *Geophys. Res. Lett.* 25(5): 703–706.

Harris, C., A. J. Erlank, A. R. Duncan and J. S. Marsh. 1987. The geochemistry of the Kirwan and other Jurassic basalts of Dronning Maud Land, and their significance for Gondwana reconstruction. In: *Geological Evolution of Antarctica* (M. R. A. Thomson, J. A. Crame and J. W. Thomson, eds.). Cambridge Univ. Press. 563–571.

Harrison, J. C. 1978. Implication of cavity, topographic and geologic influences on tilt and strain measurements. *Ninth Geodesy/Solid Earth and Ocean Physics Research conference*. Report 280, Dept. of Geod. Sci. Columbus, OH: Ohio State Univ.

Haskell, N. W. 1936. The motion of a fluid under a surface load, II. *Physics* 7: 56–61.

Haxby, W. F. and J. K. Weissel. 1986. Evidence for small scale mantle convection from Seasat altimeter data. *J. Geophys. Res.* 91(B3): 3507–3520.

Hays, J. D., J. Imbrie and N. J. Shackleton. 1976. Variations in the Earth's orbit: pacemaker of the ice ages. *Science* 194(4270): 1121–1132.

Heezen, B. C. and Marie Tharp. 1969. Map, *Pacific Ocean Floor*. Mercator projection; 5. 75 miles/inch at Equator. Washington, DC: Nat. Geogr. Soc.

Heezen, B. C. and D. J. Fornari. 1976. Pacific Ocean, in: *Geological World Atlas* (G. Choubert, ed.). Paris: UNESCO. 1976.

Heinz, D. R., R. Jeanloz and R. J. O'Connell. 1982. Bulk attenuation in a polycrystalline Earth. *J. Geophys. Res.* 7772–7778.

Heiskanen, W. 1921. Uber den einfluss der gezeiten auf die sakulare acceleration des Mondes. *Ann. Acad. Sci. Fennicae*, A18: 1–84.

Helmholtz, H. L. F. von, 1847. *Uber die Erhaltung der Kraft*. Read before Phys. Soc. of Berlin, 23 July 1847. Account and critique in Elkana 1974, 175–197.

———. 1858. On integrals of the hydrodynamical equations, which express vortex-motion. English trnsltn. by Prof. Tait, reviewed by Helmholtz: *London Edin. Dubl. Phil. Magaz. J. Sci. and J. Sci.*, Suppl. vol. XXXIII, Fourth Ser. (1867), 483–511; his concept as formulated by Kelvin is well presented in Serrin 1959, q.v.

Helmstaedt, H. H. 1995. Geotectonic controls of primary diamond deposits: implications for area selection. *J. Geochem. Explor.* 53(1–3): 125–144.

Hendershott, M. C. 1972. The effects of solid–Earth deformation on global ocean tides. *Geophys. J. Roy. Astron. Soc.* 29: 389–402.

Henderson, L. J. and R. G. Gordon. 1982. Fixed hotspots and recurrent volcanism along the Line Islands chain. *GSA Abstrs. w. Prgrm.* 14: 513.

Henriksen, S. W. 1960. *Ann. Int. Geophys. Year* 12: 86–192.

Herrick, R. R. 1994. Resurfacing history of Venus. *Geology* 22: 703–706.

Herring, T. A. 1988. VLBI studies of the nutations of the Earth. In: *The Impact of VLBI on Astrophysics and Geophysics* (J. M. Moran and M. J. Reid, eds). Dordrecht: Kluwer. 371–375.

Herring, T. A. and D. Dong. 1994. Measurement of diurnal and semidiurnal rotational variations and tidal parameters of Earth. *J. Geophys. Res.* 99(B9): 18,051–18,071.

Herring, T. A. I. I. Shapiro and T. A. Clark. 1986. Geodesy by radio interferometry. *J. Geophys. Res.* 91: 8341–8347.

Herring, T. A., J. L. Davis and I. I. Shapiro. 1990. Geodesy by radio interferometry: the application of Kalman filtering to the analysis of Very Long Baseline Interferometry data. *J. Geophys. Res.* 95(B8): 12,561–12,581.

Herron, E. M. 1972. Seafloor spreading and the Cenozoic history of the East-Central Pacific. *Bull. Geol. Soc. Amer.* 83(6): 1671–1691.

Herron, E. M. and B. E. Tucholke. 1976. Seafloor magnetic patterns and basement structure in the south-eastern Pacific, in: *Rep. Int. Seafloor Drg. Prgrm.* 35 (C. D. Hollister and C. Craddock, eds.). Pp. 263–278.

Hide, R. and S. R. C. Malin. 1970. Novel correlations between global features of the Earth's gravitational and magnetic fields. *Nature* 225: 605–609.

Hilgen, F. J., W. Krijgsman, C. G. Langereis, L. J. Lourens, A. Santarelli and W. J. Zachariasse. 1995. Extending the astronomical (polarity) scale into the Miocene. *Earth Plan. Sci. Lett.* 136: 495–510.

Hill, R. I., I. H. Campbell, G. F. Davies and R. W. Griffiths. 1992. Mantle plumes and continental tectonics. *Science* 256: 186–193.

Hinderer, J. and H. Legros. 1989. Gravity perturbations of annual period. *Proc. 11th Intern. Symp. Earth Tides, Helsinki* (Stuttgart: E. Schweizerbart'sche Verlagsbuchhandlung). 425–429.

Holmes, A. 1928a. Radioactivity and continental drift. Proc. *Glasgow Geol. Soc.* 12 January 1928. Reprinted in: *Geol. Mag.* LXV: 236–238 (1928). See also appreciation by H. D. Hedberg in: *Proc. Geol. Soc. Amer.* (1957), pp. 69–76: *Presentation of Penrose Medal to Arthur Holmes.*

———. 1928b. Continental Drift. *Nature* 122: 431–433.

———. 1944. *Principles of Physical Geology*; Edinburgh: Nelson, 532 p. Ch. XXI: *Continental Drift.*

Holzhausen, G. R. 1997. Specimen trace transmitted to RCB with letter dated January 20, 1997.

Honda, S., and S. Uyeda. 1983. Thermal process in subduction zones—a review and preliminary approach on the origin of arc volcanism. In: *Arc Volcanism: Physics and Tectonics* (D. Shimozuro and I. Yokoyama, eds.). Reidel: Terra Sci. Pubs. Pp. 117–140.

Hopkins, W. 1839. Researches in physical geology. *Phil. Trans. Roy. Soc. Lond.* 129: 81–385.

Hoskins, L. M. 1920. The strain of a gravitating sphere of variable density and elasticity. *Trans. Amer. Math. Soc.* 21: 1–43.

Hough, S. S. 1897. On the application of harmonic analysis to the dynamic theory of the tides, Pt. I. *Phil. Trans. Roy. Soc. London* A189: 201–257.

———. 1898. On the application of harmonic analysis to the dynamic theory of the tides, Pt. II. *Phil. Trans. Roy. Soc. London* A191: 40–185.

Howell, B. F. Jr. 1970. Coriolis force and the new global tectonics. *J. Geophys. Res.* 75(14): 2769–2772.

Ida, Y. 1983. Thermal and mechanical processes producing arc volcanism and back-arc spreading. In: *Arc Volcanism: Physics and Tectonics* (D. Shimozuro and I. Yokoyama, eds.). Reidel: Terra Sci. Pubs. 165–175.

Inglis, D. R. 1957. Shifting of the Earth's axis of rotation. *Revs. Mod. Phys.* 29(1): 9–19.

Irifune, T., N. Kubo, M. Isshiki and Y. Yamasaki. 1998. Phase transformations in serpentine and transportation of water into the lower mantle. *Geophys. Res. Lett.* 25(2): 203–206.

Irving, E. and W. S. Robertson. 1969. Test for polar wandering and some possible implications. *J. Geophys. Res.* 74: 026–1036.

Irving, E. 1964. *Paleomagnetism and its Application to Geological and Geophysical Problems*. New York: Wiley. 399 pp.

———. 1977. Drift of the major continental blocks since the Devonian. *Nature*, 270: 304–309.

Ito, E., and E. Takahashi. 1989. Postspinel transformations in the system Mg_2SiO_4–Fe_2SiO_4 and some geophysical implications. *J. Geophys. Res.* 94(B8): 10,637–10,646.

IUGG. 1995. IUGG *Chronicle 228*. 21st General Assembly, Plenary Session, pp. 245–252.

Ivanov, I. B. 1985. Delay in the Earth's rotation velocity. *C. R. Acad. Bulgar. Sci.* 38(1): 63 et seq.
Janse, A. J. A. 1995. Catalog of world-wide diamond and kimberlite occurrences: a selection and annotative approach. *J. Geochem. Explor.* 53(1–3): 73–111.
Jeffreys, Sir H. 1921. Tidal friction in shallow seas. *Phil. Trans. Roy. Soc. Lond.* 221A: 239–264.
_____. 1923. The movement of the Earth's surface crust. *London Edin. Dubl. Phil. Magaz. J. Sci.* v. 45, (Ser. 6): 1167–1188.
_____. 1924. *The Earth.* Cambridge Univ. Press. His 14.424, p. 225, and subsequent formulations.
_____. 1926. On Prof. Joly's theory of Earth history. *London Edin. Dubl. Phil. Magaz. J. Sci.* v. 5, 7th Ser. LXXXII, 923–931.
_____. 1928. The times of transmission and focal depths of large earthquakes. *Mon. Not. Roy. Astron. Soc.* 1: 500–521.
_____. 1929. *The Earth.* Cambridge Univ. Press. Pp. 323–325.
_____. 1939. Seismological Tables. *Mon. Not. Roy. Astron. Soc.* 99: 397–408.
_____. 1952. *The Earth.* Third Ed. Cambridge Univ. Press. Ch. VI, Stress–Differences in the Earth: 185–200.
_____. 1954. Dynamics of the Earth–Moon system, in: *The Earth as a Planet* (G. P. Kuiper, ed.). Chicago: Chicago Univ. Press. Pp. 42–56.
_____. 1962. *The Earth.* Cambridge Univ. Press. Secs. 3.09, 3.10.
_____. 1963. On the hydrostatic theory of the figure of the Earth. *Geophys. J. Roy. Astron. Soc.* 8(2): 196–202.
_____. 1972. Creep in the Earth and planets. *Tectonophysics* 13: 569–581.
_____. 1976. *The Earth.* Cambridge Univ. Press. Sec. 11. 094, pp. 458–462.
_____. 1982. Tidal friction; the core; mountain and continent formation. *Geophys. J. Roy. Astron. Soc.* 71: 555–566.
Jekeli, C. and C. K. Shum. 1998. Local gravity field determination using GPS intersatellite velocities and accelerations. *EOS (Trans. Amer. Geophys. Un.),* 79 (24, Suppl.): W14.
Jiao, W., T. C. Wallace, S. L. Beck, P. G. Silver and G. Zandt. 1995. Evidence for static displacements from the June 9, 1994 deep Bolivian earthquake. *Geophys. Res. Lett.* 22(16): 2285–2288.
Johnson, H. O. and D. C. Agnew. 1995. Monument motion and measurements of crustal velocities. *Geophys. Res. Lett.* 22: 2905–2908.
Johnston, D. E. and C. A. Langston. 1984. The effect of assumed source structure on inversion of earthquake source parameters; the eastern Hispaniola earthquake of 14 September 1981 *Earthq. Not.* 551, 15 p.
Joly, J. 1903. Radium and the geological age of the Earth. *Nature* 68(1770): 526.
_____. 1909. *Radioactivity and Geology.* London: Constable & Co. 167 pp.
Joly, J. 1925. *The Surface History of the Earth.* Oxford: Clarendon Press. Appendix to ch. 10, 180–184.
_____. 1928. The Earth's thermal history. *Phil. Magaz.* vol. V. (7th Ser.): 215–221.
Jones, G. M. and S. Gartner. 1980. Comments on "Pre-Tertiary velocities of the continents: a lower bound from paleomagnetic data" by R. G. Gordon, M. O. McWilliams and A. Cox. *J. Geophys. Res.* 85(B8): 4431–4432.
Jones, Sir H. S. 1939. The rotation of the Earth, and the secular accelerations of the sun, moon and planets. *Mon. Not. Roy. Astron. Soc.* 99: 41–558.
Jordan, T. H. 1974. Some comments on tidal drag as a mechanism for driving plate motions. *J. Geophys. Res.* 79: 2141–2142.
Jordan, T. H., P. Puster, G. A. Glatzmaier and P. J. Tackley. 1993. Comparisons between

seismic Earth structures and mantle flow modes based on radial correlation functions. *Science*, 261: 1427–1431.

Joseph, D. 1993. *A geophysical investigation of the Nova-Canton Trough; the key to the late Cretaceous evolution of the Central Pacific*. Honolulu, HI: Thes. Diss., Univ. of Hawaii at Manoa. 210 pp.

Jurdy, D. M. 1981. True polar wander. *Tectonophysics* 74: 1–16.

———. 1983. Early Tertiary subduction zones and hot spots. *J. Geophys. Res.* 88(B8): 6395–6402.

Kagan, B. A. and J. Sundermann. 1996. Dissipation of tidal energy, paleotides, and evolution of the Earth–Moon system. *Adv. Geophys.* 38: 179–266.

Kalinin, Yu. D. and V. M. Kiselev. 1980. Solar relationships and prediction of seismic activity on Earth. In: *Solar–Terrestrial Predictions Proceedings*, vol. 4. U. S. Dept. of Commerce, NOAA.

Kanamori, H. 1977. Seismic and aseismic slip along subduction zones and their tectonic implication. In: *Island Arcs, Deep–Sea Trenches and Back–Arc Basins, AGU Ewing Ser.* 1: 163–174.

———. 1978. Quantification of earthquakes. *Nature* 271: 411–414.

Kanamori, H. and F. Press. 1970. How thick is the lithosphere? *Nature* 226: 30.

Kanamori, H. and G. S. Stewart. 1976. Mode of the strain release along the Gibbs fracture zone, Mid–Atlantic Ridge. *Phys. Earth Plan. Int.* 11: 312–332.

Kant, I. 1839. "Untersuchung der Frage, ob die Erde in ihrer Umrehung um die Achse. einige Veranderung erlitten habe." In: *I. Kant's samtliche Werke*; Karl Rosenkranz and F. W. Schubeert, eds. Pt. 6, pp 3–12. (Reproduced by H. P. Munzenmayer, in: *Tidal Friction and the Earth's Rotation, II* [P. Brosche and J. Sunderman, eds.], Berlin: Springer-Verlag)

Karato, S.-I., and P. Wu. 1993. Rheology of the upper mantle: a synthesis. *Science*, 260: 771–778.

Katsura, T. and E. Ito. 1989. The system Mg_2SiO_4–Fe_2SiO_4 at high pressures and temperatures: precise determination of stabilities of olivine, modified spinel, and spinel. *J. Geophys. Res.* 94: 15,663–15,670.

Kaula, W. M. 1963. Elastic models of the mantle corresponding to variations in the external gravity field. *J. Geophys. Res.* 68: 4967–4978.

———. 1964. Tidal dissipation by solid friction and the resulting orbital evolution. *Rev. Geophys.* 2(4): 661–685.

———. 1967. Geophysical implications of satellite determinations of the Earth's gravitational field. *Space Sci. Rev.* 7: 769–794.

———. 1968. Reviewer's letter to RCB. I am greatly indebted to Prof. Kaula for formulating a separation of the interactions.

Keating, B., K. Kodama and C. E. Helsley. 1983. Paleomagnetic studies of the Bonin and Mariana island arcs. In: *Arc Volcanism: Physics and Tectonics* (D. Shimozuru and I. Yokoyama, eds.); Tokyo: Terra Sci. Pub. Co. Pp. 243–259.

Keating, B. H. 1987. Reactivation of vulcanism in the Line Islands seamount chain. *EOS (Trans. Amer. Geophys. Un.)*, 68:(44): 1451.

Kelvin, Ld. (Sir W. Thomson). 1849. On the *vis viva* of a fluid in motion. Cambridge and Dublin Math. J., iv; reprinted in *Mathematical and Physical Papers by Sir William Thomson*, vol. I, Cambridge Univ. Press, 1882, pp 107–111.

———. 1863a. On the secular cooling of the Earth. *London Edin. Dubl. Phil. Magaz. J. Sci. Ser.* 4, XXV.

———. 1863b. On the rigidity of the Earth. *London Edin. Dubl. Phil. Magaz. J. Sci. Ser.* 4, XXV: 149–151.

_____. 1869. On vortex motion. *Trans. Roy. Soc. Edin.* xxv: 217–260; his formulation is well presented in Serrin (1959), q.v.

_____. 1875a. General integration of Laplace's differential equation of the tides. *London Edin. Dubl. Phil. Magaz. J. Sci. Ser.* IV, L: 388–402.

_____. 1875b. On an alleged error in Laplace's Theory of the Tides. *London Edin. Dubl. Phil. Magaz. J. Sci. Ser.* IV, L: 227–242.

_____. 1881. On a changing tesselated structure in certain liquids. *Proc. Glasgow Phil. Soc.* 1881–1882. (Cited by Lord Rayleigh, 1916, q.v.)

_____. 1886. On the dynamical theory of the tides of long period. *Proc. Roy. Soc. Lond.* 41: 337–342.

_____. 1903. Contribution to discussion on the nature of the emanation from radium. *Brit. Assoc. Rep.* 1903: 535–537. Also: *Mathematical and Physical Papers by Sir William Thomson* (Lord Kelvin) vol. IV, Cambridge Univ. Press, paper 259: 206–209.

Kelvin, Lord, and P. G. Tait. 1895. *Treatise on Natural Philosophy.* Cambridge Univ. Press, 423–433.

Kennedy, W. Q. 1964. The structure differentiation of Africa in the pan-African (±500 m.y.) tectonic episode. In: *Leeds Univ. Research Inst. Afr. Geol. Ann. Rep. Sci. Results.* 8: 48–49.

Kent, R., W. M. Storey, A. D. Saunders, N. C. Ghose and P. D. Kempton. 1992. Origin of the Rajmahal traps and the 85 degrees E Ridge; preliminary reconstructions of the trace of the Crozet hotspot: Comment. *Geology* 20(10): 957–958.

Keondzhyan, V. P. and A. S. Monin. 1976. Calculations concerning the evolution of the interior of planets. *Bull. (Izv.) Acad. Sci. USSR, Earth Physics,* no. 4, (AGU Engl. transl.): 229–235.

Keondjian, V. P. and A. S. Monin. 1977. Calculations on the evolution of the planetary interiors. *Tectonophysics* 41: 227–242.

_____. 1977. Continental drift and large–scale wandering of the Earth's pole. *Izv. Physics of the Solid Earth* (AGU Engl. transl.): 13: 60–772.

Khramov, A. N., G. N. Petrova and D. M. Pechersky. 1981. Paleomagnetism of the Soviet Union. In: *Paleoreconstruction of the Continents* (M. W. McElhinny and D. A. Valencio, eds.), AGU Geodyn. Ser., vol. 2, 177–194.

Kilston, S. and L. Knopoff. 1983. Lunar-solar periodicities of large earthquakes in southern California. *Nature,* 304: 21–25.

King, S. D. 1994. Introduction to the special section on the transition zone. *J. Geophys. Res.* 99(B8): 15,779–15,782.

Kirschvink, J. L., R. L. Ripperton and D. A. Evans. 1997. Evidence for a large–scale reorganization of Early Cambrian continental masses by inertial interchange; True Polar Wander. *Science* 277: 541–545.

Klein, F. W. 1976. Earthquake swarms and the semidiurnal solid Earth tide. *Geophys. J. Roy. Astron. Soc.* 45: 245–295.

Klootwijk, C. T. and J. W. Peirce. 1979. India's and Australia's pole path since the late Mesozoic and the India-Asia collision. *Nature* 282: 605–607.

Knopoff, L. and A. Leeds. 1972. Lithospheric momenta and the deceleration of the Earth. *Nature* 237: 93–94.

Kohlstedt, D. L., B. Evans and S. J. Mackwell. 1995. Strength of the lithosphere: Constraints imposed by laboratory experiments. *J. Geophys. Res.* 100(B9): 17,587–17,602.

Kolsky, H. 1953. *Stress Waves in Solids.* Oxford: Clarendon Press. Ch. VII, 163–171.

Koenigsberger, L. 1906. *Hermann von Helmholtz.* Oxford: Clarendon Press; republished by Dover, New York. 1965. Excerpt from: Preface to the English Edition, by Lord Kelvin.

Koenigsberger, J. G. 1930. Ueber magnetische eigenshaft von gesteinen. *Terr. Magn.* 35: 145–148.

———. 1936. Residual magnetism and the measurement of geologic time. *Rep. Intern. Geol. Congr.* 225–231.

———. 1938. Natural residual magnetism of eruptive rocks. *Terr. Mag.* 43: 119–130.

Koppen, W. and A. Wegener. 1924. *Die Klimate der geologischen Vorzeit.* Fig. 28.5 in Holmes, A., *Principles of Physical Geology* (1978), q.v.

Kosygin, Yu. A. and L. A. Maslov. 1986. The role of lunar tides in the tectonic process. *Geotectonics* 20(6): 451–454.

———. 1989. On the physical fields of the rotating planets. *Geotectonics* 23(1): 5–9.

———. 1990. On the cosmic nature of tectonic processes. *Geotectonics* 24(5): 386–388.

Kovach, R. L. and D. L. Anderson. 1967. Study of the energy of the free oscillations of the Earth. *J. Geophys. Res.* 72(8): 2155–2170.

Kragh, H. 1993. Between Physics and Chemistry: Helmholtz's Route to a Theory of Chemical Thermodynamics. In: *Hermann von Helmholtz and the Foundations of Nineteenth Century Science*. Univ. of California Press (David Cahan, ed.), 414–415.

Kravetz, T. 1927. Ueber der Zusammenhung der Erdbeber mit den Polhohen–schwankungen. *Zeitschr. Geophys.* 3: 221–224.

Kreichgauer, D. 1926. *Die Aequatorfrage in der Geologie.* 2nd Ed. Haldenkirche. p. 55; cited in Daly (1951).

Krohn, J. and J. Sundermann. 1982. Paleotides before the Permian. In: *Tidal Friction and the Earth's Rotation, II.* (Eds. P. Brosche and J. Sundermann). Berlin: Springer. Pp. 190–209.

Krysinski, L. 1992. On the mathematical connections of planetary rotational deformations with Love's tidal problem. *Phys. Earth Plan. Int.* 72: 137–152.

Kuiper, G. P. 1955. Limits of completeness. In: *Planets and Satellites*; vol. III: The Solar System. Chicago: Chicago Univ. Press. Ch. 18.

Kukla, G. J. 1975. Missing link between Milankovitch and climate. *Nature* 253: 600–603.

Lagus, P. L. and D. L. Anderson. 1968. Tidal dissipation in the Earth and planets. *Phys. Earth Plan. Int.* 1: 505–510.

Lamb, H. H. 1970. Volcanic dust in the atmosphere; with a chronology and assessment of its meteorological significance. *Phil. Trans. Roy. Soc. London* A266: 425–533.

Lamb, Sir Horace. 1945. *Hydrodynamics.* Sixth Ed. New York: Dover. Text figures in Ch. 3, describing *irrotational* flow, pp. 31–35.

Lambeck, K. 1979. The history of the Earth's rotation. In: *The Earth: Its Origin, Structure and Evolution* (M. W. McElhinny ed.); London: Academic Press. 59–81.

———. 1988. *Geophysical Geodesy: The Slow Deformations of the Earth.* Oxford: Clarendon Press. 718 pp; esp. Ch. 11: *Tides and rotation of the Earth*, pp. 546–640.

———. 1991. A model for Devensian and Flandrian glacial rebound and sea-level change in Scotland. In: *Glacial Isostasy, Sea–Level and Mantle Rheology* (R. Sabadini et al., eds.). Kluwer. 33–60.

Lambeck, K., A. Cazenave and G. Balmino. 1974. Solid Earth and ocean tides estimated from satellite orbit analyses. *Revs. Geophys. Sp. Phys.* 12(3): 421–434.

Lambert, W. D. 1931. *Bull. U. S. Nat. Res. Council* no. 78; cited in Jeffreys, Sir H. 1976: *The Earth*, q.v.

Landau, L. D. and E. M. Lifshitz. 1959. *Fluid Mechanics.* London: Pergamon. 536 pp. Sections 7, 15, 53.

Langel, R. A. 1993. Thirty five years of satellite geomagnetism. *EOS (Trans. Amer. Geophys. Un.)*, 74(16), Suppl. p. 47.

Laplace, Marquis P.-S. de, !795. *Mécanique Céleste*. Paris. Livs. i, iv. Engl. transl.: N. Bowditch, 1829. Boston: Hilliard, Gray, Little and Wilkins.

Larsen, R. L., 1991. Latest pulse of Earth: evidence for a mid-Cretaceous superplume. *Geology*. 19: 547–550.

Laskar, J. 1989. A numerical experiment on the chaotic behavior of the solar system. *Nature* 338: 237–238.

———. 1994. Large–scale chaos in the solar system. *Astron. Astrophys*. 287: L9–L12.

Laskar, J., F. Joutel and F. Boudin. 1993. Orbital, precessional and insolation quantities for the Earth from −20 Ma to +10 Ma. *Astron. Astrophys*. 270: 522–523.

Laskar, J., F. Joutel and P. Robutel. 1993. Stabilization of the Earth's obliquity by the Moon. *Nature* 361: 15–617.

Lawver, L. A., J. G. Sclater and L. Meinke. 1985. Mesozoic and Cenozoic reconstructions of the South Atlantic. *Tectonophysics* 114: 233–254.

Le Provost, C., M. L. Genco, F. Lyard, P. Vincent and P. Canceil. 1994. Spectroscopy of the world ocean tides from a finite element hydrodynamic model. *J. Geophys. Res*. 99(C12): 24,777–24,797.

LeGrand, H. E. 1990. Rise and fall of paleomagnetic research at Carnegie's DTM. *EOS (Trans. Amer. Geophys. Un.)* 71: 043–1044.

Lenardic, A. and W. M. Kaula. 1994. Tectonic plates, D″ thermal structure, and the nature of mantle plumes. *J. Geophys. Res*. 99(B8): 15,697–15,708.

Leonard, A. 1997. Review: "Computational Methods for Fluid Dynamics" (Ferziger and Peric). *Phys. Today* 50(3): 80–84.

Levi, B. G. 1998. New measurements constrain models of mantle upwelling along a mid-ocean ridge. *Phys. Today* 51(7): 17–19.

Li, G.-Y. and H. T. Hsu. 1989. The tidal modeling theory with a lateral inhomogeneous, inelastic mantle. *Proc. Eleventh Intern. Symp. Earth Tides, Helsinki* (Stuttgart: E. Schweizerbarts'che Verlags): 601–611.

Lithgow-Bertelloni, C., M. A. Richards and Y. Ricard. 1992. Plate motions since 120 Ma: the history of subduction, net lithospheric rotation and toroidal-poloidal partitioning. *EOS (Trans. Amer. Geophys. Un.)* 73 (25, Suppl.): 66.

Lithgow-Bertelloni, C., M. A. Richards, Y. Ricard, R. J. O'Connell and D. C. Engebretson. 1993. Toroidal–poloidal partitioning of plate motions since 120 Ma. *Geophys. Res. Lett*. 20(5): 375–378.

Liu, H.-S. 1974. On the break-up of tectonic plates by polar wandering. *J. Geophys. Res*. 79(17): 2568–2572.

Liu, H.-P., D. L. Anderson and H. Kanamori. 1976. Velocity dispersion due to anelasticity: implications for seismology and mantle composition. *Geophys. J. Roy. Astron. Soc*. 47: 41–58.

Livermore, R. A., F. J. Vine and A. G. Smith. 1984. Plate motions and the geomagnetic field. *Geophys. J. Roy. Astron. Soc*. 79: 939–996.

Lliboutry, L. A. 1991. Mantle viscosity: what exactly are we looking for? In: *Glacial Isostasy, Sea–Level and Mantle Rheology* (R. Sabadini et al., eds.). Dordrecht: Kluwer. Pp. 321–341.

Lodge, Sir Oliver J. 1893. VI. Experiments on the absence of mechanical connexion between æther and matter. *Phil. Trans. Roy. Soc. London* A184: 149–166.

———. 1897. Aberration problems. A discussion concerning the motion of the æther near the Earth, and concerning the connexion between æther and gross matter; with some new experiments. *Phil. Trans. Roy. Soc. London* A189: 727–804.

Longman, I. M. 1963. A Green's function for determining the deformation of the Earth under surface mass loads; pt 2. *J. Geophys. Res*. 68(2): 485–496.

Lonsdale, P. 1988. Geography and history of the Louisville hotspot chain in the southwest Pacific. *J. Geophys. Res.* 93: 3078–3104.

Loper, D. E. and K. McCartney. 1986. Mantle plumes and the periodicity of magnetic field reversals. *Geophys. Res. Lett.* 13: 1525–1528.

Love, A. E. H. 1909. The yielding of the Earth to disturbing forces. *Proc. Roy. Soc. London* A–LXXXII: 73–88.

———. 1911. *Some Problems of Geodynamics.* Cambridge Univ. Press. Secs. 54–64.

———. 1927. *A treatise on the mathematical theory of elasticity.* (Fourth Ed. reprint: New York: Dover, 1944).

Lowe, D. R. 1994. Accretionary history of the Archean Barberton Greenstone Belt (3.5–3.22 Ga), Southern Africa. *Geology* 22(12): 1099–1102.

Lubimova, E. A. 1982. Heat flow—from the early Earth's history to the present. *Tectonophysics* 83: 143–149.

Lugt, Hans J. 1979. *Wirbelstromung in Natur und Technik*; Karlsruhe: Braun. English translation (Kuerti), 1983: *Vortex flow in Nature and Technology.* New York: Wiley & Sons. 297 pp.

Lundquist, G. M. and V. C. Cormier. 1980. Constraints on the absorption band model of Q. *J. Geophys. Res.* 85(B10): 5244–5256.

Ma, C., J. W. Ryan, D. Gordon, D. S. Caprette and W. E. Himwich. 1993. Reference frames from CDP VLBI data. In: *Contributions of Space Geodesy to Geodynamics: Crustal Dynamics. (AGU Geophysical Monographs, Geodynamics 24)*: 121–145.

MacDonald, G. J. F. 1964. Tidal friction. *Rev. Geophys.* 2(3): 467–541.

Macdonald, K. C., R. M. Haymon, S. P. Miller, J. C. Sempere and P. J. Fox. 1988. Deeptow and SeaBeam studies of dueling propagation ridges on the East Pacific Rise near 20°40′S. *J. Geophys. Res.* 93(4): 2875–2898.

MacLaurin, C. 1746. *Treatise on Fluxions.*

Mader, G. L. 1996. Calibration of GPS antennas. *EOS (Trans. Amer. Geophys. Un.)* 77(46), page F153.

Magde, L., C. Kincaid, D. W. Sparks and R. S. Detrick. 1996. Combined laboratory and numerical studies of the interaction between buoyant and plate-driven upwelling beneath segmented spreading centers. *J. Geophys. Res.* 101(B10): 22,107–22,122.

Maia, M. and M. Diament. 1991. An analysis of the altimetric geoid in various wavebands in the Central Pacific Ocean: constraints on the origin of intraplate features. *Tectonophysics* 190: 133–153.

Makada, M. and K. Lambeck. 1991. Late Pleistocene and Holocene sea-level change; evidence for lateral mantle viscosity structure. In: *Glacial Isostasy, Sea-Level and Mantle Rheology* (R. Sabadini et al., eds.). Dordrecht: Kluwer. 79–93.

Malla, R. P., S. C. Wu and S. M. Lichten. 1993. Geocenter location and variations in Earth orientation using global positioning system measurements. *J. Geophys. Res.* 98(B3): 4611–4617.

Mammerickx, J. 1992. The Foundation Seamounts: tectonic setting of a newly discovered seamount chain in the South Pacific. *Earth Plan. Sci. Lett.* 113: 293–306.

Mammerickx, J. and D. Sandwell. 1986. Rifting of old oceanic lithosphere. *J. Geophys. Res.* 91(B2): 1975–1988.

Mammerickx, J. D., F. Naar and R. L. Tyce. 1988. The Mathematician Paleoplate. *J. Geophys. Res.* 93(B4): 3025–3040.

Mammerickx, J., E. Herron and L. Dorman. 1980. Evidence for two fossil spreading ridges in the southeast Pacific. *Geol. Soc. Am. Bull.* Pt I, 91: 262–271.

Mansinha, L., D. E. Smylie and C. H. Chapman. 1979. Seismic excitation of the Chandler wobble. *Geophys. J.* 59: –17.

Manuel, F. E. 1968. *A Portrait of Isaac Newton.* Cambridge, MA: Belknap Press.

Mao, A. C., G. A. Harrison and T. Dixon. 1996. Monument stability of permanent GPS stations and tide gauges. *EOS (Trans. Amer. Geophys. Un.)* 77 (46), p. F153.

Maris. 1994. Program *"Redshift,"* vrsn. 1.2. Palo Alto, CA: Maris Media.

Marsh, J. G., F. J. Lerch and R. G. Williamson. 1985. Precision geodesy and geodynamics using Starlette laser ranging. *J. Geophys. Res.* 90(B11): 9335–9345.

Marsh, J. G., F. J. Lerch, B. H. Putney, T. L. Felsentreger, B. V. Sanchez, S. M. Klosko, G. B. Patel, J. W. Robbins, R. G. Williamson, T. L. Engelis, W. F. Eddy, N. L. Chandler, D. S. Chinn, S. Kapoor, K. E. Rachlin, L. E. Braatz and E. C. Pavlis. 1990. The GEM–T2 gravitational model. *J. Geophys. Res.* 95(13): 22,043–22,071.

Martin, A. K. and C. J. H. Hartnady 1986. Plate tectonic development of the southwest Indian Ocean: a revised reconstruction of east Antarctica and Africa. *J. Geophys. Res.* 91(B5): 4767–4786.

Martinson, D. G., N. G. Pisias, J. D. Hays, J. Imbrie, T. C. Moore, Jr. and N. S. Shackleton. 1987. *Quat. Res.* 27: 1–29.

Maruyama, T. 1964. Statical elastic dislocations in an infinite and semi-infinite medium. *Bull. Earthq. Res. Inst.* 42: 289–368.

Maslov, L. A. and H. C. Noltimier. 1993. Computational geodynamics of the Pacific Ocean and Pacific mobile belt. *Intern. Geol. Rev.* 35(6): 493–565.

Mathews, P. M. and I. I. Shapiro. 1995. Constraints on deep-Earth properties from space–geodetic data. *Phys. Earth Plan. Int.* 92: 99–107.

Mathews, P. M., B. A. Buffett and L. L. Shapiro. 1995. Love numbers for a rotating spheroidal Earth: new definitions and numerical values. *Geophys. Res. Lett.* 22(5): 579–582.

Matsumoto, K., M. Ooe, Y. Sato, Y. Tamura and Y. Imanishi. 1997. Solid–Earth Q estimated from gravity tide observations in Japan. *EOS (Trans. Amer. Geophys. Un.)* 78(46; Suppl.), Fall meeting: p. F152.

Matsumoto, K., Y. Sato, M. Ooe, Y. Tamura, Y. Imanishi and S. Okubo. 1998. A high-resolution ocean and loading tide model around Japan. *EOS (Trans. Amer. Geophys. Un.).* 79(24; Suppl.), Western Pacific meeting: p. W17.

Matyska, C. 1995. Axisymmetry of mantle aspherical structures. *Geophys. Res. Lett.* 22(4): 521–524.

Matyska, C., J. Moser and D. Yuen. 1994. The potential influence of radiative heat transfer on the formation of megaplumes in the lower mantle. *Earth Plan. Sci. Lett.* 125: 255–266.

Mauk, F. J. and M. J. S. Johnston. 1973. On the triggering of volcanic eruptions by earth tides. *J. Geophys. Res.* 78: 3356–3362.

Mayer, J. R. 1863. On celestial dynamics. VIII: The tidal wave. *London Edin. Dubl. Phil. Magaz. J. Sci.* v. XXV Ser. IV, pp 403–409.

―――. 1842. Bemerkungen ueber die Kraefte der unbelebten Natur. *Ann. Chem. Pharm. xlii,* p. 233. Engl. transl. (G. C. Foster) in *London Edin. Dubl. Phil. Magaz. J. Sci.* 1862, Ser. IV, xlviii: 371–377.

Mayes, C. L., L. A. Lawver and D. T. Sandwell. 1990. Tectonic history and new isochron chart of the South Pacific. *J. Geophys. Res.* 95(B6): 8543–8567.

McCarthy, D. D. and A. K. Babcock. 1986. The length of the day since 1656. *Phys. Earth Plan. Int.* 44: 281–292.

McElhinny, M. W. 1973. Mantle plumes, paleomagnetism and polar wandering. *Nature* 241: 522–523.

McElhinny, M. W. 1993. *Paleomagnetism from Gondwana to Rodinia.* Cox Lecture, presented at 1993 Fall Meeting, American Geophysical Union; *EOS (Trans. Amer. Geophys. Un.)* 74 no. 43 suppl., p. 213.

McGill, G. E. 1994. Hotspot evolution and Venusian tectonic style. *J. Geophys. Res.* 99(E11): 23,149–23,161.

McKenzie, D. P. 1966. The viscosity of the lower mantle. *J. Geophys. Res.* 71: 3995–4010.

———. 1968. The influence of the boundary conditions and rotation on convection in the Earth's mantle. *Geophys. J. Roy. Astron. Soc.* 15: 457–500.

McKenzie, D. P. 1972. Plate tectonics. In: *The Nature of the Solid Earth* (E. C. Robertson, ed). New York: McGraw–Hill. Pp. 323–360.

——— 1995. Lecture, May 1995. Dept. Earth Sciences, Oxford University.

McKenzie, D. and R. L. Parker. 1974. Plate tectonics in ω space. *Earth Plan. Sci. Lett.* 22: 285–293.

McKerrow, W. S. and C. R. Scotese. 1990. *Palaeozoic Palaeogeography and Biography. Geol. Soc. [London] Mem.* 12 (cover illustration).

McMurry, H. 1941. Periodicity of deep earthquakes. *Bull. Seis. Soc. Amer.* 31: 33–51.

McNish, A. G. and E. A. Johnson. 1938a. Magnetization of sediments from the bottom of the Atlantic Ocean. *Trans. Amer. Geophys. Un.* 19: 4–205.

———. 1938b. Preliminary report on measurement of magnetization of oceanic sediments. *Trans. Amer. Geophys. Un.* 19: 06.

McNutt, S. R. and R. J Beavan. 1981. Volcanic earthquakes at Pavlov volcano correlated with the solid earth tide. *Nature* 294: 615–619.

Means, W. D. 1976. *Stress and Strain.* Springer-Verlag. Chapter 24.

Melchior, Bn. P. 1983. *Tides of the Planet Earth*, 2nd Ed. Oxford: Pergamon Press.

———. 1989. The phase lag of Earth tides and the braking of the Earth's rotation. *Phys. Earth Plan. Int.* 56: 186–188.

———. 1993. The trends of earth tide research. Key-note address, in *Proc. 12th Intern. Symp. Earth Tides, Beijing, China*, 13–21.

———. 1994a. A new data bank for tidal gravity measurements [DB 92]. *Phys. Earth. Plan. Int.* 82: 125–156.

———. 1994b. Checking and correcting the tidal gravity parameters of the ICET data base. *Bull. Inf. Centr. Mar. Terr.* 119: 8979–8935.

———. 1995a. A continuing discussion about the correlation of tidal gravity anomalies and heat flow densities. *Phys. Earth. Plan. Int.* 88: 223–256.

———. 1995b. Letter to RCB. Centr. Intern. Mar. Terr. Bruxelles, 11 September, 1995.

Melchior, Bn. P. and M. De Becker. 1982. A discussion of world wide measurement of tidal gravity with respect to oceanic interactions, lithosphere heterogeneities, Earth's flattening and inertial forces. In: *Symp. Working Group I Inter-Union Commission on the Lithosphere* Tokyo, May 6 1982, pp. 9–16.

Melchior, Bn. P., B. Ducarme and M. de Becker. 1986. Correlation entre le flux de chaleur et las deformations radiales de marée terrestre en Afrique. *INQUA–Dakar Symposium Chagements Globaux en Afrique, ORSTOM, Paris, Travaux et Documents.* 197: 305–308.

Melchior, Bn. P. and B. Ducarme. 1991. Tidal gravity anomalies and tectonics. *Proc. 11th Symp. Earth Tides, Helsinki.* Stuttgart: Schweizerbart'sche Verlagsbuchhandlung. Pp. 445–454.

Melchior, Bn. P., O. Francis and B. Ducarme. 1995. Tidal gravity measurements in southeast Asia. IUGG Gen. Assembl. IAG Symp. Boulder, Colo. *Bull. Inf. Ctre. Mar. Terr.* 125: 9493–9507.

Menard, H. W. 1955. Deformation of the northeastern Pacific basin and the west coast of North America. *Bull. Geol. Soc. Amer.* 66: 1149–1198.

———. 1984. Evolution of ridges by asymmetrical spreading. *Geology* 12: 177–180.

Menard, H. W. and T. M. Atwater. 1968. Changes in direction of sea floor spreading. *Nature* 219: 463–467.

Mercier, J. C. and N. L. Carter. 1975. Pyroxene geotherms. *J. Geophys. Res.* 80: 3349–3362.

Merriam, J. B. 1986. Transverse stress Green's functions. *J. Geophys. Res.* 91: 13,903–13,913.

Merrill, R. and M. W. McElhinny. 1977. Anomalies in the time–averaged paleomagnetic field and their implications for the lower mantle. *Revs. Geophys.* 15(3): 309–323.

Michael, M. O. and D. A. Christoffel. 1975. Triggering of eruptions of Mt Ngauruhoe by fortnightly earth tide maxima, January 1972–June 1974. *N. Z. J. Geol. Geophys.* 18: 273–277.

Michelson, A. A. 1904. Relative motion of Earth and æther. *London Edin. Dubl. Phil. Magaz. J. Sci. Ser. VI*, vol. 8, No. 48: 716–719.

Michelson, I. 1974. Tides' tortured theory. *Sci. Pub. Aff.* March 1974, 31–34.

Mignard, F. 1981. The lunar orbit revisited; III. *Moon and Planets* 24: 189–207.

Milankovitch, M. 1941. Canon of insolation and the ice age problem. *Roy. Serb. Sci. Spec. Pub. 132, Sec. Math. and Natural Sci.* vol. 33; 633 pp. Engl. transl., U. S. Dept. Comm. and NSF, 1969.

_____. 1941. *Kanon der Erdbestrahlung.* Koningliche Serbische Akademie, Beograd.

Miller, G. R. 1966. The flux of tidal energy out of the deep oceans. *J. Geophys. Res.* 71(10): 2485–2489.

Miller, L., R. Cheney and J. Lillibridge. 1993. Blending ERS-1 data and tide-gauge data. *EOS (Trans. Amer. Geophys. Un.)* 74(16): 185.

Minato, M. 1968. Basement complex and Paleozoic orogeny in Japan. *Pacific Geol.* I: 85–95.

Minster, J. B., T. H. Jordan, P. Molnar and E. Haines. 1974. Numerical modelling of instantaneous plate tectonics. *Geophys. J. Roy. Astron. Soc.* 36(3): 541–576.

Minster, J. B. and T. H. Jordan. 1978. Present–day plate motion. *J. Geophys. Res.* 83: 5331–5354.

Minster, J. B. and D. L. Anderson. 1980. Dislocation and nonelastic processes in the mantle. *J. Geophys. Res.* 85(B11): 6347–6352.

MIT. 1987. Frontispiece, *"Kelvin's Baltimore Lectures and Modern Theoretical Physics"*; Boston, MA: MIT Press.

Mitchell, R. H. 1986. *Kimberlites; Mineralogy, Geochemistry and Petrology.* New York: Plenum Pr. 442 pp.

Mitchum, G. T. 1994. Comparison of TOPEX sea surface heights and tide gauge sea levels. *J. Geophys. Res.* 99(C12): 24,541–24,553.

Mitrovica, J. X., J. L. Davis, P. M. Mathews and I. I. Shapiro. 1994 Determination of tidal h Love number parameters in the diurnal band using an extensive VLBI data set. *Geophys. Res. Lett.* 21: 705–708.

Mogi, K. 1974. Active periods in the world's chief seismic belts. *Tectonophysics* 22: 265–282.

_____. 1979. Global variation of seismic activity. *Tectonophysics* 576: T47–T50.

Molines, J. M., C. LeProvost, F. Lyard, R. D. Ray, C. K. Shum and R. J. Eames. 1994. Tidal corrections in the TOPEX/POSEIDON geophysical data records. *J. Geophys. Res.* 99(C12): 24,749–24,760.

Molnar, P. and J. Stock, 1987. Relative motions of hotspots in the Pacific, Atlantic and Indian Oceans since late Cretaceous time. *Nature* 327: 587–591.

Molodenskii, S. M. (Molodensky, M. S.), and M. V. Kramer. 1980. Sur la possibilité qu'ont les hétérogénéités horizontales du manteau d'apparaître dans les observations de marées terrestres. *Fisika Zemli* 1: 3–20.

Molodenskiy, M. S. (Molodensky, S. M). 1976a. Green's functions of the elasticity equations for spheroidal elastic deformation of the Earth. *Bull. (Izv.) Acad. Sci. USSR, Physics of the Earth* no. 11.

———. 1976b. Changes in the Love numbers during variations of an Earth structure scheme. *Bull. (Izv.) Acad. Sci. USSR, Physics of the Earth* no 2.

Molodensky (Molodenskii), S. M. 1983. In: Melchior, Bn. P.: *Tides of the Planet Earth*, 2nd Ed. Pergamon. pp. 120–121.

———. 1984. Marées, nutation et structure interne de la Terre; ch. III. Dissipation de lenergie de marée dans l'envelope et dans l'ocean. *Bull. Inf. Mar. Terr.* 108: 7665–7764.

Montagner, J.-P. and B. Romanowicz. 1994. Introduction, "Symposium on GEOSCOPE Broadband Seismology." *Phys. Earth Plan. Int.* 84: p. 1.

Montagner, J.-P. and T. Tanimoto. 1991. Global upper mantle tomography of seismic velocities and anisotropies. *J. Geophys. Res.* 96(B12): 20,337–20,351.

Moon, W. M., R. Tang and B. M. Choi. 1991. Study of fluid-solid Earth coupling process using satellite altimetric data. *AGU/IUGG Geophysical Monograph* 59 (D. D. McCarthy and W. E. Carter, eds.): 85–111.

Moore, G. W. 1973. Westward tidal lag as the driving force of plate tectonics. *Geology* 1: 99–100.

Morel, P. and E. Irving. 1981. Paleomagnetism and the evolution of Pangea. *J. Geophys. Res.* 86(B3): 1858–1872.

Morelli, A. and A. M. Dziewonski. 1986. Topography of the core–mantle boundary determined with reflected and refracted rays. *EOS (Trans. Amer. Geophys. Un.)*, 67(44): 1099–1100.

Morgan, W. Jason. 1972a. Deep mantle convection plumes and plate motions. *Bull. Amer. Assoc. Petrol. Geol.* 56(2): 203–213.

———. 1972b. Plate motions and deep mantle convection. *Geol. Soc. Amer. Mem.* 22 132: 7–22.

———. 1983. Hotspot tracks and the early rifting of the Atlantic. *Tectonophysics* 94: 123–139.

Morgan, J. Phipps, W. Jason Morgan, Y.-S. Zhang and W. H. F. Smith. 1995. Observational hints for a plume-fed, suboceanic asthenosphere and its role in mantle convection. *J. Geophys. Res.* 100(B7): 12,753–12,767.

Morgan, W. J., J. O. Stoner and R. H. Dicke. 1961. Periodicity of earthquakes and the invariance of the gravitational constant. *J. Geophys. Res.* 66(11): 3831–3843.

Morley, L. W. and A. Rochelle. 1964. Paleomagnetism as a means of dating geological events. Geochronology in Canada; *Roy. Soc. Canada Sp. Publctn.* 8 (F. FitzOsborne, ed.); Toronto: Univ. of Toronto Press.

Morozov, E. G. 1995. Semidiurnal wave global field. *Deep Sea Res.* 42(1): 35–148.

Mulholland, J. D. 1980. Scientific achievements from ten years of lunar laser ranging. *Rev. Geophys. Sp. Res.* 18: 549–564.

Muller, R. D., W. R. Roest, J.-Y. Royer, L. M. Gahagan and J. D. Sclater. 1997. Digital isochrons of the world's ocean floor. *J. Geophys. Res.* 102(B2): 3211–3214.

Munk, W. H. 1968. Once again—tidal friction. *Q. J. Roy. Astron. Soc.* 9(4): 352–375.

———. 1972. Personal communication (letter to RCB).

———. 1997. Twice again—tidal friction. Presented at meeting of the Royal Society, 12 February 1997. 40 pp. I am indebted to Prof. Munk for a pre–print.

Munk, W. H. and R. R. Revelle. 1952. Sea level and the rotation of the Earth. *Am. J. Sci.* 250: 829–833.

Munk W. H. and G. J. F. MacDonald. 1960a. *The Rotation of the Earth*. Cambridge Univ. Press.

———. 1960b. Continentality and the gravitational field of the Earth. *J. Geophys. Res.* 65(7): 2169–2172.

Mysen, B. O. 1977. The solubility of H_2O and CO_2 under predicted magma genesis conditions and some petrological and geophysical implications. *Rev. Geophys. Sp. Phys.* 15(3): 351–360.

Nagata, T. 1941. *Bull. Earthq. Res. Inst.* 19: 49.

———. 1952a. Reverse thermo-remanent magnetism. *Proc. Japan Acad.* 27(10): 643–645.

———. 1952b. Self–reversal of thermo-remanent magnetism of igneous rocks. *J. Geomagn. Geoelectr.* 4(1): 22–38.

Nakanishi, M., K. Tuamaki and K. Kobayashi. 1992. Magnetic anomaly lineations from Late Jurassic to Early Cretaceous in the west-central Pacific Ocean. *Geophys. J. Intern.* 109: 701–719.

Nakiboglu, S. M. 1982. Hydrostatic theory of the Earth and its mechanical significance. *Phys. Earth Plan. Int.* 28: 302–311.

Néel, L. 1951. L'inversion de l'aimantation permanente des roches. *Ann. Geophys.* 7: 90.

Nelson, T. H. and P. G. Temple. 1972. Mainstream mantle convection: a geologic analysis of plate motion. *Bull. Amer. Assoc. Pet. Geol.* 56(2): 226–246.

Nemirovich-Danchenko, M. 1997. Stress concentration, fracture and generation of elastic waves. *Annal. Geophys.* 15 (Suppl. pt. I), p. 31.

Newsom, H. E. and S. R. Taylor. 1989. Geochemical implications of the formation of the Moon by a single giant impact. *Nature* 38: 29–34.

Newton, R. R. 1972. Astronomical evidence concerning non-gravitational forces in the Earth-Moon system. *Astrophys. Sp. Sci.* 16: 179–200.

Nicolas, A. 1990. Melt extraction from mantle peridotites: hydrofracturing and porous flow, with consequences for oceanic ridge activity. *Magma Transport and Flow* (M. P. Ryan, ed.). New York: Wiley & Sons. 159–173.

Nicolis, G., and I. Prigogine. 1989. *Exploring Complexity.* New York: W. H. Freeman. Figures 3, 30.

Niebauer, T. M., F. J. Klopping and J. E. Faller. 1995. The FG5 absolute gravimeter. *Cahiers du Centre Européen de Géodynamique et de Seismologie*, 11.

Niell, A. E. 1996. Reducing elevation-dependent errors for ground-based GPS measurements. *EOS (Trans. Amer. Geophys. Un.)* 77(46), p. F153.

Nimmo, F. and D. McKenzie. 1997. Convective thermal evolution of the upper mantles of Earth and Venus. *Geophys. Res. Lett.* 24(12): 1539–1542.

Nixon, P. H. and E. Condliffe. 1989. Tanzania kimberlites: a preliminary heavy-mineral study. In: *Kimberlites and Related Rocks* (J. Ross, ed.), *G. S. A. Spec. Pub.* #14: 407–416.

Noether, E., 1918. Invariante Variationsprobleme. *Nachr. v. d. Gesellsch. d. Wissensch. z. Göttingen*, 235–257. Engl. summary by K. Uhlenbeck (1983): Conservation laws and their application in global differential geometry, in: *Emmy Noether in Bryn Mawr* (B. Srinavasan and J. D. Kelly, eds.). New York: Springer-Verlag. 103–115.

Nolet, G. 1994. Oceans in the upper mantle. *EOS (Trans. Amer. Geophys. Un.)* 75(16, Suppl.): 232.

Nolet, G. and A. Zielhaus. 1994. Low S-velocities under the Tornquist-Teissayre zone: evidence for water injection into the transition zone by subduction. *J. Geophys. Res.* 99(B8): 15,813–15,820.

Nolet, G. S., P. Grand and B. L. N. Kennett. 1994. Seismic heterogeneity in the upper mantle. *J. Geophys. Res.* 99(B12): 23,753–23,766.

O'Connell, R. J. and A. M. Dziewonski. 1976. Excitation of the Chandler wobble by large earthquakes. *Nature* 262(5566): 259–262.

O'Connell, R. J. and B. Budiansky. 1977. Viscoelastic properties of fluid-saturated cracked solids. *J. Geophys. Res.* 82(36): 5719–5735.

O'Connell, R. J. and B. H. Hager. 1980. On the thermal state of the Earth. In: *Physics of the Earth's Interior* (A. M. Dziewonski ed.), *Proc. Intern. Sch. Phys. "E. Fermi," Crse.* LXXVIII, pp. 270–317.

O'Connell, R. J., C. W. Gable and B. H. Hager. 1991. Toroidal-poloidal partitioning of lithospheric plate motions. In: *Glacial Isostasy, Sea–Level and Mantle–Rheology* (R. Sabadini, K. Lambeck and E. Boschi, eds). Dordrecht: Kluwer. 535–551.

O'Connell, R. J. and B. Steinberger. 1998. True polar wander and hot-spot motion during the Cenozoic from mantle flow models. *Annal. Geophys. Suppl.* v. 16. (Abstr.); *News Letter, Eur. Geophys. Soc.*, no. 66, p. 72.

Odling, N. W. A. 1995. An experimental replication of upper-mantle metasomatism. *Nature* 373: 58–60.

O'Neill, B., J. D. Bass and G. R. Rossman. 1993. Elastic properties of hydrogrossular garnet and implications for water in the upper mantle. *J. Geophys. Res.* 98(B11): 20,031–20,037.

Ohtani, E., T. Shibata, T. Kubo and T. Kato. 1995. Stability of hydrous phases in the transition zone and the uppermost part of the lower mantle. *Geophys. Res. Lett.* 22(19): 2553–2556.

Okal, E. I. and R. Batiza. 1987. Hotspots: the first 25 years. In: *Seamounts, Islands and Atolls* (B. H. Keating, P. Fryer, R. Batiza and G. W. Boehlert, eds.). *AGU Geophys. Monograph* 43: 1–11.

Okubo, S. 1982. Theoretical and observed Q of the Chandler wobble; Love number approach. *Geophys. J.* 71: 647–665.

Okubo, S., S. Yoshida, T. Sato, Y. Tamura and Y. Imanishi. 1997. Verifying the precision of a new generation absolute gravimeter FG5–comparison with superconducting gravimeters and detection of oceanic loading tide. *Geophys. Res. Lett.* 24(4): 489–492.

Oliver, J. and B. Isacks 1967. Deep earthquake zones, anomalous structures in the upper mantle, and the lithosphere. *J. Geophys. Res.* 72 (16): 4259–4275.

Olivet, J.-L., J. Bonnin, P. Beuzart and J.-M. Auzende. 1982. Cinematique des plaques et paleogeographie: une revue. *Bull. Soc. Geol. France* XXIV, 875–892.

Olson, P. P., G. D. Silver and R. W. Carlson. 1990. The large–scale structure of convection in the Earth's mantle. *Nature* 344: 209–214.

O'Neill, H. St C. 1991. The origin of the Moon and the early history of the Earth's chemical model. *Geochim. Cosmochim. Acta*, 55(4): 1136–1157.

Ooe, M. 1978. An optimal complex AR, MA model of the Chandler wobble. *Geophys. J. Roy. Astron. Soc.* 53: 445–457.

Orowan, E. 1958. Lecture, *Mechanical problems of Geology*, at Pasadena: Calif. Inst. Technol. Cited by Elsasser (1969), q.v.

Oxburgh, E. R. and D. E. L. Turcotte. 1974. Membrane tectonics and the East African Rift. *Earth Plan. Sci. Lett.* 22: 133–144.

Pacheco, J. F. and L. R. Sykes. 1992. Seismic moment catalog of large shallow earthquakes, 1900 to 1989. *Bull. Seis. Soc. Amer.* 82(3): 1306–1349.

Packham, G. H. 1982. Foreword to papers on the tectonics of the Southwest Pacific Region. *Tectonophysics* 87: 1–10.

———. 1993. Plate tectonics and the development of sedimentary basins of the dextral regime in Southeast Asia. *J. Southeast Asian Earth Sci.* 8: 497–511.

Pan, C. 1985. Polar instability, plate motion, and geodynamics of the mantle. *J. Phys. Earth* 33: 411–434.

Pang, K. D., K. Yau and H.-H. Chou. 1995. The Earth's paleorotation, post-glacial rebound and lower mantle viscosity from analysis of ancient Chinese eclipse records. *Pageophys.* 145(3/4): 459–485.

Pang, K. D. and K. Yau. 1994. Ancient and modern eclipses and a 4000–year geodetic baseline. *EOS (Trans. Amer. Geophys. Un.)*, 75(Suppl.): 75.

Pari, G. and W. R. Peltier. 1996. The free–air gravity constraint on subcontinental dynamics. *J. Geophys. Res.* 101(12): 28,105–28,132.

Pariiskii, N. N., M. V. Kuznetsov and L. V. Kuznetsova. 1972. On the influence of ocean tides on the secular deceleration of the Earth's rotation. *Fiz. Zemli*, 2: 3–12.

Parke, M. E. and M. C. Hendershott. 1980. M2, S2, K1 models of the global ocean tide on an elastic Earth. *Mar. Geod.* 3: 379–407.

Pavoni, N. 1969. Zonen lateraler horizontaler Verscheibung in der Erdkruste und darauableiitbare Aussagen zur globalen Tektonik. *Geol. Rundschau*, 59: S. 56–77.

———. 1981. A global geotectonic reference system inferred from Cenozoic tectonics. *Geol. Rundsch.* 70: S. 189–206.

———. 1985. Pacific/anti-Pacific bipolarity in the structure of the Earth's mantle. *EOS (Trans. Amer. Geophys. Un.)*, 66: 497.

———. 1991. Bipolarity in structure and dynamics of the Earth's mantle. *Eclogae geol. Helv.* 84(2): 327–343.

———. 1993. Pattern of mantle convection and Pangæa break–up, as revealed by the evolution of the African plate. *J. Geol. Soc.* [London] 150: 953–964.

Peacock, S. M. 1990. Fluid processes in subduction zones. *Science* 248: 329–337.

Pekeris, C. L. 1935. Thermal convection in the interior of the Earth. *Mon. Not. Roy. Astron. Soc. Geophys. Suppl.* 3: 342–367.

Peltier, W. R. 1986. Slow changes in the Earth's shape and gravitational field: Constraints on the glaciation history and internal viscosity stratification. In: *Space Geodesy and Geodynamics* (A. J. Anderson and A. Cazenave, eds.). London: Academic Press. 75–109.

———. 1989. Mantle viscosity, in: *Mantle Convection, Plate Tectonics and Global Dynamics* (New York: Gordon and Breach). Ch. 6, 389–478.

Philippot, P. 1993. Fluid-melt-rock interaction in mafic eclogites and coesite-bearing metasediments: constraints on volatile recycling during subduction. *Chem. Geol.* 108: 93–112.

Phillips, R. J., R. E. Grimm and M. C. Malin. 1991. Hotspot evolution and the global tectonics of Venus. *Science* 252: 651–658.

Phillipsen, S. 1996. A long–term comparison of continuous GPS and relative gravity. *EOS (Trans. Amer. Geophys. Un.)*, 77(46, Suppl.): p. F154.

Platzman, G. W. 1984. Planetary energy balance for tidal dissipation. *Rev. Geophys. Plan. Phys.* 22(1): 73–84.

Pockalny, R. A., P. J. Fox, D. J. Fornari, K. C. MacDonald and M. R. Perfit. 1997. Tectonic reconstruction of the Clipperton and Siqueiros fracture zones: evidence and consequences of plate motion change for the last 3 Myr. *J. Geophys. Res.* 102(B2): 3167–3181.

Pollack, H. N. 1973. Longman tidal formulas: resolution of horizontal components. *J. Geophys. Res.* 78(14): 2598–2600.

Pollack, H. N., S. J. Hurter and J. R. Johnson. 1993. Heat flow from the Earth's interior. *Rev. Geophys.* 31(3): 267–280.

Powell, C. McA. and B. D. Johnson. 1980. Constraints on the Cenozoic position of Sundaland. *Tectonophysics* 63: 1–109.

Powell, C. McA., S. R. Roots and J. J. Veevers. 1988. Pre–breakup continental extension in

East Gondwanaland and the early opening of the eastern Indian Ocean. *Tectonophysics* 155(1–4): 261–283.
Press, F. 1965. Displacements, strains and tilts at teleseismic distances. *J. Geophys. Res.* 70(10): 2395–2412.
Priestley, K. F., G. Zandt and G. E. Randall. 1988. Crustal structure in western Kazakh, USSR, from teleseismic receiver functions. *Geophys. Res. Lett.* 15(6): 613–616.
Prigogine, I. 1995. Why irreversibility? The formulation of classical and quantum mechanics for non-integrable systems. *Intern. J. Bifurc. Chaos Appl. Sci. Eng.* 5(1): 3–16.
———. 1997. Non-linear science and the laws of nature. *J. Frank. Inst.* 334B, no. 5–6, 745–758.
Prothero, W. A. Jnr. and J. M. Goodkind. 1972. Earth–tide measurements with the superconducting gravimeter. *J. Geophys. Res.* 77: 926–932.
Proudman, J. 1960. The condition that a long–period tide shall follow the equilibrium-law. *Geophys. J. Roy. Astron. Soc.* 3: 244–249.
Puster, P. B., H. Hager and T. H. Jordan. 1995. Mantle convection experiments with evolving plates. *Geophys. Res. Lett.* 22(16): 2223–2226.
Ranalli, G. and H. H. Schloessin. 1989. Role of episodic creep in global mantle deformation. In: Slow Deformation and Transmission of Stress in the Earth; *Geophys. Monogr. AGU* no. 49, (S. C. Cohen and P. Vanicek, eds.), 55–63.
Rankine, W. J. M. 1853. Address *"On the general law of the transformation of energy"* before Phil. Society of Glasgow; reproduced in Rankine, W. J. M. 1880, *Miscellaneous Scientific Papers.*
Rapp, R. H. 1988. Second-order radial gradient of the Earth's disturbing gravitational potential at an elevation of 160 km above the reference ellipsoid. Computed by Prof. Rapp, in benefit of predicting returns from an orbital gradiometer; data transmitted with letter 621, NASA/Goddard from J. Marsh to RCB, May 27, 1988; image constructed by Karen Settle, published in *EOS (Trans. Amer. Geophys. Un.)*, vol. 69, no. 18, May 3 1988.
Rapp, R. H. and N. K. Pavlis. 1990. Development and analysis of geopotential coefficient models to spherical harmonic degree 360. *J. Geophys. Res.* 95(B13): 21,885–21,911.
Rautenberg, V., H.-P. Plag, M. Burns, G. E. Stedman and H.-U. Juttner. 1997. Tidally-induced Sagnac signal in a ring laser. *Geophys. Res. Lett.* 24(8): 893–896.
Ray, R. D. 1994. Tidal energy dissipation: observations from astronomy, geodesy and oceanography. In: *The Oceans: Physical-Chemical Dynamics and Human Impact* (S. K. Majumdar et al., eds.), Pennsylv. Acad. Sci. Ch. 11: 171–185.
———. 1995. Letter to RCB. Goddard Space Flight Center, NASA, 1995 23 Nov.
Ray, R. D. and B. V. Sanchez. 1989. Radial deformation of the Earth by oceanic tidal loading. *NASA Techn. Mem.* 100743; 50 pp.
Ray, R. D., S. Bettadpur, R. J. Eanes and E. J. O. Schrama. 1994. Geometrical determination of the Love number h2 at four tidal frequencies. *Geophys. Res. Lett.* 16: 2175–2178.
Ray, R. D. and G. T. Mitchum. 1996. Surface manifestation of internal tides generated near Hawaii. *Geophys. Res. Lett.* 23(16): 2101–2104.
Ray, R. D., R. J. Eanes and B. F. Chao. 1996. Detection of tidal dissipation in the solid Earth by satellite tracking and altimetry. *Nature* 381: 595–597.
Rayleigh, Ld. (R. J. Strutt). 1906. On the distribution of radium in the Earth's crust, and the Earth's internal heat. *Proc. Roy. Soc. London* A, LXXVII, 472–485.
———. 1916. On convection currents in a horizontal layer of fluid, when the higher temperature is in the underside. *London Edin. Dubl. Phil. Magaz. J. Sci.* Ser. VI, LIX, 529–546.

Reif, F. 1965. *Fundamentals of Statistical and Thermal Physics*. New York: McGraw-Hill. 651 pp.
Renne, P. R. and A. R. Basu. 1991. Rapid eruption of the Siberian traps: flood basalts at the Permo-Triassic boundary. *Science* 253: 176–179.
Revenaugh, J. S. and T. H. Jordan. 1989. A study of mantle layering beneath the western Pacific. *J. Geophys. Res.* 94: 5787–5813.
Revuzhenko, A. F. 1991. Tidal mechanism of mass transfer. *Izvestiya, Earth Physics (AGU Engl. transl.)*, 27(6): 445–451.
Revuzhenko, A. F., A. P. Bobryakov and E. I. Shemyakin. 1983. A potential mechanism for mass transfer in the Earth. *Dokl. Akad. Nauk CCCR*, 272(5): 1087–1099.
Ribe, N. M. 1992. The dynamics of thin shells with variable viscosity and the origin of toroidal flow in the mantle. *Geophys. J. Intern.* 110: 537–552.
Ricard, Y., C. Doglioni and R. Sabadini. 1991. Differential rotation between lithosphere and mantle: a consequence of lateral mantle viscosity variations. *J. Geophys. Res.* 96(B5): 8407–8415.
Ricard, Y., R. Sabadini and G. Spada. 1992. Isostatic deformations and polar wander induced by redistribution of mass within the Earth. *J. Geophys. Res.* 97(10): 14,223–14,236.
Ricard, Y., G. Spada and R. Sabadini. 1993. Polar wandering of a dynamic Earth. *Geophys. J. Intern.* 113: 284–298.
Rice, A. R. 1971. Mechanism of dissipation in mantle convection. *J. Geophys. Res.* 76(5): 1450–1459.
———. 1972. On the stagnation temperature in the computation of natural convective flows. (Abstr.) *EOS (Trans. Amer. Geophys. Un.)*, 53(4): 385.
Richards, M. A. 1991. Hotspots and the case for a high viscosity lower mantle. In: *Glacial Isostasy, Sea–Level and Mantle Rheology* (R. Sabadini et al., eds.); Dordrecht: Kluwer. Pp. 571–587.
———. 1997. Rotational dynamics and mantle convection. *EOS (Trans. Amer. Geophys. Un.)*, 78 (46; Suppl.): F183.
Richards, M. A., R. A. Duncan and V. E. Courtillot. 1989. Flood basalts and hotspot tracks: Plume heads and tails. *Science*, 245: 103–107.
Richards, M. A. and D. C. Engebretson. 1992. Large–scale mantle convection and the history of subduction. *Nature* 355: 437–440.
Richards, M. A., B. H. Hager and N. H. Sleep. 1988. Dynamically supported geoid highs over hotspots: observation and theory. *J. Geophys. Res.* 93: 7690–7708.
Richards, M. A., R. A. Duncan and V. E. Courtillot. 1989. Flood basalts and hot–spot tracks: plume heads and tails. *Science* 246: 103–107.
Richter, F. M. and B. Parsons. 1975. On the interaction of two scales of convection in the mantle. *J. Geophys. Res.* 80(17): 2529–2541.
Robertson, D. S. 1987. Radio Interferometry. *Rev. Geophys.* 25(5): 867–870.
Robertson, D. S., J. R. Ray and W. E. Carter. 1994. Tidal variations in UT1 observed with very long baseline. *J. Geophys. Res.* 99(B1): 621–636.
Rock, N. M. S. 1989. Kimberlites as varieties of lamprophyres: implications for geological mapping, petrological research and mineral exploration. In: *Kimberlites and Related Rocks*, (J. Ross, ed.). *Geol. Soc. Austral. Sp. P.* no. 14: Blackwell Sci. Pubs.: 46–59.
Roeder, D. H. and T. H. Nelson. 1971 Subduction, orogeny and mainstream mantle convection. *Bull. Amer. Asoc. Petrol. Geol.* 55, p. 361.
Romanowicz, B. 1993. Spatiotemporal patterns in the energy release of great earthquakes. *Science* 260: 1923–1926.

———. 1994. On the measurement of anelastic attenuation using amplitudes of low-frequency surface waves. *Phys. Earth Plan. Int.* 84: 179–191.

Royer, J. Y. and D. Sandwell. 1989. Evolution of the eastern Indian Ocean since the late Cretaceous: constraints from Geosat altimetry. *J. Geophys. Res.* 94(B10): 13,755–13,782.

Rudnicki, J. and H. Kanamori. 1981. Effects of fault interaction on moment, stress and strain energy release. *J. Geophy Res.* 86(3): 1785–1793.

Ruff, L. and H. Kanamori. 1983. Seismic coupling and uncoupling at subduction zones. *Tectonophysics* 99: 99–117.

Rumble III, D. 1994. Water circulation in metamorphism. *J. Geophys. Res.* 99(B8): 15,499–15,502.

Rummel, R. 1996. Expected geoid precision from planned ESA Explorer mission GOCE. *EOS (Trans. Amer. Geophys. Un.)*, 77(46, Suppl.): page F136.

Runcorn, S. K. 1957. Convection currents in the mantle and recent developments in geophysics. In: *Gedenboek F. A. Vening Meinesz.* (The Hague: Mouton and Co.), 271–277.

———. 1962a. Towards a theory of continental drift. *Nature* 193 (4813): 311–314.

———. 1962b. Drawing given to RCB by Keith Runcorn at AGU Fall Meeting, 1990, on presenting "Systematics of continental motion: N and S hemispheres" *(EOS (Trans. Amer. Geophys. Un.)*, 71[43]: p. 1941). An almost identical figure was published earlier by Runcorn in: *Research* (1962). London, 16: 103–108.

———. 1968. Polar wandering and continental drift. In: *Continental Drift, Secular Motion of the Pole, and Rotation of the Earth*, (Eds. W. Markowitz and B. Guinot). Dordrecht: Kluwer. 80–85.

———. 1973a. In report: "Mantle plumes get cool reception." *Nature* 243, p. 192.

———. 1973b. Polar wandering and continental drift. In: *Implications of Continental Drift to the Earth Sciences* (D. H. Tarling and S. K. Runcorn, eds.). London: Academic Press. 995–997.

———. 1982. The role of the core in irregular fluctuations of the Earth's rotation and excitation of the Chandler wobble. *Phil. Trans. Roy. London* A306: 261–270.

Runcorn, S. K., A. C. Benson, A. F. Moore and D. H. Griffiths. 1951. Measurements of the variation with depth of the main geomagnetic field. *Phil. Trans. Roy. Soc. London* A244: 113–151.

Russell, C. T. 1993. Magnetic fields of the terrestrial planets. *J. Geophys. Res.* 98(E10): 18,681–18,695.

Rutherford, E. 1906. The mass and velocity of the α particles expelled from radium and actinium. *London Edin. Dubl. Phil. Magaz. J. Sci.* XLI: 348–371.

Ryan, J. W. 1997. Letter to RCB, Jan. 22, 1997.

Ryan, J. W., T. A. Clark, R. J. Coates, C. Ma, W. T. Wildes, C. R. Gwinn, T. A. Herring, I. I. Shapiro, B. E. Corey, C. C. Counselman, H. F. Hinteregger, A. E. E. Rogers, A. R. Whitney, C. A. Knight, N. R. Vandenberg, J. C. Pigg, B. R. Schupler and B. O. Roennaeng. 1986. *J. Geophys. Res.* 91(2): 1935–1946.

Ryan, J. W., T. A. Clark, C. Ma, D. Gordon, D. S. Caprette and W. E. Himwich. 1993. Global scale tectonic plate motions measured with CDP VLBI data. In: *Contributions of Space Geodesy to Geodynamics: Crustal Dynamics (AGU Geophysical Monographs, Geodynamics* 23): 37–49.

Rydelek, P. A., P. M. Davis and R. Y. Koyanagi. 1988. Tidal triggering of earthquake swarms at Kilauea Volcano, Hawaii. *J. Geophys. Res.* 93(B5): 4401–4411.

Rydelek, P. A., W. Zurn and J. Hinderer. 1991. On tidal gravity, heat flow and lateral homogeneities. *Phys. Earth Plan. Int.* 68: 215–229.

Sabadini, R. and D. A. Yuen. 1989. Mantle stratification and long-term polar wander. *Nature* 339: 373–375.

Sabadini, R. D., A. Yuen and P. Gasperini. 1989. Viscoelastic deformation and temporal variations in the geopotential. In: *Slow Deformation and Transmission of Stress in the Earth*; Geophys. Monogr. AGU no. 49, (S. C. Cohen and P. Vanicek, eds.), 115–123.

Sager, W. W. and B. H. Keating. 1984. Paleomagnetism of Line Islands seamounts: evidence for late Cretaceous and early Tertiary volcanism. *J. Geophys. Res.* 89(B13): 11,135–11,151.

Sager, W. W. and U. Bleil. 1987. Latitudinal shift of Pacific hotspots during the late Cretaceous and early Tertiary. *Nature* 326: 488–490.

Sager W. W. and M. S. Pringle. 1988. Mid–Cretaceous to early Tertiary apparent polar wander path of the Pacific plate. *J. Geophys. Res.* 93(B10): 11,753–11,771.

Sagnac, G. 1913. L'éther lumineux démontré par l'effet du vent relatif d'éther dans un interféromètre en rotation uniforme. *Contes Rend. Acad. Sci.* (Paris) 157: 708–710. (Reference fr. Anderson, Bilger and Stedman [1994]).

Sanchez, B. and N. K. Pavlis. 1995. Estimation of main tidal constituents from Topex/Poseidon altimetry using a Proudman function expansion. *J. Geophys. Res.* 100(C12): 25,229–25,248.

Sandwell, D. T., E. L. Winterer, J. Mammerickx, R. A. Duncan, M. A. Lynch, D. A. Levitt and C. L. Johnson. 1995. Evidence for diffuse extension of the Pacific plate from Pukapuka ridges and cross-grain gravity lineations. *J. Geophys. Res.* 100(B8): 15,087–15,099.

Sandwell, D. T. and Walter H. F. Smith. 1996. *Marine Gravity Anomaly from Satellite Altimetry.* Geol. Data Center, MC–0223, Scripps Inst. Oceanog., Univ. of California at San Diego, La Jolla, California; http://www.noaa.gov/mgg/announcements/announce_predict.html.

Sato, H. I., S. Sacks, T. Murase, G. Muncill and H. Fukuyama. 1989. Q_p–melting temperature relation in peridotite at high pressure and temperature: attenuation mechanism and implications for the mechanical properties of the upper mantle. *J. Geophys. Res.* 94(B8): 10,647–10661.

Satyabala, S. P. and H. K. Gupta. 1996. Is the quiescence of major earthquakes (M ≥ 7.5) since 1952 in the Himalaya and northeast India real? *Bull. Seis. Soc. Amer.* 86(6): 1983–1986.

Saunders, A. D., M. Storey, R. W. Kent and M. J. Norry. 1992. Consequences of plume–lithosphere interactions. In: *Magmatism and the Causes for Continental Break–Up; Geological Soc.* [London] *Spec. Pub.* no. 68, (B. C. Storey, T. S Alabaster and R. J. Pankhurst, eds.): 41–60.

Savostin, L. A., J.-C. Sibuet, L. P. Zonenshain, X. LePichon and M.-J. Roulet. 1986. Kinematic evolution of the Tethys belt from the Atlantic Ocean to the Pamirs since the Triassic. *Tectonophysics* 123: 1–35.

Schlanger, S. O., M. O. Garcia, B. H. Keating, J. J. Naughton, W. W. Sager, J. A. Haggert, J. A. Philpotts and R. A. Duncan. 1984. Geology and geochronology of the Line Islands. *J. Geophys. Res.* 89(B13): 11,261–11,272.

Schmeling, H. and G. Y. Bussod. 1996. Variable viscosity convection and partial melting in the continental asthenosphere. *J. Geophys. Res.* 101(B3): 5411–5423.

Schrama, E. J. O. and R. D. Ray. 1994. A preliminary tidal analysis of Topex/Poseidon altimetry. *J. Geophys. Res.* 99(C12): 24,799–24,808.

Schrama, E. J., O. R. Rummel and N. Sneeuw. 1993. The role of STEP as a gravity field mapping mission. *EOS (Trans. Amer. Geophys. Un.)* 74 (Suppl.): p. 97.

Schuh, H. 1989. Earth's rotation measured by VLBI. In: *Earth's Rotation from Eons to Days* (P. Brosche and J. Sundermann, eds.). Berlin: Springer-Verlag. 3–11.

Schuh, H. and W. Schwegmann. 1998. First steps towards real-time VLBI. *Newsletter, Europ. Geophys. Soc.* Number 66 March 1998 (Sci. Prgrm. Nice 1998), p. 135.

Schum, C. K., P. L. Woodworth, O. B. Andersen, G. Egbert, O. Francis, C. King, S. Klosko, C. LeProvost, X. Li., J. Molines, M. Parke, R. Ray, M. Schlax, D. Stammer, C. Tierney, P. Vincent and C. Wunsch. 1996. *J. Geophys. Res.*, in press.

Schwiderski, E. W. 1979. *Global ocean tides, Pt. II: the semidiurnal principal lunar tide M2 atlas of tidal charts and maps.* TR 79–414, 22448. Dahlgren, VA· Naval Surface Weapons Center.

_____. 1980. On charting global ocean tides. *Revs. Geophy. Sp. Phys.* 18(1): 243–268.

_____. 1985a. On tidal friction and the deceleration of the Earth's rotation and Moon's revolution. *Mar. Geod.* 9: 399–450.

_____. 1985b. Energy balance of the Moon-Ocean-Earth tidal system and the braking of Earth's rotation and Moon's revolution; inc. discussion. In: *Proc. Tenth Symp. Earth Tides*, (R. Vieira, ed.). Madrid: Consejo Sup. Inv. Cient., 449–455.

Scotese, C. R. and S. F. Barrett. 1990. Gondwana's movement over the South Pole during the Paleozoic: evidence from lithological indicators of climate. In: *Palaeozoic Palaeogeography and Biogeography* (W. S. McKerrow and C. R. Scotese, eds). *Geol. Soc. Mem.* no. 12: 75–85.

Scotese, C. R. and W. S. McKerrow. 1990. Revised world maps and introduction. In: *Palaeozoic Palaeogeography and Biogeography* (W. S. McKerrow and C. R. Scotese, eds.). *Geol. Soc. Mem.* no. 12: 1–21.

Scrutton, C. T. 1978. Periodic growth features in fossil organisms and the length of the day and month. In: *Tidal Friction and the Earth's Rotation* (P. Brosche and J. Sundermann, eds.). Springer-Verlag. Pp. 154–196.

Seiler, U. 1990. Variations of the angular momentum budget for tides of the present ocean. In: *Tidal Friction and the Earth's Rotation* (P. Brosche and J. Sundermann, eds.); Springer-Verlag. Pp. 81–94.

Sengor, A. M. C. and K. Burke. 1978. Relative timing of rifting and volcanism on Earth and its tectonic implications. *Geophys. Res. Lett.* 5(6): 419–421.

Serrin, J. 1959. *Mathematical Principles of Classical Fluid Mechanics. Handbuch der Physik*, vol. VIII/1 (S. Flugge, ed.). New York: Springer-Verlag. 125–262; esp. III, §25.

Sharpe, H. N. and W. R. Peltier. 1979. A thermal history model for the Earth with parametrized convection. *Geophys. J. Roy. Astron. Soc.* 59: 171–203.

Shaw, H. R. 1970. Earth tides, global heat flow, and tectonics. *Science*, 168(3935): 1084–1087.

Shaw, H. R., R. W. Kistler and J. F. Evernden. 1971. Sierra Nevada plutonic cycle: Pt. II, Tidal energy and a hypothesis for orogenic–epeirogenic periodicities. *Geol. Soc. Amer. Bull.* 82: 869–896.

Shaw, H. R. and J. G. Moore. 1988. Magmatic heat and the El Nino cycle. *EOS (Trans. Amer. Geophys. Un.)*, 69(45): 1553, 1564–1565.

Shimozuru, D. 1987. Tidal effects in Hawaiian volcanism. In: *Volcanism in Hawaii, vol. 2* (R. W. Decker, T. L. Wright and P. H. Stauffer, eds.), *U. S. Geol Surv. Prof. Paper* 1350: 1337–1342.

Shirokov, V. A. 1983. The influence of the 19-year tidal cycle on large–scale eruptions and earthquakes in Kamchatka, and their long–term prediction. In: *The Great Tolbachik Fissure Eruption; Geological and Geophysical Data 1975–1976*. Cambridge Univ. Press. 232–242.

Shuhei, O., S. Yoshida, T. Sato, Y. Tamura and Y. Imanishi. 1997. Verifying the precision of a new generation of absolute gravimeter FG5; comparison with superconducting gravimeters and detection of oceanic loading tide. *Geophys. Res. Lett.* 24(4): 489–492.

Sjoberg, B. and A. Stigebrand. 1992. Computations of the geographical distribution of the energy flux to mixing processes via internal tides and the associated vertical circulation in the ocean. *Deep–Sea Res.* 39(2): 269–291.

Skinner, E. M. W. 1989. Contrasting Group I and Group II kimberlite petrology: towards a model for kimberlites. In: *Kimberlites and Related Rocks*, (J. Ross, ed.), v. 1, G. S. A. Spec. Pub. #14, 528–544.

Sleep, N. H. 1990. Monteregian hotspot track: a long–lived mantle plume. *J. Geophys. Res.* 95(B13): 21,983–21,990.

Slichter, L. B. 1963. Secular effects of tidal friction upon the Earth's rotation. *J. Geophys. Res.*, 68(14): 4281–4288

Smith, A. G. and R. A. Livermore. 1987. Hotlines and Cenozoic volcanism in East Antarctica and eastern Australia. In: *Geological Evolution of Antarctica* (M. R. A. Thomson, J. A. Crame and J. W. Thomson, eds.); Cambridge Univ. Press. 593.

Smith, M. L. 1974. The scalar equations of motion of infinitesimal elastic–gravitational motion for a rotating, slightly elliptical Earth. *Geophys. J. Roy. Astr. Soc.* 37: 491–526.

———. 1977. Wobble and nutation of the Earth. *Geophys. J.*, 50: 103–140.

Smith, W. and D. Sandwell. 1997. Map: *Measured and estimated seafloor topography.* Vrsn. 4. 2. Nat. Geophys. Data Center, Boulder, Co.

Snider-Pelligrini, A. 1858. *La Création et Ses Mystères.* Paris. Cited by W. Sullivan. 1976, q.v.

Solomatov, V. S. and L.-N. Moresi. 1996. Stagnant lid convection on Venus. *J. Geophys. Res.* 101(E2): 4737–4753.

———. 1997. Three regimes of mantle convection with non-Newtonian viscosity and stagnant lid convection on the terrestrial planets. *Geophys. Res. Lett.* 24(15): 1907–1910.

Sommerfeld, A. 1950. *Mechanics of Deformable Bodies.* (Engl. transl.:) New York: Academic Press. Chapter IV.

Sovers, O. J. 1994. Vertical ocean loading amplitudes from VLBI measurements. *Geoph. Res. Lett.* 21(5): 357–360.

Sovers, O. J., C. S. Jacobs and R. S. Gross. 1993. Measuring rapid ocean tidal earth orientation variations with very long base-line interferometry. *J. Geophys. Res.* 98(B11): 19,959–19,971.

Sovers, O. J., J. B. Thomas, J. L. Fanselow, E. J. Cohen, G. H. Purcell, D. H. Rogstad, L. J. Skjerve and D. J. Spitzmesser. 1984. Radio interferometric determination of intercontinental baselines and Earth orientation utilizing deep space network antennnas, 1971–1980. *J. Geophys. Res.* 89(9): 7597–7607.

Spada, G. 1993. True polar wander and long-wavelength dynamic topography. *Tectonophysics* 223: 3–13.

Spada, G., Y. Ricard and R. Sabadini. 1992. Excitation of true polar wander by subduction. *Nature* 360: 452–454.

Spencer, R. and J. M. Vassie. 1985. Comparisons of sea–level measurements obtained from deep pressure sensors. *Adv. Underw. Techn. Offsh. Engnrg.* 4: 183–207.

Spitaler, R. 1930. Die Achsenschwankungen der Erde und ihre Folgen. *Gerlands. Beitr. Geophys.* 26: 94–97.

Stacey, F. D. 1977. A thermal model of the Earth. *Phys. Earth Plan. Int.* 15: 341–348.

Steensrup, S. and L. Gerward. 1996. Becquerel's discovery of radioactivity—a centenary. *Phys. Tr.* 34(5): 286–287.

Stein, C. A. and S. Stein. 1994. Comparison of plate and asthenospheric flow models for the thermal evolution of oceanic lithosphere. *Geophys. Res. Lett.* 21(8): 709–712.

_____. 1994. Constraints on hydrothermal heat flow through the oceanic lithosphere from global heat flow. *J. Geophys. Res.* 99(B20): 3081–3095.
Steketee, J. A. 1958. On Volterra's dislocations in a semi-infinite medium. *Can. J. Phys.* 36: 192–205.
Stetson, H. T. 1929. The correlation of deep–focus earthquakes with lunar hour angle and declination. *Science*, 69: 523–524.
Stevens, C., et al. 1999. Rapid rotations about a vertical axis in a collisional setting revealed by the Palu fault, Sulawesi, Indonesia. *Geophys. Res. Lett.* 26(17): 2677–2680.
Stevenson, D. J. 1983. Anomalous bulk viscosity of two–phase fluids and implications for planetary interiors. *J. Geophys. Res.* 88:(B3): 2445–2455.
_____. 1993a. Why is true polar wander so slow? *EOS (Trans. Amer. Geophys. Un.)* 74(43 Suppl.), p. 212.
_____. 1993b. Does mantle convection know about Earth rotation? In: *Geodynamics workshop in the Czech Republic* (O. Cadek and D. A. Yuen, eds.). *Terra Nova* 5: 589–590.
Stevenson, F. R. 1978. Pre-telescopic astronomical observations. In: *Tidal Friction and the Earth's Rotation* (P. Brosche and J. Sundemann, eds.); Springer-Verlag. Pp. 5–21.
Sterling, A. and E. Smets. 1971. Study of earth tides, earthquakes and terrestrial spectroscopy by analysis of the level fluctuations in a borehole at Heibaart (Belgium). *Geophys. J. Roy. Astron. Soc.* 23: 225–242.
Stiller, H. 1971. Zur Bedeutung der Arbeiten von H. v. Helmholtz für die geophysikalische Hydrodynamik und für die Physik des Erdinnern. Gedanken von Helmholtz über schöpferische Impulse und über das Zusammenwirken verschiedener. *Wissen–schaftszweige Sutz. d. Plen. U. Klassen, Akad. Wissen. DDR* 1. 1972: 45–48.
Storey, B. C. 1995. The role of mantle plumes in continental breakup: case histories from Gondwanaland. *Nature* 377: 301–308.
Strom, R. G., G. G. Schaber and D. D. Dawson. 1994. The global resurfacing of Venus. *J. Geophys. Res.* 99(E5): 10,899–10,926.
Su, W. R., L. Woodward and A. M. Dziewonski. 1994. Degree 12 model of shear velocity heterogeneity in the mantle. *J. Geophys. Res.* 99(B4): 6945–6980.
Suess, E. 1909. *Das Antlitz der Erde.* Prague: F. Tempsky.
Sugisaki, R. 1981. Deep-seated gas emission induced by Earth tide: a basic observation for geochemical earthquake prediction. *Science*, 212: 1264–1266.
Sullivan, Walter 1974. *Continents in Motion: The New Earth Debate.* New York: McGraw-Hill.
Sundvik, M. T. and R. L. Larson. 1988. Seafloor spreading history of the western North Atlantic Basin derived from the Keathley sequence and computer graphics. *Tectonophysics* 155: 49–71.
Sutherland, W. 1903. The cause of the Earth's magnetism. *Terr. Magn.* 8: 49–52.
Sykes, L. R. 1967. Mechanism of earthquakes and nature of faulting on the mid–ocean ridges. *J. Geophys. Res.* 72: 2131–2150.
Tackley, P. J., D. J. Stevenson, G. A. Glatzmaier and G. Schubert. 1993. Effects of an endothermic phase transition at 670 km depth in a spherical model of convection in the Earth's mantle. *Nature* 361: 699–704.
Takahashi, E. 1986. Melting of a dry peridotite KLB-1 up to 14 GPa: Implications on the origin of peridotitic upper mantle. *J. Geophys. Res.* 91: 9367–9382.
Takeuchi, H. 1951. On the Earth tide in the compressible Earth of varying density and elasticity. *J. Fac. Sci. Un. of Tokyo, Sectn. II*, Vol. VII, Pt. II, 1–118.
Tanimoto, T. 1989. Moment of inertia of three-dimensional models of the Earth. *Geophys. Res. Lett.* 16(5): 389–393.
Tanimoto, T. and Y.-S. Zhang. 1990. Lithospheric thickness and thermal anomalies in the

upper mantle inferred from the Love wave data. *Geophys. Res. Lett.* 17(13): 2405–2408.

Tanimoto, T. and Y.-S. Zhang. 1992. Cause of low-velocity anomaly along the South Atlantic hotspots. *Geophys. Res. Lett.* 19(15): 1567–1570.

Tapley, B. D., W. G. Melbourne and C. Reigber. 1996. GRACE: a satellite-to-satellite geopotential mapping mission. *EOS (Trans. Amer. Geophys. Un.)*, 77(46, Suppl.): page F136.

Taylor, F. B. 1910a. Sliding continents and tidal and rotational forces. In: *The Theory of Continental Drift.* (Tulsa: American Association of Petroleum Geologists): 158–177.

_____. 1910b. Bearing of the Tertiary mountain belt on the origin of the Earth's plan. *Geol. Soc. Amer. Bull.* 21: 179–226.

Taylor, G. I. 1920. I, Tidal friction in the Irish Sea. *Phil. Trans. Roy. Soc. London* 220A: 1–33.

Thomas, J. B., J. L. Fanselow, P. F. MacDoran, L. J. Skjerve, D. J. Spitzmesser and H. F. Fliegel. 1976. A demonstration of an independent-station radio interferometry system with 4-cm precision on a 16-km base line. *J. Geophys. Res.* 81(5): 995–1005.

Thompson, A. B. 1992. Water in the Earth's upper mantle. *Nature* 358: 295–302.

Thomson, Sir W. (later Ld Kelvin), 1867. On vortex motion. *Trans. Roy. Soc. Edinb. v.* XXV: 217–260.

Todhunter, I. 1876. William Whewell, D. D. *Master of Trinity College, Cambridge.* London: MacMillan and Co. Chapter 6.

Tomaschek, R. 1957. Tides of the solid Earth. *Handbuch der Physik* (Berlin: Springer-Verlag), v. XLVIII, II: 775–845.

Tonn, R., and J. Zschau. 1989. On the figures of the Earth. In: *Slow Deformation and Transmission of Stress in the Earth* (S. C. Cohen and P. Vanicek, eds.). *Geophys. Monograph* 49, *IUGG,* Vol. 4, 7–15.

Touma, J. and J. Wisdom. 1994. Evolution of the Earth-Moon system. *Astron. J.* 108(5): 1943–1961.

Tozer, D. C. 1965. Heat transfer and convection currents. *Phil. Trans. Roy. Soc. London* A258: 252–271.

Truesdell, C. 1954. The *Kinematics of Vorticity.* Indiana Univ. Press. See also Serrin (1959), *supra.*

Turcotte, D. L. 1974. Membrane tectonics. *Geophys. J. Roy. Astron. Soc.* 36: 33–42.

_____. 1993. An episodic hypothesis for Venusian tectonics. *J. Geophys. Res.* 98(E9): 17,061–17,068.

_____. 1996. Magellan and comparative planetology. *J. Geophys. Res.* 101(E2): 4765–4773.

Turcotte, D. L. and E. R. Oxburgh. 1973. Mid-plate tectonics. *Nature* 244: 337–339.

Turnbull, H. W. 1960. *The Correspondence of Isaac Newton,* II. Cambridge Univ. Press. 417 pp.

Turner, G. J. and F. D. Stacey. 1981. Frequency dependence of Q for rock stressed near the breaking point. In: *Anelasticity in the Earth.* F. D. Stacey, M. S. Paterson and A. Nicholas, eds. Geodynamics Ser. v. 4. (Washington DC: AGU) p. 83–85.

Um, J. and F. A. Dahlen. 1992. Normal mode multiplet coupling on an aspherical, anelastic Earth. *Geophys. J. Intern.* 111(1): 11–31.

Uyeda, S. 1955. Magnetic interaction between ferromagnetic materials contained in rocks. *J. Geomag. Geoelectr.* VII, 9–36.

Uyeda, S. 1988. Implications: Geodynamics. In: *Handbook of Terrestrial Heat-Flow Density Determination* (R. Haenel et al., eds.). Dordrecht: Kluwer. Pp. 317–346.

Uyeda, S. and H. Kanamori. 1979. Back-arc opening and the mode of subduction. *J. Geophys. Res.* 84(B3): 1049–1061.
vanDam, T. M. 1997. Personal communication, 97–02–05.
vanDam, T. M. and T. A. Herring. 1994. Detection of atmospheric pressure loading using very long baseline interferometry measurements. *J. Geophys. Res.* 99(B3): 4505–4517.
vanDam, T. M., G. Blewitt and M. B. Heflin. 1994. Atmospheric pressure loading effects on global positioning system coordinate determination. *J. Geophys. Res.* 99(B12): 23,939–23,950.
VanDecar, J. C., D. E. James and M. Assumpcaon. 1995. Seismic evidence for a fossil mantle plume beneath South America and implications for plate driving forces. *Nature* 378: 25–31.
Van der Gracht, W. A. and J. M. van Waterschoot. 1928. Remarks concerning the papers offered by the other contributors to the Symposium. In: *The Theory of Continental Drift*. (Tulsa: American Association of Petroleum Geologists). Pp. 197–226.
Van der Voo, R. 1994. True polar wander during the middle Paleozoic? *Earth Plan. Sci. Lett.* 122: 239–243.
Vanicek, P. R., B. Langley, D. E. Wells and D. Delikaraoglou. 1984. Geometrical aspects of differential GPS positioning. *Bull. Geod.* 58: 37–52.
Varga, P. and E. Grafarend. 1996. Distribution of the lunisolar tidal elastic stress tensor components within the Earth's mantle. *Phys. Earth. Plan. Int.* 93: 285–297.
Vasco, D. W., L. R. Johnson and J. Pulliam. 1994. Lateral variations in mantle velocity structure and discontinuities determined from P, PP, S, SS and SS–Sd S travel time residuals. *J. Geophys. Res.* 100(B12): 24,037–24,059.
Vening Meinesz, F. A. 1944. Der verdeeling van continenten en oceanen over het aardopperviak. *Proc. Ned. Akad. Wet.* 53: no. 4.
———. 1947. Shear patterns of the Earth's crust. *Trans. Amer. Geophys. Un.* 28: 1–61.
———. 1948. Major tectonic phenomena and the hypothesis of convection currents in the Earth. *Q. J. Geol. Soc.* [London] 103: 191–207.
———. 1958. Convection currents in the Earth. Origin of continents and oceans. Ch. 11, p. 397–441, in: *The Earth and its Gravity Field* by W. A. Heiskanen and F. A. Vening Meinesz (New York: McGraw–Hill); 470 pp.
Vermeersen, L. L., A. R Sabadini, G. Spada and N. J. Vlaar. 1994. Mountain building and Earth rotation. *Geophys. J. Intern.* 117: 610–624.
Vetter, U. R. and R. O. Meissner. 1977. Creep in geodynamic processes. *Tectonophysics* 42: 37–54.
Vidale, J. E., S. Goes and P. G. Richards. 1995. Near-field deformation seen on distant broadband seismograms. *Geophys. Res. Lett.* 22(1): 1–4.
Vigny, C., Y. Ricard and C. Froidevaux. 1991. The driving mechanism of plate tectonics. *Tectonophysics* 187: 345–360.
Vincent, P. and J. R. Rundle. 1998. Synthetic aperture radar interferometry capability now available to universities. *EOS (Trans. Amer. Geophys. Un.)* 79 (3): p. 34.
Vine, F. J. and D. H. Matthews. 1963. Magnetic anomalies over oceanic ridges. *Nature* 199: 947–949.
Wadati, K. 1928. Shallow and deep earthquakes. *Geophys. Magaz.* 1: 162–202.
Wagner, C. A. 1991a. A refined M2 tide from GeoSat altimetry. *NOAA Techn. Rep. NOS* 136 NGS 48 71.
———. 1991b. How well do we know the deep ocean tides? *Mar. Geod.* 16: 18–140.
Wahr, J. M. 1981. Body tides on an elliptical, rotating, elastic and oceanless Earth. *Geophys. J. Roy. Astron. Soc.* 64(3): 677–703.

———. 1983. The effects of the atmosphere and oceans on the Earth's wobble and on the seasonal variations in the length of day–II. Results. *Geophys. J.* 74: 451–487.
Wahr, J. and Z. Bergen. 1986. The effects of mantle anelasticity on nutations, earth tides, and tidal variations in rotation rate. *Geophys. J. Roy. Astron. Soc.* 87: 633–668.
Ware, R. H. and C. Rocken. 1986. A global positioning system baseline determination including bias fixing and water vapor radiometer corrections. *J. Geophys. Res.* 91(B9): 9183–9192.
Watts, A. B., J. K. Weissel, R. A. Duncan and R. L. Larson. 1988. Origin of the Louisville Ridge and its relationship to the Eltanin fracture zone system. *J. Geophys. Res.* 93: 3051–3077.
Weems, R. E. and W. H. Perry. 1989. Strong correlation of major earthquakes with solid–Earth tides in part of the eastern United States. *Geology*, 17: 661–664.
Weertman, J. 1975. High temperature creep produced by dislocation models. In: *Plastic Deformation of Materials* (American Soc. of Metals), pp. 315–336.
Wegener, Alfred. 1915. *Die Enstehung der Kontinente und Ozeane*. First Ed.
———. 1928. Two notes concerning my theory of continental drift. In: *The Theory of Continental Drift*. (Tulsa: American Association of Petroleum Geologists). 97–103.
Weissel, J. K. and R. N. Anderson. 1978. Is there a Caroline plate? *Earth Plan Sci. Lett.* 41(2): 143–158.
Wells, J. W. 1963. Coral growth and geochronometry. *Nature* 197: 448–450.
Wells, W. C. 1984. *Spaceborne gravity measurements*. NASA Conference Publication 2305.
Wessel, P., D. Bercovici and L. W. Kroenke. 1994. The possible reflection of mantle discontinuities in Pacific geoid and bathymetry. *Geophys. Res. Lett.* 18: 1943–1946.
White, D. A., D. H. Roeder, T. H. Nelson and J. C. Crowell. 1970. Subduction. Term for abrupt descent of a segment of lithosphere; advocate use because of plate tectonics applicability. *Geol. Soc. Amer. Bull.* 81(11): 3431–3432.
White, R. S. and D. McKenzie. 1995. Mantle plumes and flood basalts. *J. Geophys. Res.* 100(B9): 17,543–17586.
Whitehead, J. A. and R. F. Gans. 1974. A new, theoretically tractable earthquake model. *Geophys. J. Roy. Asron. Soc.* 39: 11 28.
Widmer, R., G. Masters and F. Gilbert. 1991. Spherically symmetric attenuation within the Earth from normal mode data. *Geophys. J. Intern.* 104: 541–553.
Wiechert, E. 1907. Die Erdbebensforschung, ihre Hilfsmittel und ihre Resultate fur die Geophysik. *Phys. Zeitschr.* 9: 36–47.
Wiens, D. A. and S. Stein. 1985. Implications of oceanic intraplate seismicity for plate stresses, driving forces and rheology. *Tectonophysics* 116, 143–162.
Wilcock, W. S., D. S. C. Solomon, G. M. Purdy and D. R. Toomey. 1992. The seismic attenuation structure of a fast-spreading mid-ocean ridge. *Science*, 258 1470–1474.
Williams, G. E. 1989. Late pre-Cambrian tidal rhythmites in South Australia and the history of the Earth's rotation. *J. Geol. Soc.* [London] 146: 97–111.
———. 1993. History of the Earth's obliquity. *Earth-Science Revs.* 34(1): 1–45.
———. 1994. Resonances of the fluid core for a tidally decelerating Earth: cause of increased plume activity and tectonothermal reworking events? *Earth Plan. Sci. Lett.* 128: 155–167.
———. 1997. Precambrian length of day and the validity of tidal rhythmite paleotidal values. *Geophy. Res. Lett.* 24(4): 421–424.
Wilson, C. R. and R. O. Vicente. 1980. An analysis of the homogeneous ILS polar motion series. *Geophys. J. Roy. Astron. Soc.* 62: 605–616.
Wilson, J. T. 1963. Evidence from islands on the spreading of ocean floors. *Nature* 197: 536–538.

_____. 1965. A new class of faults and their bearing on continental drift. *Nature* 207: 43–347.

_____. 1970. Some possible effects if North America has overridden part of the East Pacific Rise. *Abstr. w. Programs–Geol. Soc. Amer.* 2; 7 722–723.

_____. 1973. Mantle plumes and plate motions. *Tectonophysics* 19: 149–164.

_____. 1976. *Continents Adrift and Continents Aground*. S. Francisco: W H Freeman and Co., pp. 230.

Woodhouse, H. 1988. The calculation of eigenfrequencies and eigenfunctions of the free oscillations of the Earth and the Sun. *Seismological Algorithms*. London: Academic Press. Pp. 322–370.

Woodward, R. L., A. M. Dziewonski and A. R. Peltier. 1994. Comparisons of seismic heterogeneity models and convective flow calculations. *Geophys. Res. Lett.* 21:(5): 325–328.

Wooley, A. R. 1989. The spatial and temporal distribution of carbonatites. In: *Carbonatites: Genesis and Evolution*. (K. Bell, ed.) Allen and Unwin. ch. 2, 15–21.

Wunsch, C. 1975. Internal tides in the ocean. *Rev. Geophys. Sp. Phys.* 13(1): 167–182.

Wunsch, J. 1990. Astrometric observations of Hevelius and derived values of ΔT (Dynamical time–Universal time). In: *Earth's Rotation from Eons to Days* (P. Brosche and J. Sundermann, eds.). Berlin: Springer-Verlag. Pp. 21–26.

Wyllie, P. J. 1987. Transfer of sub-cratonic carbon into kimberlites and rare earth carbonatites, in: *Magmatic Processes: Physiochemical Principles* (B. O. Mysen, ed.); *Geochemical Society Spec. Pub.* no. 1: 107–119.

_____. 1989. The genesis of kimberlites and some low-SiO_2, high–alkali magmas. In: *Kimberlites and Related Rocks*, Vol. 1, *G.S.A. Spec. Pub.* 14: 603–615.

_____. 1995. Experimental petrology of upper mantle materials, processes and products. *J. Geodynamics* 20(4): 429–468.

Wyllie, P. J., M. B. Baker and B. S. White. 1990. Experimental boundaries for the origin and evolution of carbonatites. *Lithos* 26: 3–19.

Wyss, M. and J. N. Brune. 1967. The Alaska earthquake of 28 March 1964; a complex multiple rupture. *Bull. Seis. Soc. Amer.* 57: 1017–1023.

Xiao, Q. and H. T. Hsu. 1989. The impulse response of a PREM–Zschau Earth. *Proc. 11th Symp. Earth Tides, Helsinki*. (Stuttgart: E. Schweizerbartsche Verlag): 493–501.

Yanshin, A. L., P. Melchior, V. I. Keilis-Borok, M. de Becker, M. Ducarme and A. M. Sadowsky. 1986. Global distribution of tidal anomalies and an attempt of its geotectonic interpretation. In: *Proc. Tenth Symp. Earth Tides, Madrid*, (R. Vieira, ed.). Madrid: Consejo Sup. Inv. Cient. pp. 731–755.

Yoshioka, N. 1994. The role of plastic deformation in normal loading and unloading cycles. *J. Geophys. Res.* 99(B8): 15,561–15,568.

Zahel, W. 1978. The influence of solid Earth deformations on semidiurnal and diurnal oceanic tides. In: *Tidal Friction and the Earth's Rotation* (P. Brosche and J. Sundemann, eds.). Berlin: Springer-Verlag. Pp. 98–124.

Zahel, Wilfried. 1991. Modeling ocean tides with and without assimilating data. *J. Geophys. Res.* 96(B12): 20,379–20,391.

Zhang, C. Z. and K. Zhang. 1995. On the internal structure and magnetic fields of Venus. *Earth, Moon and Planets*, 69: 237–247.

Zhang, Y.-S., T. Tanimoto and E. M. Stolper. 1984. S-wave velocity, basalt chemistry and bathymetry along the Mid-Atlantic Ridge. In: *Ten Years of GEOSCOP Broadband Seismology* (Montagner, J. P. and B. Romanowicz, eds.) *Phys. Earth Plan. Int.* 84(1–4): 79–93.

Zhang, Y. and J. T. Kuo. 1989. Time–domain earth tides for a rotating Earth. *Proc. 11th Symp. Earth Tides, Helsinki*. (Stuttgart: E. Schweizerbart'sche Verlag). Pp. 273–279.

Zhang, Y.-S. and T. Tanimoto. 1991. Global Love wave phase velocity variation and its significance to global tectonics. *Phys. Earth Plan. Int.* 66(3–4): 160–202.

Zhang, Y.-S. and T. Tanimoto. 1992. Ridges, hotspots and their interaction as observed in seismic velocity maps. *Nature* 355: 45–49.

Zhang, Y.-S. and T. Tanimoto. 1993. High-resolution global upper mantle structure and plate tectonics. *J. Geophys. Res.* 98(B6): 9793–9823.

Zhao, L. and F. A. Dahlen. 1995. Asymptotic normal modes of the Earth, II: Eigenfunctions. *Geophys. J. Intern.* 121: 585–626.

Ziegler, A. M., C. R. Scotese and S. F. Barrett. 1982. Mesozoic and Cenozoic paleogeographic maps. In: *Tidal Friction and the Earth's Rotation, II.* (P. Brosche and E. Sundemann, eds.). Berlin: Springer-Verlag. Pp. 240–252.

Zschau, J. 1978. Tidal friction in the solid Earth: loading tides versus body tides. In: *Tidal Friction and the Earth's Rotation* (Eds. P. Brosche and J. Sundemann). Berlin: Springer-Verlag. Pp. 62–94.

―――. 1983. Rheology of the Earth's mantle at tidal and Chandler wobble periods. In: *Proc. Ninth Symp. Earth Tides* (Stuttgart: E. Schweizerbart'sche Verlags). Pp. 605–629.

―――. 1986a. Tidal friction in the solid Earth as deduced from the Chandler wobble period. In: *Proc. Tenth Symp. Earth Tides, Madrid*, (Ed. R. Vieira). Madrid: Consejo Sup. Inv. Cient. Pp. 433–448.

―――. 1986b. Tidal friction in the solid Earth: constraints from the Chandler wobble period. In: *Space Geodesy and Geodynamics* (A. J. Anderson and A. Cazenave, eds.). London: Academic Press. Pp. 315–344c.

Zschau, J. and R. Wang. 1986. Imperfect elasticity in the Earth's mantle; implications for Earth tides and long period deformations. *Proc. Tenth Symp. Earth Tides, Madrid*, (R. Vieira, Ed.), Madrid: Consejo Sup. Inv. Cient. Pp. 1986–1995.

Zurn, W., D. Emter and H. Otto. 1991. Ultra-short strain meters: tides are in the smallest cracks. *Bull. Inf. Centr. Mar. Terr.* 109: 7912 7921.

Index

"absolute" motion of lithosphere plates
  relative to deep mantle objects, 144, *145*
  relative to mantle plumes, *108*
absorption band (Q/frequency)
  of Liu, Anderson, and Kanamori, *46*
  of Zschau, *46*, 47
Africa, tectonics under TPW, *173*, 174
altimetric radar satellites, 24
  Topex/Poseidon, 24, 206
Americas
  motion of, 127, *130, 151*
Ampferer, O.
  on asymmetry of Verschluckung-Zonen, 124
  on Atlantic Ocean opening, 21
Amstutz, A., *124*
antisymmetric part of tidal stress tensor, 52
  dissipation attributable to, 59
asthenosphere
  autonomous convection in, *72*, 73
  Barrell's, 69
  phase transitions in, 23, *73*, 92
asymmetry
  of EPR (Project MELT), 150
  of mantle convection, *83*, 126–127, 149
  "mainstream" m. c., 58
  of seafloor spreading, 129–132
  of subduction zones, 124, *125*
Asynchronous rotation of Earth
  rotation of tidal potential, *28*, 29

Atlantic Ocean
  asymmetry of development, *132*
  fracture zones, 141
  hotspot (plume) contribution to development, *130*
  Mid-Atlantic Ridge (MAR), *140*, 141

Becquerel, A. H.
  activité radiante spontanée, 38
biota, terrestrial; dependence upon mobile lithosphere, 220
"bulges," bodily tidal, 39
  modeled as geostationary, 29, 79
  represent mobile waves, 29–34, 80

Centre International des Marées Terrestres, 8, 25, 190–195, 207
CMB (Core-Mantle Boundary), 148
  role of TVI in planetary magnetization, 82, 153–154
continental drift
  early views on, 21
  hypothesized by Joly, 59
  non-acceptance, 21–22
Convection (bulk flow) in mantle
  Gasparik representation, *92*
  hypothesis traced by Sullivan, 21
  as origin of global tectonics, 21
  reconstruction by Vigny et al., *119*
  slab-pull as driving force, 150

viscosity dissipation number, 86–88
whole-mantle vs. two-layer, 71–72, *72*
cotidal seismicity
  Du's 18.6-yr correlation, 100
  far-field displacement, 98–99
  inadequacy to explain tidal dissipation, 100
  of Klein, 96–99, *97*
couples (force-couples), internal
  under wave tides, 49–52
creep, as defined by microstructural dislocation, 59, *77*

Darwin, Sir G. H., portrait, *35*
  on crustal displacement under retarding couple, 19
  on isostasy, 22
  on tides in viscous spheroid, 17–19
  on true polar wander, 19
delta (δ) gravimetric tidal factor of Dehant, 187
displacement, cumulative, 58
dissipation, (tidal energy contribution)
  marine, 42–43, 61, 182
    due ocean floor heaving (Hendershott), 61
    model of Le Provost et al., 98, 98–102
    model of Miller, 188
    via internal (baroclinic) waves, 99
  relative to viscous portion in convection, 88–90, 218
  in shallow seas
    reservations of Cartwright, Ray, 43, 189
    reservations of Munk, 13
  solid Earth, 78–82
    in Earth critically pre-stressed, 90, 103
    for values of Q, 44–46
    in subduction zones, 93–95

Earth/Moon double planet (Kuiper), 6, 29, 153, 240
  duality intrinsic in global tectonics, 152–153, 220
Earth tides (bodily tides)
  Darwin's dissipation assumption, 31–32
  Love's geostationary (irrotational) assumption, 182–185
  models of, 24–33

East Pacific Rise (EPR)
  asymmetry, 144, *149*
    (Project MELT), 150
  maintaining location, 149–150
  stepping or "jumping," 126–127, *128, 137*
Eddington, Sir Arthur, *36*
  reservations in respect to tidal theory, 19, 36, 38, 219
  tidal braking, 33, 36
El Niño, incidence of, 213–214

flattening
  excess, of Earth's figure of rotation, 121
  as demonstrating strength of Earth, 24
force-couples generated by rotating potential, *28*

"geohedron," 147
geoid
  actual (referred to estimated rotation-figure) *118*
  "best-fitting" (observed-flattening removed), 117
  flow models based on, 118
GPS (Global Positioning System), 199
Gravity
  gradient at high elevation (Rapp's projection), 134
growth-lines, tidal
  in organisms and rhythmites, 42

heat efflux, terrestrial
  correlation with seismic wave velocity, 112
  over subduction zones, 95
  total, 116
heaving of ocean floor, under bodily tide (Hendershott), 61
Helmholtz, H. L. F. von
  acquaintance with Kelvin, 32
  energy conservation, *34*
  portrait, *34*
  vortex flow, 79–80
    identification of, 31–32
Holmes, A.
  on continental drift, 22
  on opening of the Atlantic, 21

264   Index

hotspots
    as "absolute motion" reference, 133
    as convection-roll expression, 139
    gravity expression, *135, 138, 140*
    "standard set," 134

Indian Ocean region,142

Jeffreys, Sir H.
    portrait, *37*
    seismologic tables, 20
Joly, John
    on mantle fusion, convection, 59
    portrait, *35*

Lambeck, K.
    apportionment of tidal dissipation, 182, 219
Laplace, Marquis P.-S. de
    tide species, *15*, 16
Lithosphere
    mobility, 144, 152, 220
    net rotation, 109–110
    as plates; identification by J. T. Wilson, 23
load tide (oceanic), *203*
Love, A. E. H.
    bodily tide assumption, 183
    portrait, *36*
Love numbers, 40–41, 182–185, 187

magnetization, planetary
    TVI as driver, 82, 153–154
        (Bostrom 1998), 225
marine tides
    equilibrium assumption of Laplace, 15
McKenzie, D. P., 117, 159, 219
Melchior, Bn. P.
    Earth-Tides Center, Brussels (CIMT), 25
    gravimetric observation of earth-tides, 190 et seq.
    ICET Databank DB92, 190, 194 (Table 7.3, n.)
    reference standard, bodily tides, 63–64, 192, *194*
    relationship tidal anomalies/tectonics? 64, 96, 203
    on tiltmeters, 207

metastability in asthenosphere, 168
    phase multiplicity, 23
mobilist vs. fixist tectonics
    perceptions of, 21
mobility of Earth's lithosphere, 144, 152, 220
models
    of mantle bulk flow
        based on geoid, 113 et seq.
        numerical, 70, 217–218
        viscosity-based, 115, 116
    of tidal Earth, 25 et seq., 181
motion, of plates
    "absolute," 110
    asymmetric, 150
    random, 144
    relative to
        to deep mantle objects, 150
        hotspots, *108*
    toroidal, 58
Munk and MacDonald (*The Rotation of the Earth*)
    tidal dissipation, 67

nucleation of strain (seismic energy release), 76, 96

observation techniques, earth tides
    GPS (Global Positioning System), 199–201
    gravimetric, 190–195, *193,194*
    tiltmeter (geophysical clinometer), 207–210
    VLBI (Very Long Baseline Interferometer), 195–199
oceanic effect, on bodily tides
    attraction and loading, 37, 61, 100, 202–204
orbital elements
    long-term indeterminacy, 214–217
    role in "ice ages," 214–216

Pacific region
    tidal input in flow system, 83
    western Pacific tectonics, 122, *123*
        deep seismicity concentration, 122
        zonal-flow model of Crook, 122
paleomagnetic record
    durability, *22*

plate rotation displayed by, *114–115*
use by Runcorn, 22–23, *1i3*
passive Earth, 40, 47
phase-lag
   of gravimetric tide, 45, 47
   of tidal bulges observed, 45
phase-transitions, P/T, *73, 75, 77*
plate tectonics
   recognition of, 23
Platzmann's formulation (tidal dissipation), 62, 80–81
plumes, in mantle
   distortion during uprise, 148
   as mantle flow component, *132*
   unresolved by tomography, *69*
polar wander. *See also TPW*
   Darwin's perception, 19
   mechanism of Goldreich and Toomre, 24, 160–161
polarity in global tectonics (Pavoni), *177*
potential, tidal
   geostationary, of G. H. Darwin, Love, 27, 31–33, 35
   rotating, 48, 79–81
pre-stress, by mantle convection, 90
principal axes of moment, *158*
   analysis of Goldreich and Toomre, 120, 160
      effects of exchange, 180

Q (elastic quality factor)
   based on VLBI, 66
   frequency-dependence, 59
   in respect to Chandler wobble, 45
   in respect to elastic wave decay 44

radioactivity
   energy source in global tectonics, 21
Rayleigh number
   under joint TVI/ buoyancy convection, 218
retarding couple, lunar
   displacement under, 53
rheology, of mantle
   creep, diffusion- and dislocation-, 74–75
   power-law, 59–60
rotationality measure, Truesdell's, 103, 105
Runcorn, S. K.
   paleomagnetism, 23

portrait, *37*
   toroidal vs. poloidal velocity components, 58

Schwiderski, E. W.
   marine tidal dissipation, 64, 202, 254
seafloor spreading
   a passive response? 127, 137, 150
secular term, in tidal stress, 52
   cumulative displacement under, 58
      dissipation associated with, 59, 88–90
      vorticity induction by, *81*
self-organization of flow, 103, 218
seismicity, cotidal
   Du's 18.6-yr correlation, 100
   far-field displacement, 98–99
   inadequacy to explain tidal dissipation, 100
   of Klein, 96–99, *97*
self-organization
   intrinsic in rotational flow, 103, 218
shallow seas
   as seat of tidal energy dissipation, 17
      reservations of Cartwright, Ray 189
slab-pull vs. ridge-push, 150
state-change, in volcanism
   cascading effect of depressurization, 78, 93
strain ellipsoid
   rotation relative to particle lines, 53
stress
   diffusion, 84
      rate, distance, 84–86
   ellipsoid, tidal, 28, 49
subduction (downflow) regions
   compression regime in, *94*
   Gasparik representation of, 92
   heatflow anomaly (Ida's paradoxx), 95
   tidal dissipation in, 94

Takeuchi, H.
   internal stress under stationary tidal bulge, 50
tidal action
   observation using radar altimetry, 42–43
   in seafloor spreading, *83*
   rotational asymmetry, 58
tidal gravimetric factor (Dehant), 47

tidal potential
  geostationary vs. rotating, 28, 79, *184*
tide-gauge records, 43
tiltmeter
  geophysical, 207
  signal relative to Earth rotation mode, 48
  to monitor vorticity induction, 209
tomographic images of mantle, *69, 71, 81*
toroidal plate motion
  identification by Runcorn, 58
  observed, 107–109
  separation by Lithgow-Bertelloni et al., *87*
Torques, body
  internal
    under tidal bulge passage, 50–52
    modeled as reconfiguration, 28
TPW (true polar wander)
  figure adjustment, 166–169
  geological record, 174
  Goldreich and Toomre's analysis, 24, 160
  inhibition of, 175
  transitions, discontinuous
    phase-, *73, 75, 77*
    state-, *169, 173*
    dissipation due to (non-linearity), 90
TRM (thermal remanent magnetism)
  durability of, 22
TVI (tidal vorticity induction)
  as agency in planetary magnetization, 82, 153–154
  convection modulation (flow as resultant), 93
  dissipation attributable to, *65, 66,* 218

Vening Meinesz, F.
  on mantle convection, 23
    bulge migration stress, *178*

Venus
  CO_2 locus, 152
    Bostrom (1998), 225

  resurfacing, 6
  rotation rate, 6, 151
    'stagnant' lithosphere, 151
  subduction absent, 154
    topography, vs. Earth's, *152*
  tectonics, contrasting geotectonics, 152–153
    convection but no CMF, 153
viscosity, 7, 11, 17–19, 24
  role of volatiles content, 74
  upper mantle, 21, 47
  variation with depth, *117*
volatiles (H_2O, CO_2)
  hydration-dehydration cycle, 90, *92*
    internal preservation by subduction, 152
  mantle's water content, 90
  role in convection, 74
volcanicity; association with tides, 8–9, 210–212
vortex motion (elemental rotational flow)
  identification by Helmholtz, 31, 32
  induction by wave tides, 79–82, 103
  as sketched by Lamb, *55*
  self-organization is intrinsic, 218
vorticity induction, tidal
  as driver of planetary magnetization, 153–154
    (Bostrom 1998), 225
  minimal in Venus, 152
  Pacific/Atlantic region, *83*

Wadati, K.
  identification of deep seismicity, 124

yield stress
  as impediment to mantle convection, 24

Zschau, J.
  estimation of tidal dissipation, 45–47, 65
  gravimetric, as against physical, tidal "bulge," 45